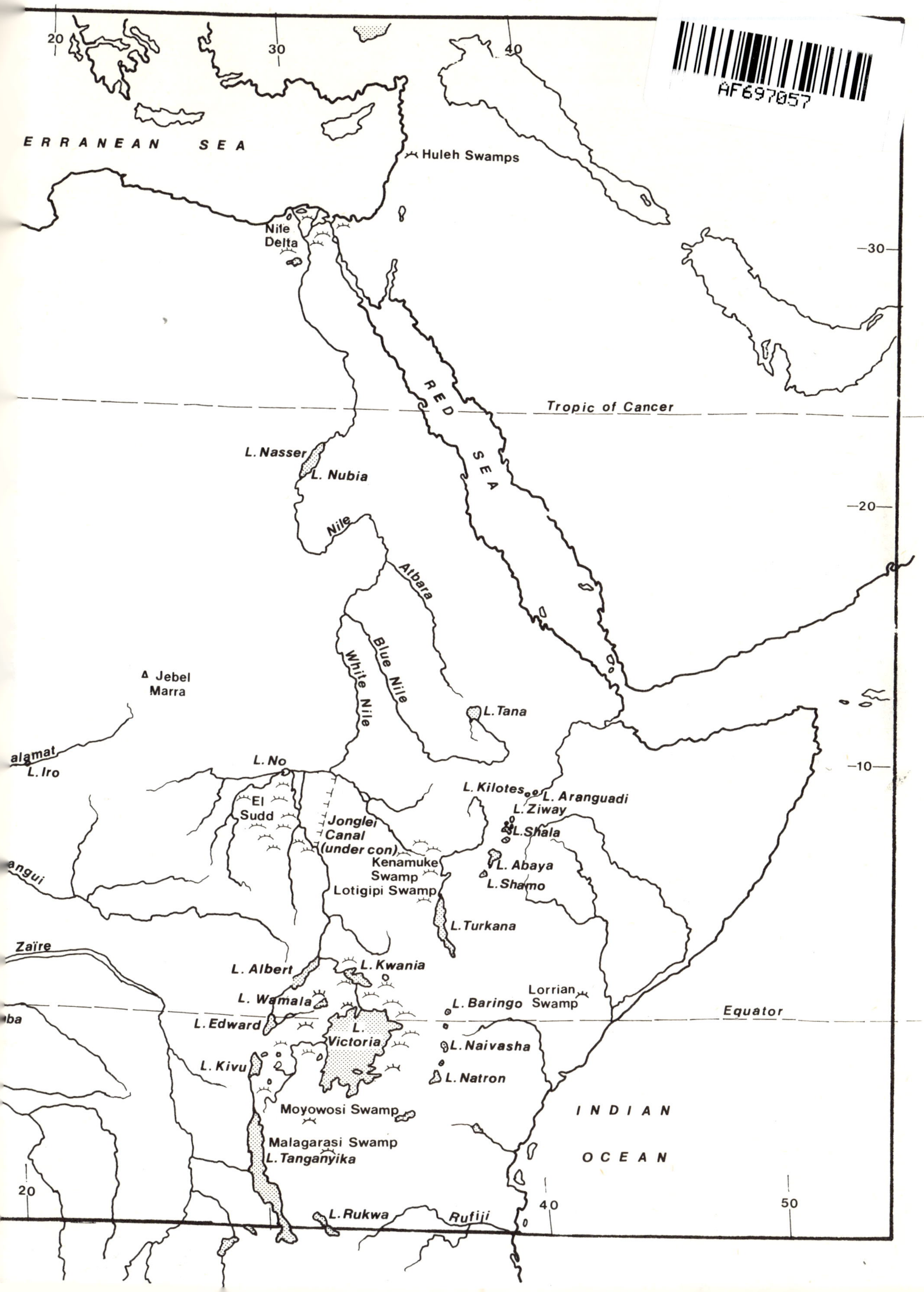
20
30
40
ERRANEAN SEA
Huleh Swamps
Nile Delta
30
RED SEA
Tropic of Cancer
L. Nasser
L. Nubia
Nile
20
Atbara
Blue Nile
White Nile
Jebel Marra
L. Tana
alamat
L. Iro
L. No
10
El Sudd
Jonglei Canal (under con)
Kenamuke Swamp
Lotigipi Swamp
L. Kilotes
L. Aranguadi
L. Ziway
L. Shala
L. Abaya
L. Shamo
angui
L. Turkana
Zaïre
L. Albert
L. Kwania
L. Wamala
L. Baringo
Lorrian Swamp
Equator
ba
L. Edward
L. Victoria
L. Naivasha
L. Kivu
L. Natron
Moyowosi Swamp
INDIAN OCEAN
Malagarasi Swamp
L. Tanganyika
20
40
50
L. Rukwa
Rufiji

The ecology and management of African wetland vegetation

Geobotany 6

Series Editor

M. J. A. WERGER

The ecology and management of African wetland vegetation

A botanical account of African swamps and shallow waterbodies

edited by

PATRICK DENNY

1985 DR W. JUNK PUBLISHERS
A MEMBER OF THE KLUWER ACADEMIC PUBLISHERS GROUP
DORDRECHT / BOSTON / LANCASTER

Distributors

for the United States and Canada: Kluwer Boston, Inc., 190 Old Derby Street, Hingham, MA 02043, USA
for the UK and Ireland: Kluwer Academic Publishers, MTP Press Limited, Falcon House, Queen Square, Lancaster LA1 1RN, England
for all other countries: Kluwer Academic Publishers Group, Distribution Center, P.O. Box 322, 3300 AH Dordrecht, The Netherlands

Library of Congress Cataloging in Publication Data

Main entry under title:

The Ecology and management of African wetland vegetation.

(Geobotany ; 6)
Bibliography: p.
Includes indexes.
1. Wetland flora--Africa--Ecology. 2. Wetland conservation--Africa. I. Denny, Patrick, 1939-
II. Series.
QK381.E26 1985 574.5'26325'096 84-21305

ISBN 90-6193-509-1 (this volume)
ISBN 90-6193-895-3 (series)

Copyright

PRINTED IN THE NETHERLANDS

The magnitude of the task. A satellite-view of the African Continent.

This picture was transmitted by the European Space Agency's geostationary weather satellite Meteosat-1 at 1155 GMT on 31 May 1978, stationed over the Greenwich meridian at the equator. Meteosat-1 orbits at 36 000 kilometres above the earth's surface; it remains motionless above a fixed point on the equator and scans the same large area continuously.

Contents

A table on pp. 206–211 shows the enormous extent of wetlands in Africa, but it shows also how few of them, as yet, have any conservation status. It is concluded that 'the best management strategy to maintain African wetlands at their natural level of productivity and to retain their services as erosion barriers and water storage sites is to leave them alone'. But reasons are advanced that often this is impossible, for there are many ways in which these natural assets can be turned to human benefit. However, now that the many types of wetland ecosystems can be identified, it is of high importance that good examples of each are demarcated for permanent conservation – as bench-marks so to speak – so that effects of development can be assessed. The comprehensive indexes and coss-referenced bibliography (Chapter 10) provide the much-needed additional reference material essential for the sound management and future survival of African wetlands.

In the main text, references to literature mainly cite work published during the past two decades when many specialists of diverse nationalities visited Africa, and a growing number of indigenous scientists became interested in the environment. But it would be wrong to deduce from this that the principles of environmental management and conservation were not recognised and stated during the colonial period, albeit with rather less evidence. In fact, colonial scientists were in one way at an advantage in that the problems of different countries could more easily be considered in collaboration. The 'wind of change' tended to compartmentalise African research but this book should go some way to redress the balance.

A final note must be made on the excellent photographs, drawings and diagrams which bring the wetland vegetation alive. Some, such as the view through the canopy of papyrus swamp (page 126) have a rare artistic as well as scientific merit.

Sussex, December 1984 E. Barton Worthington

Preface

In December 1979 limnologists from all over Africa gathered in Nairobi to attend the SIL-UNEP (International Association of Limnology – United Nations Environmental Programme) Workshop on African Limnology. In the course of the meeting the lack of a text-book on African Wetland Vegetation was bemoaned and the concept of this book was born. Dr. John Gaudet agreed to be Editor, and much of the initial impetus stems from his enthusiasm. Unfortunately for us, John was soon appointed to the US-AID Office in Nairobi and owing to the additional demands this entailed he felt he could not continue. I succeeded him as editor and am grateful for his guidance and considerable assistance in the period of transition.

The book is part of the SCOPE (Scientific Committee on Problems of the Environment) Continental Wetlands Project on African Wetland and Shallow Waterbodies and is designed to integrate with other publications. It is intended for those with an interest in wetland ecology, management and conservation and is especially suitable for government organisations, research personnel, and students of hydrobiology, limnology and freshwater biology.

The early chapters provide an introduction to the biology of plants from the four groups – emergent, floating-leaved, submerged and surface-floating types – and discuss them in relation to vegetation dynamics. The structure and functioning of wetland vegetation is considered with special reference to zonation, distribution, primary production, decomposition, nutrition and mineral cycling. The authors then consider waterweed problems, and other aspects of management and conservation. The serious and urgent need for the conservation of African Wetlands is made clear, but proposals are sympathetic to sociological problems. The final chapter has a full, cross-indexed bibliography of over 800 inclusions.

I have been fortunate to have substantial assistance and goodwill from many friends and colleagues. Although it is impossible to mention them all individually, their kindnesses are no less appreciated. A special thanks must go to the authors who have tolerated my impossible demands with cheerfulness, and to the publishers who accepted unkept deadlines philosophically. The gestation period for one chapter was far beyond my wildest estimates! Clive Howard-Williams and David Mitchell have both been very much part of the book from its conception, and its overall content and balance have evolved through their comments and suggestions. The painstaking efforts of Clive Hobday and John Lynn warrant special mention. Clive has been through all the manuscripts to amend some of the more glaring linguistic idiosyncrasies and has compiled the geographical index. He, together with Alison Evans, kindly prepared the endpaper maps. John compiled the author and the systematic indices and lent his hand to drawings and diagrams. The editorial task was simplified by the good auspices of the Westfield College Computer Unit who trained and assisted me in the art of word processing. Brian Place nursed me through many a processing hiccup, and Jenny Allen cheerfully copy-typed and formatted several of the manuscripts. I am grateful to those who have provided photographs, water (shallow lake) in a wetland, but these are generally less than the areas of vegetation (see also the definition in section 3.1). This book is restricted to the wetlands of African athalassic (non-marine) waters and excludes estuarine and marine systems.

1.2 Life-form classification

For many purposes it is convenient to group wetland plants into categories according to their life-forms, and we should consider first the classic work of the Danish botanist, Raunkiaer. Readers are referred to Raunkiaer's book (translated edition, 1937) for full coverage; only an outline can be provided here. In his terminology most aquatic macrophytes are *Cryptophytes, i.e.* plants in which the dormant buds survive periods unfavourable for active growth either under the ground or in the water. Within the cryptophytes, Raunkiaer recognises three groups, *viz. Geophytes, Helophytes,* and *Hydrophytes*. The geophytes are predominantly terrestrial plants and need not be considered further. Helophytes are cryptophytes that grow in soil saturated with water, or in the water itself, from which both leaf- and flower-bearing shoots emerge. Raunkiaer describes hydrophytes as follows: '. . . the vegetative shoots are sunk in the water and only the flowers or inflorescences rise above the surface for pollination. The leaves are either completely submerged or the leaf laminae rest on the surface of the water (*i.e.* floating-leaves) . . .'

There are serious shortcomings in Raunkiaer's classification, especially as he restricts his attention to flowering plants. Freshwater macro-algae such as the charophytes, the bryophytes *e.g. Fontinalis*, and pteridophytes, *e.g. Equisetum* and *Isoetes*, are all excluded. Further, within his strict definitions, not all marsh or swamp plants are cryptophytes, *e.g. Mentha aquatica* and most surface-floating plants such as *Pistia* and *Eichhornia crassipes*. Thus, whereas helophytes and hydrophytes may initially appear to include all aquatic macrophytes, this is not so. Braun-Blanquet (1932) amended Raunkiaer's classification so that the term hydrophyte included all waterplants other than plankton, whilst geophytes covered the terrestrial perennials. He conveniently glossed over the troublesome helophyte (swamp and marsh) species. Within the hydrophytes, Braun-Blanquet recognised:

(a) *Hydrophyta natantia* – free-floating plants such as *Lemna, Riccia* and *Ceratophyllum*;
(b) *Hydrophyta adnata* – waterplants anchored to the substratum, *e.g. Fontinalis*;
(c) *Hydrophyta radicantia* – aquatics with roots. In this sub-division are included most of the common submerged and floating-leaved aquatic macrophytes such as *Nymphaea, Potamogeton* and *Vallisneria*.

To provide a classification specifically for waterplants, Den Hartog and Segal (1964) revised and extended Raunkiaer's system. However, it is unsuitable for our purposes since they excluded: (a) those plants which can remain submerged but are unable to reproduce in this state (*i.e. pseudo-hydrophytes*), (b) emergent plants and (c) plants floating on the water surface, *e.g. Eichhornia crassipes* but not *Salvinia natans*, which they consider to be an *acropleustophyte* (see Chapter 4 for an explanation of this term). Cook (1968) favoured a slightly modified Hejný system (1960). Based on a Raunkiaer-type classification, it has the advantage of including swamp and marsh species. The three main divisions are:

(a) *Hydatophytes* – obligate waterplants;
(b) *Tenagophytes* – plants that may, or may not, require submergence but do require the aerial phase for sexual reproduction, *i.e.* amphibious plants of wet or regularly flooded areas;
(c) *Trichophytes* – essentially terrestrial plants that can tolerate a degree of flooding.

This system is practical and covers most of the needs of botanists working in the aquatic environment. Within the hydatophytes are divisions and subdivisions which separate plants attached to rocks *etc. (Podostemophytes)* from totally submerged plants *(Euhydatophytes)* and those with vegetative parts coming into contact with the air *(Hydatoaerophytes)*. This last category is subdivided into surface-floating and bottom rooted species, but as these groups have very different growth habits and occupy distinct ecological niches, they should be in entirely independent categories.

In Britain, Clapham, Tutin and Warburg (1962) loosely interpreted helophytes as swamp and marsh plants and hydrophytes as waterplants. This can lead to confusion, especially with regard to surface-floating plants. Spence (1964) preferred to separate swamps from marshes (swamps being wetter) and confined helophytes largely to marshes. He grouped swamp and open water plants together under hydrophytes and included emergent and non-rooting (surface-floating?) plants, as well as floating-leaved and submerged plants. This is very different from Raunkiaer's original conception and goes further than Braun-Blanquet by taking into consideration the swamp-marsh plant anomalies. Hydrophytes so defined would be approximately equivalent to Hejný's combined hydatophytes and tenagophytes and would have the merit of being close to the literal Greek translation: *waterplant*.

Ambiguity can thus arise from apparently obvious terms used to describe aquatic plants. To a degree this is understandable. Some biologists take an ecological viewpoint and wish to use terms which categorise plants into communities. Others have a greater interest in the comparative biology of various life-forms. It so happens that certain life-forms can predominate in particular communities (emergent plants in swamps, for example), but the association of a life-form with a specific community is not rigid. Hutchinson (1975, pp. 118–123) reviews the multiplicity of terms that have been used to overcome this vexing problem, but I suspect he may only have achieved greater confusion.

1.2.1 *Definitions*

Although there is the obvious danger of proliferating the confusion by suggesting yet another system, the classification adopted in this book is both simple and convenient. Terms are kept to a minimum and are described briefly below. Fuller descriptions may be found in the appropriate chapters and a summary appears in Fig. 1.1.

1.2.1.1 *Hydrophyte*

Waterplant (*sensu* Spence, 1964). To be used in the literal and widest sense to describe those plants which grow in water or are found in predominantly wet places. Clearly the limit of the range towards terrestrial plants cannot be exact. Swamp and marsh plants are hydrophytes, and one can also include those of the floodplains as they are flood-tolerant species occurring in seasonal wetlands.

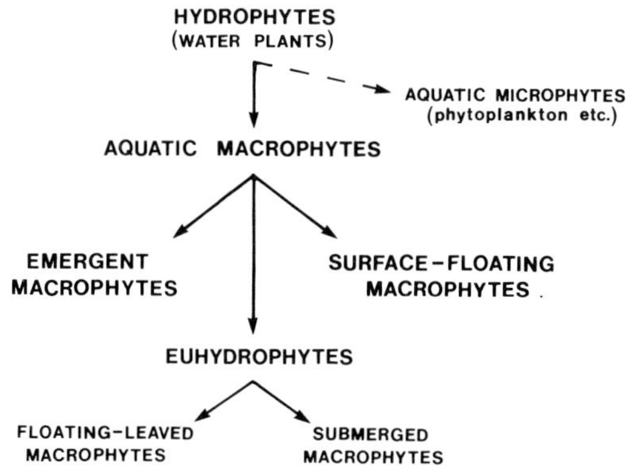

Fig. 1.1. A flow diagram showing the relationship of the terms for waterplants used in this book.

1.2.1.2 *Freshwater macrophyte*

The definition in Cook *et al.* (1974), with slight modification to emphasise the inclusion of emergent plants, can be used. The term comprises all charophytes (stoneworts), bryophytes (mosses and liverworts), pteridophytes (ferns and fern allies), and spermatophytes (seed-bearing plants) with vegetative parts that are permanently or seasonally submerged in or emerge from freshwater, or float on the water surface. The broader term, *aquatic macrophyte*, includes marine and brackish water plants.

1.2.1.3 *Emergent macrophyte* (Plate 1.1)

These aquatic macrophytes (Chapter 3) are mainly rooted or anchored in the substratum and produce emergent vegetative shoots. Emergent plants predominate in many swamps and marshes and often form a fringe of vegetation around lake shores and along rivers. The edges of the swamp can break away and form rafts or islands of floating vegetation. Typical emergent plants include *Cyperus papyrus* (papyrus), *Phragmites* (reed), *Typha* (cattail), *Equisetum* and *Marsilea*. Associated swamp plants such as *Ipomoea aquatica* and *Cayratia ibuenis* are often included, as are grasses from the seasonally flooded plains.

1.2.1.4 *Surface-floating macrophytes*
These aquatic macrophytes (Chapter 4) normally float on the water surface. Plants particularly adapted to the surface-floating habit are termed *obligate acropleustophytes* (Mitchell, Chapter 4) to separate them from species which are not so restricted. The location of surface-floating macrophytes is largely dependent upon wind and water movement, and they can become densely packed to form extensive rafts of vegetation. The more modest components of the group include members of the Lemnaceae (duckweeds) such as *Lemna minor*, *Spirodela*, and *Wolffia*, but in the tropics and sub-tropics obnoxious weeds such as *Eichhornia crassipes, Salvinia, Pistia* and *Azolla* come most readily to mind. Floating islands or rafts of emergent plants are not included. (Plate 4.4).

1.2.1.5 *Floating-leaved macrophytes*
These aquatic macrophytes (Chapter 2) are rooted or anchored in the substratum and produce coriaceous leaves which float on the surface of the water. Some submerged leaves are often present, and flowers or inflorescences usually emerge; examples include members of the Nymphaeaceae (water-lilies) and some potamogetons, *e.g. Potamogeton thunbergii*. (Plate 1.3).

1.2.1.6 *Submerged macrophytes*
The term is synonymous with the term *submersed* plants and is defined as aquatic macrophytes with vegetative parts below the surface of the water at all times, although flowers and inflorescences may emerge (Chapter 2). They are normally rooted in, or anchored to, the substratum for a large part of their life cycle, although, at certain times of the year, shoots may fragment and float away to form new plants. A few species included here, such as *Lemna trisulca* and some other members of the Lemnaceae, remain in suspension below the water surface. Submerged aquatic macrophytes are an assorted group of plants from major taxonomic groups including the Characeae, aquatic mosses and liverworts, and various angiosperms such as *Hydrilla, Ceratophyllum, Vallisneria, Najas* and *Potamogeton pectinatus*.

Plants such as *Trapa natans* (water chestnut), *Callitriche*, and *Potamogeton octandrus* have floating-leaves and submerged leaves. Into which category do we put these? A general term for these polymorphs, and the floating-leaved and submerged life-forms, would be a great help. They frequently occur together in plant communities and whilst one can talk conveniently of surface-floating or emergent vegetation, there is no term to cover the plants of the floating-leaved/submerged community. To rectify this ommission, the term *euhydrophyte* is proposed and is defined as follows:

1.2.1.7 *Euhydrophyte*
(*Gr. Eu* = true or well; *hydrophyte* = waterplant). Floating-leaved and submerged aquatic macrophytes. The plants are completely submerged or anchored to the substratum with floating leaves, or have both floating and submerged leaves. Flowers and inflorescences of angiosperms may emerge above the water surface. Euhydrophytes are commonly found lakeward of the emergent plant community and occupy the body of water in the euphotic zone not taken over by emergent plants or covered by the more permanent surface-floating vegetation. Typical examples include: *Chara, Fontinalis, Nymphaea, Ceratophyllum and Vallisneria*. Free-floating submerged plants such as *Lemna trisulca* and *Wolffiopsis* are included as well.

In this book the terms will be used in accordance with the definitions given above, and the text is sub-divided along the lines: emergent macrophytes, surface-floating macrophytes and euhydrophytes. Of course, there will be some plants that still defy satisfactory classification!

1.3 Wetland vegetation

1.3.1 *Zonation*

The aquatic macrophyte communities of lakes and rivers are part of a vegetation continuum from land

Plate 1.1. Kabanyolo Swamp, one of the numerous papyrus valley swamps in southern Uganda formed as a result of tectonic tilting. Most lie within the Lake Victoria forest belt and the *Raphia* palm (above) is an indicator of this. (Photo: K. Thompson)

to below water, down to the depth at which no further plant growth is possible. Frequently, there are distinct zones of vegetation along the continuum, each dominated by a different plant life-form. The transition from dry land to the water's edge is often diffuse. It may be occupied by an extensive floodplain with flood-tolerant species grading into a permanent swamp of emergent plants; or there may be a clear demarcation between dry land and swamp. Towards deeper water, the emergent plant zone is successively replaced by other zones dominated by floating-leaved and submerged plants. Within these zones and further out in the lake, surface-floating plants may drift about.

An idealised plant zonation, published by Van der Ben (1959), for the vegetation of Lake Kivu, provides an excellent example. On dry land elephant grass, *Pennisetum purpureum*, is abundant. At the edge of the water the emergent plant zone is occupied by dense stands of *Phragmites mauritianus*. Creepers (*Vigna luteola* and *Ipomoea cairica*) grow over the wave-washed zone between the *Pennisetum* and *Phragmites*. In the water below a metre the *Phragmites* is replaced by a mixed vegetation, the upper zone of which is dominated by *Scirpus subulatus* and the lower, by the aquatic, rhizomatous grass, *Paspalidium geminatum*. Beds of floating-leaved, rooted water-lilies (*Nymphaea calintha* and *N. mildbraedii*) also occur in the mixed zone, and in the shelter of their leaves submerged

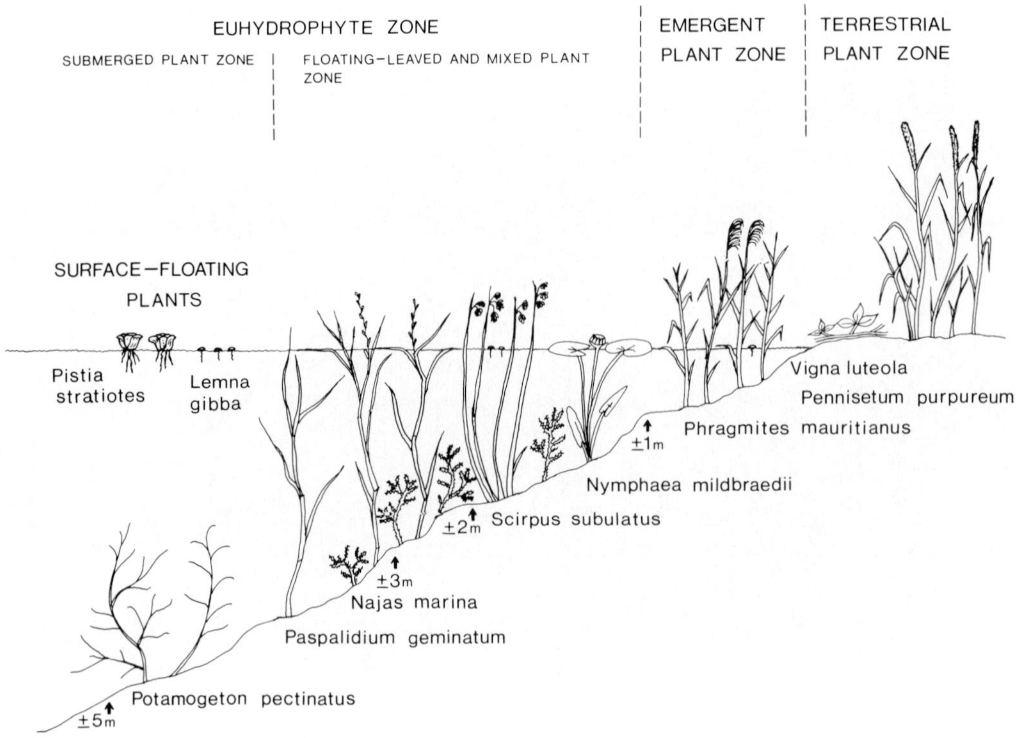

Fig. 1.2. Schematic representation of a typical wetland zonation of vegetation. (Redrawn and modified from Van der Ben, 1959, Fig. 21.)

plants such as *Najas marina* subsp. *armata* can be found. Below 3 metres the vegetation is dominated by submerged species such as *Potamogeton pectinatus* and *Najas marina*. The *Najas* appears unable to flourish in depths greater than about 3.5 m and *P. pectinatus* therefore survives as the dominant species to approximately 5 m where colonisation ceases. This is diagramatically represented in Fig. 1.2. Other examples can be found in later chapters.

1.3.2 *Succession*

The continuum is a *Hydrosere*, and the observed zones may indicate plant succession. The classic concept of the hydrosere as a succession is described by Tansley, (1965). He based his text largely on the studies of Pearsall (1917, 1920) in the English Lake District, and put forward the view that the submerged plant zone is a pioneer colonisation stage. Colonisation of the substrate by such plants, and the resultant reduction in water turbulance, encourages sedimentation, so in time the water becomes shallower. This allows submerged plants to be succeeded by floating-leaved plants which cannot survive in the deeper water. The succession continues: floating-leaved plants are replaced by swamp emergents, and finally by terrestrial plants. Thus, a zonation of vegetation may be a horizontal representation of succession.

This rather simple concept of succession, however, is not now so widely supported. Walker (1970) points out that Tansley's examples are restricted to autogenic changes within lakes and do not represent the overall aquatic environment. Walker shows that not only are successions in a hydrosere predominantly the result of allochthonous deposition (*i.e.* the deposition of substances introduced from outside the system), but the classic succession from submerged → floating-leaved → emergent plants, although common, is by no means universal. The actual sequence of change of vegeta-

Plate 1.2. The ground flora of papyrus swamp is typically poor: the canopy is dense and dead culms accumulate below it. Although shade-tolerant, ferns (above) are still more common in swamp fringes (Plate 3.5) and on nutrient-poor sites where papyrus is stunted (above). Floodplain swamp, Ankoro, Zaire. (Photo: K. Thompson)

Plate 1.3. Where the emergent canopy is broken other communities develop in the swamp. Above, a pool in a *Typha* swamp is dominated by the waterlily *Mymphaea alba.* Okavango Swamps, Botswana. (Photo: K. Thompson)

tion depends upon many factors including local topography and water balance. Van der Valk (1981) has proposed a model which may make the prediction of the vegetation sequence in an allogenic succession more reliable. He reiterates that changes in floristic composition result from a multiplicity of factors including those mentioned above, and, in addition, biological pressures such as the destruction of plants by pathogens, herbivores or man, plant interaction, and invasion by new species. Changes will also depend upon the life histories of the species involved, their life-spans, their propagule longevity and their propagule establishment requirements. If this information can be obtained, then the model may predict likely changes. This would be of especial importance in the management of newly-flooded (or re-flooded) areas when effects of plant colonisation and weed development need to be anticipated.

Successions from largely allogenic processes result from organic and inorganic deposition. The two processes are not mutually exclusive, but inorganic deposition is more common. Allogenic organic deposition is mainly associated with organic input into the system from surrounding grasslands and forests as well as from beds of vegetation upstream. Allogenic inorganic sedimentation is a particular feature of deltas. Suspended silts carried by a river are deposited at the entrance to a lake where the water flow is reduced. Vegetation reduces water movement further and the finest particles can also settle (see Chapter 5). Similar processes take place on lake shores, in the floodplain zones, and along rivers where areas of erosion and sedimentation proliferate. Colonisation by plants, and a succession of vegetation, can quickly develop where sedimentation occurs.

Autogenic organic deposition is associated with a rapid accumulation of detritus in conditions unfavourable for decomposition. In the warm tropics these conditions arise in waters and swamps which support a great biomass of vegetation. If there is a lack of water circulation anaerobic conditions prevail, aerobic decomposition is inhibited by oxygen depletion and partially decomposed or undecomposed litter accumulates. A classic example of this is Tasek Bera, a tropical freshwater peat swamp in

Plate 1.4. Trapa natans showing floating leaves (the small mass of leaves) and submerged leaves. The large leaves are those of waterlilies. (Photo: K. Thompson)

the Malaysian Peninsula (Furtado and Mori, 1982). In Africa, a raft of vegetation composed of, say, *Cyperus papyrus*, may totally deoxygenate water many metres below and detritus from the bottom of the raft will perpetually 'rain down' through the water column to accumulate on the lake floor. (This phenomenon is considered in detail in Chapter 6.) In the shallower waters the litter from euhydrophytes and emergent swamp vegetation accumulates *in situ*, new plants develop, and ultimately the vegetation may become raised on a bed of peat above the water level.

Atypical sequences of plants occur in successions associated with newly formed man-made lakes. Initially, plants may grow so rapidly that a weed problem results (see Chapter 7), but the weeds are generally composed of either surface-floating or emergent plants. Rarely do we hear of submerged weed problems. The reasons are manifold (Chapter 7): surface-floating plants, especially *Salvinia, Eichhornia* and *Pistia,* may be carried down river into a lake where they become established; emergent plants, such as *Typha,* produce large quantities of viable, wind-borne seeds; and inundated swamp vegetation provides a rich source of propagules for freshly flooded, shallow areas. The numbers of euhydrophyte propagules, on the other hand, are relatively fewer and, as the turbidity of

Plate 1.5. Variety of aquatic plants ranging from rooted, with floating leaves (*Nymphaea*), and surface-floating (*Pistia*), to emergents such as *Vossia, Cyperus papyrus* (background), *Typha* (centre) and *Limnophyton* (foreground, right). Part of convoluted shore of the shallow Lake Kioga, Uganda. (Photo: K. Thompson)

the water initially can be high (due largely to algal blooms), they have to compete with surface-floating and emergent vegetation for light and space. Large changes in water level may make colonisation within the euhydrophyte zone impossible. Where the water is clearer and less prone to fluctuations, euhydrophytes will colonise swamp pools, and will establish themselves beyond the emergent plant zone. Nyumba ya Mungu reservoir may be cited as an example of this type of succession (Welsh and Denny, 1978). Eight years after inundation of the Pangani River valley, the vegetation of the reservoir was dominated by emergent plants. At the inflow, a very extensive *Typha* swamp developed, and euhydrophytes were restricted to channels in the swamp and around its fringe. Surface-floating weeds, fortunately, were absent from the rivers upstream. The shores of the reservoir were fringed with extensive beds of emergent plants, but euhydrophytes were absent, even

Plate. 1.6. Boro River, Okavango Swamps. Note vegetation differences on opposite sides of curve: Deep-rooted *Phragmites* and *Miscanthidium* dominate outer fringe, where levée is higher; *Ficus* and papyrus are adapted to more permanent water on inside curve. (Photo: K. Thompson)

though conditions appeared suitable. This is illustrated in Fig. 1.3.

There are many notorious examples where surface-floating vegetation has been a rapid coloniser of man-made lakes: the *Salvinia* invasion of Lake Kariba, and the *Eichhornia* infestation of Hartebeespoort Dam in South Africa, to mention just two.

1.3.2.1 *Rates of succession*

Rates of succession are regulated by climatic and environmental factors such as temperature, water balance, nutrient availability, and rates of sedimentation. The succession can be held at a subclimax indefinitely, or it can be reversed if one or another of these factors so determines. A succession may take many hundreds of years to evolve, and examination of soil profiles below swamps, lakes and rivers can provide much valuable information about past history. In the profile the surface soil or peat is the most recent, and successive layers below the surface represent deposition from suc-

Plate. 1.7. A seasonal swamp in the Okavango Delta. Stable zones of *Phragmites* (background) *Typha* (centre) and SICA – *Scirpus inclinatus* (stem tips in foreground) / *Cyperus articulatus* – evolve in response to complete annual flood / drought régimes. *Typha* and SICA both also have greater drought tolerance than *P. australis*. See Chapter 3. (Photo: K. Thompson)

cessively earlier times. As deposits are laid down or peat develops, plant material is trapped and some, especially pollen grains, remains undecomposed. The different layers can be dated using isotope techniques, and a vertical profile through the strata provides a botanical palimpsest. As pollen grain wall structure is frequently characteristic for particular specis or genera, a picture of changes in vegetation through successive climatic changes can be built up. A good example of this is given by Morrison (1968) for Muchoya Swamp, Uganda, an upland swamp (*alt.* 2256 m) just west of Lake Bunyonyi, dominated by large tussocks of the sedge *Pycreus*, together with various swamp herbs. Towards the swamp edge and in many small bays the robust swamp herb, *Gunnera perpensa*, predominates. Morrison took a sediment core from the centre of the swamp and constructed a vertical profile for pollen from the different layers. The

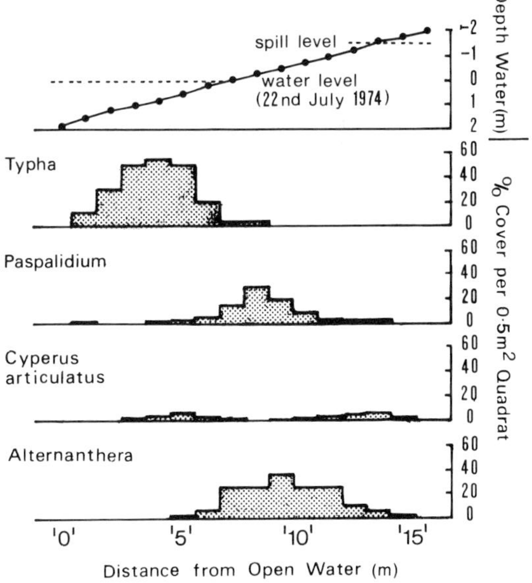

Fig. 1.3. A transect through a bed of vegetation that has developed in the recently formed man-made lake. Nyumba ya Mungu, Tanzania. Note the absence of euhydrophytes in the zonation. (After Welsh and Denny, 1978, transect 2.)

Plate 1.8. Scirpus littoralis colonising open water in Lake Chilwa. (Photo: C. Howard-Williams)

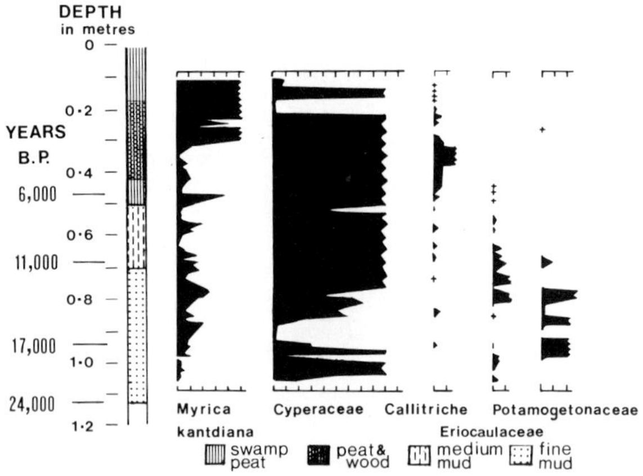

Fig. 1.4. Pollen diagram vertical profile through a sediment core taken from the centre of Muchoya Swamp, south-western Uganda, showing the development of swamp vegetation from a former lake. Only the hydrophytes have been selected from Morrison's pollen data to show plant succession from submerged plants, to floating-leaved and, finally, to emergent plants. The pollen total upon which the percentages are based includes trees, herbs and grasses. (Redrawn from Morrison, 1968, Fig. 5.)

sediments at the bottom of the profile were composed of fine lake muds containing substantial amounts of euhydrophyte pollen from the Potamogetonaceae. In the overlying, coarser muds the *Potamogeton* pollen had been replaced largely by pollen from swamp emergents of the Cyperaceae and Eriocaulaceae families. The coarse muds gave way to an upper peat layer containing plant remains and pollen of swamp shrubs, especially *Myrica*. The data clearly show that Muchoya valley began as a shallow lake which became filled in and supported a well-developed *Myrica kantdiana* – dominated swamp forest. The swamp forest finally degenerated into the *Pycreus* swamp of the present day, the whole process taking about 24 000 years (Fig. 1.4).

Plate 1.9. Polygonum senegalense invading a rotting mat of *Salvinia* on the shore of Lake Kariba following a fall in waterlevel during the 1983–1985 drought. (Photo: P. Denny)

1.4 Methods of vegetation analysis

1.4.1 *Classical (descriptive) methods*

There are many ways in which aquatic vegetation can be studied and analysed. The traditional methods, such as those employed by Pearsall (1917), Tansley (1965) and Spence (1964), tend to divide vegetation into units which are then considered separately. The initial division is often according to habitat or plant life-form and, as shown in the previous section, a plethora of descriptive terms has evolved. Terms such as *Marsh, Fen, Carr,* and *Meadow* are conventionally classified according to particular types of plant associations from temperate climate vegetation and are not normally used for tropical environments. The term *marsh* was used by Howard-Williams and Walker (1974) in Malawi for vegetation dominated by sedges on waterlogged soils (the water levels never being much above the soil surface), but the discrimination between marsh and seasonal swamp is indistinct. In tropical and sub-tropical Africa, the continuum can be separated conveniently into *Floodplain, Swamp, Euhydrophyte* and *Surface-floating* vegetation. (This book is mainly concerned with the wetter end of the hydrosere whilst the floodplains, *i.e.* the seasonally flooded grasslands around lakes and rivers, are well discussed by Welcomme, 1979.) Swamps are either permanently flooded areas, or remain flooded for most of the year, and are dominated by emergent plant life-forms, generally greater than a metre high. They may be distinguished from *bogs* which tend to have *short* (<1 metre) herbaceous vegetation. Fragments of swamp vegetation can break away to form floating islands and, conversely, surface-floating plants can conglomerate to form impenetrable swamp. (Per-

Plate 1.10. A mature stand of *Polygonum,* 4 m tall, further back in the succession on the shore of Lake Kariba. The level of the lake had fallen by several metres due to drought. (Photo: P. Denny)

haps the most notable example of this habit is the Sudd vegetation of the Nile.) Euhydrophyte vegetation is truly aquatic and consists of submerged and floating-leaved life-forms. Plants can survive drying out of the habitat through specialised structures such as fruits or seeds, but few of these can tolerate long periods of desiccation. Surface-floating vegetation is unaffected by water depth but is at the mercy of the wind, and is generally unstable.

The basic workable unit is the plant community; this is defined as a group of plants which normally grow together in a particular habitat and form a recognisable type of vegetation. Communities are identified by dominant species or by groups of species which form discrete associations. For example, swamp vegetation may include communities dominated by *Typha, Cyperus papyrus* or *Phragmites*. Also, some swamps are typically *Cyperus papyrus:Phragmites* associations; others may be *Cyperus papyrus:Typha* associations. Each community has its own environmental and floristic character. What constitutes a community is somewhat arbitrary, and ecologists will vary in their choice of unit size, but this does not really matter as use of the term is merely one of convenience. The survey by Lind and Visser (1962) of Namiro Swamp at the north-west end of Lake Victoria is one such example. The authors made a belt transect from land to water across a corner of the swamp (see Chapter 2). In so doing, they recognised six vegetational zones which they treated as separate communities, *viz:*

(a) A shrubby area dominated by *Kotschya africana*;
(b) *Miscanthidium violaceum:Loudeia phragmatoides* zone;
(c) *Miscanthidium* as dominant; with other grasses;
(d) *Miscanthidium:Cyperus papyrus* co-dominants;
(e) *Cyperus papyrus* dominant;
(f) Floating rhizomes of *Cyperus papyrus*, with euhydrophytes.

Lists of species and habitat factors were recorded from each zone. By comparing habitats and vegetation zones, the authors provided a description of each community and indicated possible limiting environmental factors for each zone.

1.4.2 *Phytosociological methods*

An alternative approach to the description of vegetation is provided by phytosociological analyses for which Braun-Blanquet and the Montpellier school are the greatest exponents (Braun-Blanquet, 1932). The Braun-Blanquet system has been adopted largely by workers from Europe. For those with a limited knowledge of the German language, the reviews and assessments by Poore (1955*a, b, c*) and Moore (1962) are admirable. Up to date and comprehensive accounts are given by Westhoff and Van der Maarel (1973) and Mueller-Dombois and Ellenberg (1974). Basically, the Braun-Blanquet system consists of a hierarchical classification and nomenclature of plant communities. The fundamental unit of vegetation is the *Association*, which is defined on floristic composition, although environmental characteristics may influence ultimate categories. This contrasts with the British/American system which categorises vegetation on habitat and floral composition equally. Each association is characterised by a few species of high *'fidelity'* to that particular association, *i.e.* the complete or partial (local) restriction of certain species to one particular association. Associations are biased towards *'faithful'* or *'indicator'* species which may be low in dominance, whereas the British/American system puts a greater emphasis on species dominance within a vegetation unit, though these differences may be more apparent than real (Moore, 1962). Associations are arranged in an ascending hierarchy of *Alliances, Orders, Classes etc.* based on their floristic composition and in a similar way to the taxonomic classification of plants into Species, Genera, Families, *etc.* The hierarchical classification makes no claim to be natural: it is merely a convenience (Moore, 1962). The method by which fidelities of species can be assessed, and associations built up, are quantitative. Uniform sites of vegetation are chosen, and within the sites, equivalent areas (stands) of homogeneous vegetation are selected for scrutiny. Every species present in the stand is recorded and is given an index of *cover-abundance* and *sociability* (see Poore, 1955*a*, p. 236). Association tables are then constructed from lists of stands which have similar species composition. A study of these lists enables an association to be described. Poore states: 'These tables are never final; for, as additional lists are made or as more detailed ecological information is accumulated, it may be found necessary to divide the tables of what was originally one association into two, or to combine two which were originally thought separate.' Ultimately, it is hoped that, if the Braun-Blanquet system is widely accepted, there can be a world-wide classification of plant communities and, as nomenclature is standardised, comparisons of vegetation from various parts of the world can be made. It is of interest to note that, at the 1979 SIL-UNEP Workshop on African Limnology held in Nairobi, it was recommended that the Braun-Blanquet system should be adopted generally for African aquatic vegetation (Symoens, Burgis and Gaudet, 1981).

Particularly noteworthy phytosociological studies in Africa are those carried out in the former Belgian Congo which was inspired by Lebrun (1947) and developed by Léonard (1950), Van der Ben (1959), and Van Meel (1952) on the African Great Lakes bordering that territory. The aquatic vegetation of the area belong to the Order, *Nymphaeetalia Loti* Lebrun, (1947), (*i.e.* the vegetation of tropical and sub-tropical freshwaters in the Old World), and the Alliance, *Nymphaeion Loti*. In the alliance, 4 associations were recognised: (a) *Lemneto-Pistietum*; (b) *Utricularieto-Nymphaeetum*; (c) *Eichhornieto-Ranalismetum*; (d) *Nymphaeetum Afro-orientale*.

The amphibious association (roughly equivalent to floating swamp) was the *Jussieueto-Enhydretum*. Each association was carefully described, together with environmental characteristics such as exposure and water depth, and associations were divided into smaller groups for detailed analysis. By this means a comprehensive picture of the aquatic vegetation was built up. Lock (1973), in his study of the vegetation of Lake George, integrated the British/American ecological approach with the Braun-Blanquet phytosociological approach by first describing the vegetation communities and then comparing them with the approximately equivalent phytosociological alliances and associations.

1.4.3 *Multivariate analytical methods*

Other methods of vegetation analysis which have gained favour over the last twenty years or so include multivariate classification techniques and ordination of plant communities. Both require quantitative data on floral composition of a sufficient number of stands with homogeneous vegetation. As a minimum requirement, data must provide reliable information on presence and frequency of species within sample areas. By acceptable mathematical manipulation of the data, the importance of species, the association of species with each other within sample plots and the relationships between species and habitats can be statistically analysed. In multivariate classificatory analyses, stands of similar floristic structure can be grouped together, using cluster analyses based on inter-stand correlation coefficients. The construction of dendrograms gives a picture of the clustering of vegetationally similar stands into groups, and the degree of similarity between groups can be seen. Vegetation is thus classified according to the similarity of sample stands; environmental patterns may then be sought.

This method of vegetation analysis was used by Howard-Williams and Walker (1974) to study the vegetation of Lake Chilwa, Malawi. To classify the vegetation, the authors obtained data on species' presence and frequency from 73 stands selected randomly within the lake area. A cluster analysis dendrogram of the stands resulted in the identification of a major dichotomy: Division A – those stands in shallow water which dried out for much of the year, and Division B – those stands which were flooded for all or most of the year. Within the two Divisions, at a higher clustering level, 7 vegetation types were apparent (see Fig. 1.5). The flora and habitat of each vegetation type was of distinctive character, and each could be fully described and classified.

We have already established that aquatic and wetland areas comprise a hydrosere which is part of a vegetation continuum. In order to determine the nature and relative importance of environmental gradients within the continuum, quantitative data from stands must be arranged in a way for trends to

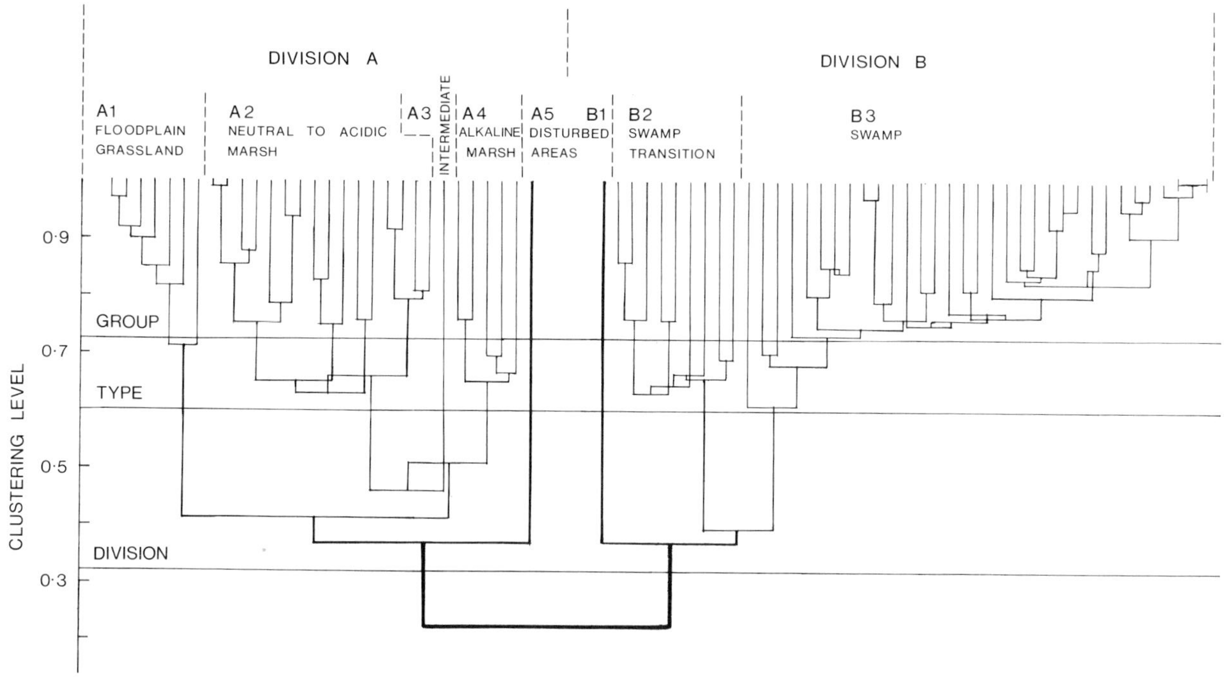

Fig. 1.5. A dendrogram constructed from cluster analysis of 73 stands of homogeneous vegetation from Lake Chilwa, Malawi. In the hierarchy, note the first dichotomy into: (A) areas which are dry for much of the year and (B) permanently wet areas. At a higher level of separation, 7 types of vegetation are identified. The ultimate level of separation is into individual stands. (Redrawn from Howard-Williams and Walker, 1974, Fig. 3.)

be highlighted. Of course, species:species and species:habitat relationships are complex and multivariate. The art is to display the data in a few dimensions, rather than in the numerous dimensions which actually exist in a plant community, and yet represent the original data as fairly as possible. In other words, species or stand data need to be ordinated along major axes. By superimposing environmental information on these axes, environmental gradients, which impose structure upon a community, may become apparent. Howard-Williams and Walker (1974) subjected their data to a principle components analysis. The first five components accounted for about 70% of total variance, and the authors ordered species on the first three of these. Stand ordinations derived from the species data were arranged on two axes, and environmental data were superimposed on the ordinations. Ordination of stands on the first axis showed a close link with the water table and with soil conductivity. The conductivity relationship may indicate that soil salinity exerts some control on the vegetation type. The second axis showed a relationship with soil pH, with high pH favouring a *Typha domingensis* swamp.

1.5 Conclusion

In this chapter an attempt has been made to provide a background for vegetation studies described in detail later. It must be remembered that information is lacking from large parts of Africa; and in other areas, because of the difficulty of the terrain, the information gleaned is of a very basic nature. Workers of many different interests and disciplines have contributed to our knowledge, but they have used methods most suitable for their own needs; it is not easy to make comparisons from such varied techniques. In the following chapters these limitations may well become apparent, but it is hoped they will not detract from the intrinsic value of this attempt to describe African aquatic and wetland vegetation.

1.6 References

Braun-Blanquet, J., 1932 English translation by G.D. Fuller, and H.S. Conard, 1965. Plant Sociology. Hafner Publishing Company, New York and London.

Clapham, A.R., Tutin, T.G. & Warburg, E.F., 1962 (2nd edit.). Flora of the British Isles. Cambridge University Press.

Cook, C.D.K., 1968. The vegetation of the Kainji Reservoir site in Northern Nigeria. Vegatatio. 15:225–243.

Cook, C.D.K., Gut, B.J., Rix, E.M., Schneller, J. & Seitz, M., 1974. Water plants of the world. Dr. W. Junk, The Hague.

Den Hartog, C. & Segal, S., 1964. A new classification of the water-plant communities. Acta Botanica Neerlandica 13: 367–393.

Denny, P., 1973. Lakes of South-western Uganda ii: Vegetation studies on Lake Bunyonyi. Freshwater Biology 3:123–135.

Furtado, J.I. & Mori, S. (eds.), 1982. Tasek Bera. The ecology of a freshwater swamp. Dr. W. Junk, The Hague.

Hejný, S., 1960. Okologische Characteristik der Wasser- und Sumpfpflanzen in den slowakischen Tiefebenen (Donau- und Theissgebiet). Bratislava.

Howard-Williams, C. & Walker, B.H., 1974. The vegetation of a tropical African lake: classification and ordination of the vegetation of Lake Chilwa (Malawi). Journal of Ecology 62:831–854.

Hutchinson, G.E., 1975. A treatise on Limnology. Volume III, Limnological Botany. John Wiley and Sons, New York, London, Sydney, Toronto.

Lebrun, J., 1947. La végétation de la plaine alluviale au sud du Lac Edouard. Institut des Parcs Nationale du Congo Belge. Bruxelles. 2 volumes.

Léonard, J., 1950. Botanique du Congo Belge. Cahiers d'Office de la Recherche Scientific et Technique Outre-Mer, 1:345–389.

Lind, E.M. & Visser, S.A., 1962. A study of a swamp at the north end of Lake Victoria. Journal of Ecology 50:599–613.

Lock, J.M., 1973. The aquatic vegetation of Lake George, Uganda. Phytocoenologia 1:250–262.

Moore, J.J., 1962. The Braun-Blanquet System: A reassessment. Journal of Ecology 50:761–769.

Morrison, M.E.S., 1968. Vegetation and climate in the uplands of South Western Uganda during the later Pleistocene period. Journal of Ecology 56:363–384.

Mueller-Dombois, D. & Ellenberg, H., 1974. Aims and methods in vegetation ecology. John Wiley and Sons, New York.

Pearsall, W.H., 1917. The aquatic and marsh vegetation of Esthwaite water. Journal of Ecology. 5:180–202.

Pearsall, W.H., 1920. The aquatic vegetation of the English Lakes. Journal of Ecology. 8:163–201.

Poore, M.E.D., 1955*a*. The use of phytosociological methods in ecological investigations. I: The Braun-Blanquet system. Journal of Ecology 43:226–244.

Poore, M.E.D., 1955*b*. The use of phytosociological methods in ecological investigations. II: Practical issues involved in an attempt to apply the Braun-Blanquet system. Journal of Ecology 43:245–269.

Poore, M.E.D., 1955*c*. The use of phytosociological methods in ecological investigations. III: Practical application. Journal of Ecology 43:606–651.

Raunkiaer, C., 1937. (Translated by Gilbert-Carter, H.) Plant Life Forms. Clarendon Press, Oxford.

Spence, D.H.M., 1964. The macrophytic vegetation of lochs, swamps and associated fens. In: J.H. Burnett, (ed.) The vegetation of Scotland. Oliver and Boyd, Edinburgh: 306–425.

Symoens, J.J., Burgis, M., & Gaudet, J.J., (eds.) 1981. The ecology and utilization of African inland waters. UNEP Report and Proceedings, Series 1. Nairobi.

Tansley, A.G., 1965 (reprinted). The British Isles and their vegetation. Cambridge University Press. 2 volumes.

Van der Ben, D., 1959 IV. La végétation des rives des lacs Kivu, Edouard et Albert. In: Exploration hydrobiologique des lacs Kivu, Edouard et Albert, 1952–1954. Institut Royal des Sciences Naturelles de Belgique, Bruxelles.

Van der Valk, A.G., 1981. Succession in wetlands: a Gleasonian approach. Ecology 62:688–696.

Van Meel, L., 1952. Le Milieu Vegetal. In: Exploration Hydrobiologique du Lac Tanganika (1946–1947). Institut Royal des Sciences Naturelles de Belgique, Bruxelles: 51–68.

Walker, D., 1970. Direction and rate in some British postglacial hydroseres. In: Walker, D. & West, R.G. (eds.), Studies in the vegetational history of the British Isles. Cambridge University Press: 117–139.

Welcomme, R.L., 1979. Fisheries ecology of floodplain rivers. Longmans, London.

Welsh, R.P.H. & Denny, P., 1978. The vegetation of Nyumba ya Mungu reservoir, Tanzania. Biological Journal of the Linnean Society. 10:67–92.

Westhoff, V. & Van der Maarel, E., 1973. The Braun-Blanquet approach. In: Whittaker R.H., (ed.), Ordination and classification of communities. Handbook of vegetation science, Vol. 5. Dr. W. Junk, The Hague: 617–726.

Worthington, E.B., 1976. The conservation of wetlands in Africa. Proceedings of the Symposium on the Okavango Delta and its future utilization. Published by the Botswana Society, Gaborone: 61–66.

Author's address:
P. Denny
School of Biological Sciences
Queen Mary College, The University of London
London EI 4NS, England

CHAPTER 2

Submerged and floating-leaved aquatic macrophytes (euhydrophytes)

PATRICK DENNY

Detailed contents

Euhydrophytes, *i.e.* floating-leaved rooted, and submerged plants (see Chapter 1), rarely attract the same attention or have the same environmental impact in African waterways as emergent and surface-floating species. The problems with papyrus swamps in East Africa and the Sudd in the Sudan, and the water-weed menaces of *Pistia stratiotes, Eichhornia crassipes* and *Salvinia molesta* come immediately to mind (see Chapter 7). However, euhydrophytes are not without importance as they provide breeding and feeding grounds for fish, and valuable niches for many invertebrates (Chapter 5). The submerged plants, through their photosynthetic activity, release oxygen into the water for subsequent use by the aquatic fauna. On the detrimental side, they encourage the spread of water-borne diseases by supplying suitable habitats for invertebrates such as snails, mosquitoes, and *Simulium* larvae, all of which are intermediate hosts for some of the more serious parasites affecting man. Additionally, euhydrophytes can grow profusely, thereby impairing the free flow of water in canals and irrigation schemes, impeding navigation and restricting fishing.

Any discussion of African euhydrophytes is somewhat limited by the shortage of data on their ecology and distribution. Few geographical areas have been studied sufficiently, and floral lists are incomplete. In addition, there is often a lack of detail from vegetation surveys and insufficient environmental data to make meaningful comparisons between various bodies of water. In this chapter, the taxonomy and life-forms of euhydrophytes, their biogeography, and the major factors which may determine the presence or absence of a taxon in a flora, will be considered. Their biology and ecology will be explored from descriptive accounts of floras, whilst the dynamics and functioning of euhydrophyte vegetation will be considered in detail in Chapter 5. Apparent omissions and thinly presented arguments will, perhaps, provide an indicator and stimulus for further studies.

Patrick Denny (ed.), The ecology and management of African wetland vegetation.
 ISBN 90 6193 509 1

2.1 General biology

2.1.1 *Classification*

Euhydrophytes include all anchored or rooted aquatic macrophytes of various taxonomic groups which are either totally submerged (except for flowering shoots) or have their leaves floating on the water surface (see Table 2.1 and Chapter 1). Free-floating submerged plants such as *Ceratophyllum demersum* and *Lemna trisulca* are included in this group. A glance through Wild's (1961) and Cook *et al.*'s (1974) books shows that these plants belong to a wide variety of families which have one to several representatives. The most familiar floating-leaved plants are the waterlilies (Nymphaeaceae). The families Aponogetonaceae and members of the Potamogetonaceae are also very abundant, together with *Trapa*, of the monogeneric family, Trapaceae. The Hydrocharitaceae has one of the widest selections of common submerged genera in Africa, including *Ottelia, Vallisneria* and members of the tribe Hydrilleae (*Hydrilla, Egeria, Lagarosiphon*, and *Elodea*). The Lemnaceae and Lentibulariaceae are well represented, and the Podostemaceae has no fewer than fifteen endemic genera (Cook *et al.*, 1974). The potamogetons and the genera *Ceratophyllum* and *Najas* include relatively few species, but are extremely abundant and can be found in most bodies of water colonised by plants.

2.1.2 *Identification*

The identification of euhydrophytes is hindered by the absence of accurate drawings and descriptions of many species, a problem compounded by the variability and plasticity of morphological charac-

Table 2.1. Examples of the range of taxonomic groups and plant life-forms in the euhydrophytes.

Plant	Taxonomic group	Description
(1) Floating-leaved, rooted macrophytes		
Trapa natans	angiosperm	Submerged and coriaceous floating leaves. Straggly stem with 'water' roots developing from leaf scars and 'anchor' roots growing into the mud. Can become detached and free-floating.
Nymphaea caerulea	angiosperm	Rhizomatous plant firmly rooted into mud. Large coriaceous floating leaves, and, sometimes, delicate submerged leaves.
Potamogeton thunbergii	angiosperm	Rhizomatous plant rooted in mud. Floating leaves and occasional non-persistent submerged leaves.
Potamogeton octandrus	angiosperm	Rhizomatous plant rooted into mud. Coriaceous floating leaves and submerged leaves present.
P. × *bunyonyiensis*	angiosperm	Rhizomatous plant rooted into mud. Coriaceous floating leaves and submerged leaves. Phenotypes ranging from predominantly floating-leaved to predominantly submerged-leaved forms.
(2) Submerged macrophytes		
Chara sp.	alga	Anchored into mud by the high density of their cell contents.
Ricciocarpus natans	liverwort	Free-floating just below the water surface.
Fontinalis sp.	moss	Anchored onto substratum by rhizoids.
Isoetes sp.	pteridophyte	Anchored to substratum by numerous branched roots.
Lemna trisulca	angiosperm	Plant with delicate roots; free-floating below the surface.
Hydrilla verticillata	angiosperm	Anchored to the substratum by roots.
Vallisneria sp.	angiosperm	Anchored to the substratum by roots.
Potamogeton pectinatus	angiosperm	Anchored to the substratum by roots.
Ceratophyllum demersum	angiosperm	Rootless. Free-floating or anchored in mud.

Plate 2.1. A typical mixed euhydrophyte community with a floating-leaved potamogeton and submerged *Vallisneria* and *Lagarosiphon* in Lake Kariba. (Photo: P. Denny)

ters (Arber, 1920; Sculthorpe, 1967). Irradiation, water temperature, oxygen content, water depth and flow rate may all influence gross morphology and produce variable phenotypes, whilst hybridisation between species produces a diversity of polymorphs. The variation found within the *Potamogeton schweinfurthii* complex shows this well (Fig. 2.1). *Potamogeton schweinfurthii* A. Benn. exhibits considerable plasticity depending upon environment, the shape of both floating and submerged leaves being very variable. A totally submerged form has been shown to hybridise with *P. thunbergii* (a floating-leaved species) to produce F1 hybrids with a range of floating and submerged leaf shapes, some forms of which are indistinguishable from *P. schweinfurthii* (Denny and Lye, 1973). At present, the potamogetons in this complex are variously named: *P. schweinfurthii* A. Benn.; *P.* × *bunyonyiensis* Denny and Lye; and *P. lucens* L.

The Hydrocharitaceae, and especially members of the Hydrilleae, can also cause taxonomic difficulties, but the current revisions will largely overcome these (Lowden, 1982; Cook and Luond, 1982*a*, *b* and *c*, 1983). Cook *et al.*'s *Waterplants of the World* (1974) is an invaluable guide to the identification of most genera, and the species of a few families have been excellently described in regional floras, *e.g.* the Potamogetonaceae in *Flora d'Afrique Centrale* (Lisowski *et al.*, 1978). The *Flora of West Tropical Africa* (2nd. edition revised by Hepper) and the published parts of the *Flora of Tropical East Africa* are also useful. *The Flowering Plants of the Sudan* by Andrews (3 volumes, 1950–56) is of limited value, but the *Students' Flora of Egypt* by Tackholm (2nd. edition 1974) is a good companion. An introductory guide and key to waterplant genera can be found in *Flore et Faune aquatiques de l'Afrique Sahelo-Soudanienne*, by

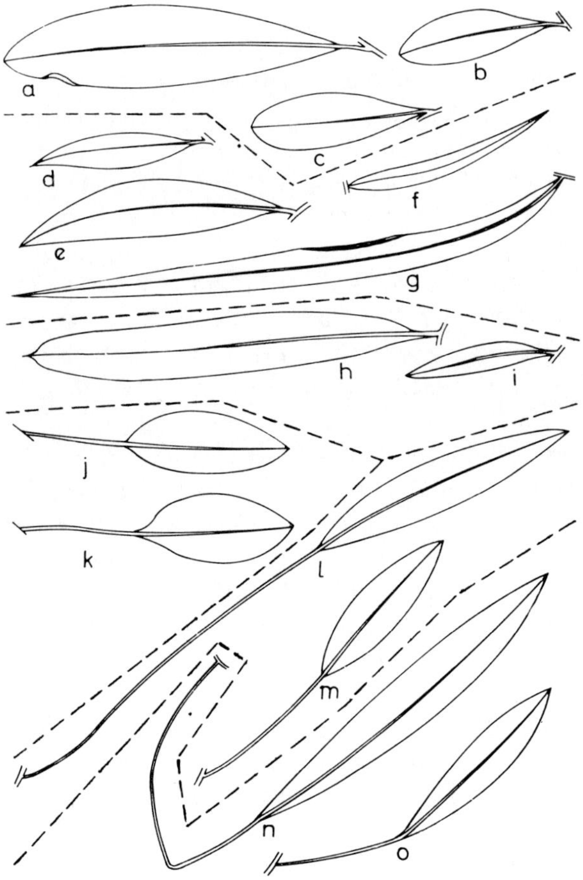

Fig. 2.1. Ranges of leaf morphology found within the *Potamogeton schweinfurthii* A. Benn. complex in Uganda. *a–c,* Broad, totally submerged leaves from the Lake Mutanda population (= *P. lucens*). *d–g,* leaves from the totally submerged population of Lake Bunyonyi grown in ponds at Makerere University, Kampala. *h–i,* leaves from the submerged population in Lake Bunyonyi. *j–k,* coriaceous floating leaves of *P. thunbergii* from Lake Mutanda. *l–m,* floating leaves of the hybrid, *P.* × *bunyonyiensis* (*P. thunbergii* × *P. schweinfurthii*) from Lake Bunyonyi. *n–o,* submerged and floating leaves respectively from other hybrid populations in Lake Bunyonyi. (From Denny and Lye, 1973, Fig. 1.)

Durand and Lévêque (1980). Descriptions of southern African species may be found in *Flora Capensis* or *Flora Zambeziaca*, while Musil (1973) has written a guide to the waterplants of Natal. Gibbs Russell (1975) has provided a valuable bibliography to taxonomic studies of aquatic macrophytes in southern Africa and, more recently (1977), has produced an excellent key to the aquatic plants of Rhodesia (now Zimbabwe). Until suitable keys and floras for all African waterplants are produced, accurate identification of many species has to remain the prerogative of experts.

2.1.3 *Propagation and dispersal*

The modes of propagation of plants are relevant to their geographical distribution. Plants have two possible mechanisms of reproduction: sexual reproduction with the formation of fruits and seeds, and asexual reproduction by vegetative propagules. This is discussed with reference to aquatic plants in Lind and Morrison (1974, pp. 120–122). Many aquatic angiosperms produce numerous viable seeds which provide the main mechanism of propagation and dispersal of the species. In *Nymphaea*, the absorption of water by the mucilage within the fruit capsule causes the pericarp to split and release clusters of seeds. The seeds have arils which trap air, and these act as float chambers to aid distribution. In Uganda I observed that seeds sank after a few days, germination being encouraged by anoxic conditions. Thus the reducing muds prevalent in the warm, shallow waters in and around swamps provide an ideal environment for their development. The dispersal of *Trapa* is curious; it is an annual, producing woody, spiny nuts which are full of starch and fat (they are widely collected for food in Asia and India). After floating for a short while, they sink into the mud where the seedlings become firmly attached to the substratum. The young plants grow rapidly; when the surface leaves have developed, the lower parts of the shoots decay and the plant becomes free-floating (Sculthorpe, 1967). Some of the potamogetons produce good viable fruits, but others apparently rely largely upon vegetative means for dispersal (Denny and Lye, 1973). *Potamogeton schweinfurthii* and *P. thunbergii* both produce viable fruits which float for long periods. Indeed, laboratory experiments have shown that ripe fruits of *P. schweinfurthii* remain green, and cells of the mesocarp continue to photosynthesise, for up to 18 months. The gas produced from this photosynthesis is thought to maintain the buoyancy of the fruits (Spence *et al.*, 1971). After they have sunk and the endocarps have rotted away, the embryos require a stimulation by light to trigger germination. Consequently, fruits that fall into deep, dark water, or settle amid the dense shade of vegetation have little chance of successful germination.

Amongst submerged life-forms vegetative propagation is the prevalent form of dispersal. Plants such as *Vallisneria spiralis* and *Potamogeton schweinfurthii* produce runners from which young plants can become detached and distributed by the water. *Ceratophyllum demersum*, *Hydrilla verticillata*, and similar growth forms have brittle stems which fragment at certain times of the year (usually after periods of rapid growth); the fragments float around for a while, and then they become established as new plants. Resting and dormant stages enable some plants to survive adverse conditions – for example, rhizomes in the Nymphaeaceae, tubers in the Alismataceae and some Potamogetonaceae (*P. pectinatus*), turions in other potamogetons (*P. panormitanus, P. trichoides*, and *P. crispus*) and turion-like structures in some of the Hydrilleae. The formation of dormant (overwintering) propagules are of particular significance to the survival of plants in the cooler regions prevalent at higher altitudes and greater latitudes in Africa. However, few are resistant to periods of desiccation; one of the main hazards to the continued existence of euhydrophytes in tropical regions.

Clearly, water is the main agent for the dispersal of fruits, seeds and asexual propagules through wind action at the water surface and through water flow, but now, the activity of man is increasingly more significant (see Chapter 7). Although many of the species are endemic, others are introduced and some, such as *Elodea* from the New World, have become weeds.

The influence of the various Palaearctic-African bird migrations on the distribution of species is interesting. Several authors think that waterfowl may be important long-range agents (Good, 1974; El-Hadidi, 1965). Many Palaearctic water-birds overwinter in the northern tropics in a zone roughly between 10° and 15°N (Moreau, 1972). Large expanses of open water, such as Lake Chad, are little frequented, but the shore areas, and areas which become seasonally flooded during the summer rains, attract large numbers of migrants, especially in western Africa. In Senegal, these areas are invaded by hundreds of thousands of water-birds every autumn – dominated by Garganey ducks (*Anas querquedula*), Ruffs (*Philomachus pugnax*) and Black-tailed Godwits (*Limosa limosa*). In the great floodplains of the upper Niger, in the floodplains south of Lake Chad, and, to a lesser extent, in the Sudan, similar patterns occur. The Mediterranean Sea and the Sahara Desert appear to be no barrier to the migrations, and, surprisingly, birds do not concentrate at oases or in the Nile Valley but fly indiscriminately over the desert, covering at least 1500 km at one go (Moreau, 1967). Some birds fly directly south from Europe, but many migrations originate to the east and beyond the Caspian Sea, following a south-westerly route over the Sahara. Birds from the central and eastern Palaearctic zone have also been recorded from eastern Africa. El-Hadidi (1965) noticed a close relationship between the distribution of *Potamogeton trichoides* in Africa and the so-called 'Caucaso-Zambian' fan of bird migration. Likewise, the distribution of *Potamogeton perfoliatus* in Egypt could be correlated with bird migration from the Mediterranean, over the Sinai, onto the Nile at Allaqui. He concluded that these species are probably distributed by water-fowl along their migration routes. Moreau (1972) estimates that the autumn migration from the Palaearctic zone to south of the Sahara involves in excess of 5000 million birds – excluding water-birds which have not been estimated yet! It may be anticipated, therefore, that occasional transplanting of waterplant propagules by so many birds is inevitable.

2.2 Factors affecting biogeographical distribution

Climate and physico-chemical characteristics of water bodies restrict the ability or otherwise of euhydrophytes to colonise a region. The physical geography of an area, its leeward or windward aspect, and features such as sediment and the undulations of lake beds may affect vegetation either directly or indirectly through wave action and water turbulence. The depth of water limits colonisation and is a major regulator in the zonation of species. It is not possible to cover all aspects in detail here: a good, general account of factors affecting distribution can be found in Sculthorpe (1967), whilst aspects pertinent to the zonation and distribution of euhydrophyte vegetation within particular African waterbodies are discussed fully in Chapter 5. Emphasis, therefore, is given to those, such as salinity and alkalinity, which are not normally appreciated sufficiently.

2.2.1 *Radiation*

Total available radiation and daylength vary according to geographical and climatic factors. African floating-leaved plants tend to thrive in full sunlight, but a few species, such as *Nymphaea maculata* which is found in raffia swamps (Cook, 1968), apparently require shade. In contrast, submerged macrophytes have rather particular light requirements, and the underwater radiation may affect their distribution significantly (Chapter 5).

2.2.2 *Temperature*

There is good evidence that temperature is a major factor in limiting the distribution of some of the more obnoxious tropical surface-floating waterweeds (see Chapter 4). The temperature optimum for *Eichhornia crassipes*, for example, is 28–30°C; growth is high over the range 22–35°C, whilst below about 12°C it is inhibited (Knipling, West and Haller, 1970); but similar information for euhydrophytes is limited. Different species seem to have different temperature optima; some, such as *Potamogeton schweinfurthii*, are restricted to the warmer regions, whilst others (*e.g. P. lucens*) occur only in colder waters. However, correlations between water temperatures and distribution cannot be considered properly until there is more information and wider collections are made.

2.2.3 *Nutrition*

Similar comments are applicable to data on nutrient requirements of aquatic plants. Some species have relatively low nutrient demands and occur in the more oligotrophic waters, whilst others have high requirements and, indeed, may serve as indicators of eutrophication or salinity (*e.g. Najas marina*). Euhydrophytes have two potential sources of nutrient supply: either the bottom substrate or the surrounding water (see Chapter 5). Both have to be considered in relation to the distribution of plants, although not all species respond equally to the two nutrient environments. *Nymphaea lotus* for example, favours organic soils but occurs in a variety of waters. In Sierra Leone I found it growing in extremely oligotrophic waters where the water conductivity, K_{25}, is less than 8 μS cm^{-1} (Gerrath and Denny, 1979), and here it must obtain nourishment from the substrate *via* the roots. *Lemna trisulca*, on the other hand, has insignificant roots and is rarely in intimate contact with the substratum. Its distribution is unaffected by the nutrient status of the sediment, but it does favour nutrient-rich water.

Levels of nutrients in waters are dependent largely upon catchment areas (type of soil, land-use, *etc.*), and rates of cycling within the water body. As microbial activity and mineralisation processes are a function of water temperature, mineral cycling in the warm waters of tropical Africa is generally rapid. Flowing water provides a continuous replacement of nutrients, which in nutrient-poor areas is particularly advantageous, whereas standing waters may become depleted. Shorelines of natural and man-made lakes with changing water levels receive a nutrient boost when water levels rise (McLachlan, 1974; Viner, 1977), which encourages macrophytic and algal growth (Denny *et al.*, 1978). Nutrients in the bottom soils depend upon regional lithology, surrounding soils and shore exposure. Strong wave action, currents and fast-flowing waters sift out fine particulate matter, leaving a coarse gravel or sandy sediment low in nutrients (Chapter 5). Plants found there must be physiologically adapted to nutrient-poor sediments and morphologically adapted to resist washing away. In areas of sedimentation, fine, particulate silt and clay soils enriched with organic and inorganic matter commonly develop swamp vegetation which itself contributes a substantial amount of autochthonous organic matter.

2.2.4 *Carbon dioxide supply*

Carbon dioxide availability also affects the distribution of plants. There are two potential sources of carbon dioxide for photosynthesis in submerged plants: free dissolved CO_2, prevalent in low pH waters mainly of low conductivity and bicarbonate ions prevalent in alkaline waters. Submerged taxa can be divided into bicarbonate- and non-bicarbonate- users (Ruttner, 1963), but this division is not clear-cut. *Potamogeton schweinfurthii* has a mechanism for actively absorbing and utilising bicarbonate ions despite a preference for free dissolved CO_2 (Denny and Weeks, 1968; 1970). Some plants appear not to have the bicarbonate uptake mechanism at all (*Fontinalis antipyretica* is a much-quoted example), and are often restricted in distribution to waters of high free-CO_2 content. Members of the Podostemaceae occur in rapids and waterfalls where aeration ensures a continuous dissolved-CO_2 supply. Other plants such as *Potamogeton pectinatus* thrive in waters of high pH values (>9.5) where free dissolved CO_2 is minimal, the main carbon source being bicarbonate ions.

2.2.5 *Salinity, alkalinity and pH*

A great diversity of salinities can be found within the African inland water system. Waters from catchments of hard, well-weathered rocks such as those of Lake Tibi, Sierra Leone (Gerrath and Denny, 1979), or those that have passed through swamps (see Chapter 6) may have conductivity values of less than 10 $\mu S\ cm^{-1}$. Conversely, inland soda lakes like Lake Nakuru, Kenya, and coastal lagoons of brackish waters such as Lake Swartvlei in South Africa, are in excess of 160 000 $\mu S\ cm^{-1}$. Characteristics of African athalassic (non-marine) waters, particularly those from central and eastern Africa, are fully described by Talling and Talling (1965), whilst additional limnochemical information can be found in Beadle (1981). The latter has especially interesting sections on Lakes Turkana and Chad.

Total ionic concentration (*i.e.* salinity) is closely correlated with conductivity. The principal ions in most African waters are sodium and bicarbonate (in contrast to temperate region waters which are predominately calcium bicarbonate) and, in these waters, there is a strong correlation between conductivity and total alkalinity (HCO_3^- + CO_3^{2-} + OH^-). pH generally rises with increased alkalinity. Chloride ion equivalent concentrations are lower than total alkalinity and usually rise proportionally with increased alkalinity, though there are exceptions. Lakes Malawi and Chad, for example, have unusually low proportions of chloride, whilst Lake Turkana has a higher than average amount, and lagoons under the influence of seawater are dominated by sodium chloride. Inorganic nitrogen concentrations are generally very low (below 30 μg NO_3-N dm^{-3}), but phosphorus concentrations are variable (up to 150 μg P dm^{-3}, or more) and, in the more saline lakes, such as Lake Turkana, can exceed 1000 $\mu g\ dm^{-3}$ (Talling and Talling, 1965). The proportions of the major cations – sodium, potassium, calcium and magnesium – vary more than the anions, but sodium generally predominates with increasing proportion in highly saline lakes.

Sculthorpe (1967) highlights the difficulty of separating salinity from other variables when considering the distribution of freshwater macrophytes. He believed salinity to be important and provided examples of saline-tolerant euhydrophytes including *Potamogeton pectinatus* and *Najas marina*. The vegetation in the Maputaland coastal lakes in South Africa, which range from freshwater (<1000 $\mu S\ cm^{-1}$) to varying concentrations of seawater (>10000 $\mu S\ cm^{-1}$), convinced Howard-Williams (1980*b*) that the salinity factor was the overriding regulator in species occurrence.

The effect of salinity is emphasised by arranging a variety of athalassic lakes and coastal lagoons along a salinity gradient (conductivity gradient) and noting the occurrence of some of the more common African euhydrophytes (Table 2.2). *Zostera capensis, Ruppia maritima, R. spiralis* (= *R. cirrhosa*) and *Lamprothamnum papulosum* are obligate marine and estuarine species; indeed, they are abundant in many coastal lakes of South Cape (Howard-Williams, 1980*b*). It can be seen that most other species become intolerant of salinities >1000 $\mu S\ cm^{-1}$ and some, such as *Potamogeton thunbergii*, are not recorded from lakes with >800

Table 2.2

LAKES	Chad south	6	Vic	Mal	Bun	7	Naiv	NYM NE	Tang	8	Alb	4	10	Edw	NYM NW	Chad north	Chil	Kivu	Turk	12	16	3	Swart Lake	Swart est.	1	13
Chemistry (References)	u	h	m	m	c	h	fm	e	m	ah	m	wh	xh	m	e	u	tr	m	s	xh	h	h	i	r	h	xh
Conductivity μS cm^{-1}	50–120	100	105	210	262	300	363	540	610	630	800	800–1057	900	1000	1000	800–1100	1250	1360	3200–3600	8000	8000	10× 10^3	13× 10^3	20× 10^3	20× 10^3	20× 10^3
pH	7–8		7.1	8.0	8.5	7.3	8.8	7.5		8.3	9.0	8.0	8.3	9.1	8.2	8–9	8.5	9.1	9.3	8.1			7.9	8.2		8.2
Alkalinity meq dm^{-3}	0.5–1.2		0.9	2.4	2.1	1.2	3.3	5.6	6.5	2.5	7.3	5.4		9.9	9.0	8–10.5	7–8	16.0	20–25				3.2	4.1	4+	4+
Sodium mg dm^{-3}	6		10	21	20	88	45	75	60	86	91	100–200	270	110	150	80	324	130	850	2377	5889	2970	3863	5940	5940	5940
Chloride mg dm^{-3}	0		4	4	29	106	14	18	28	140	33	191–245	475	36	40	low	264	35	450	4220	4438	5277	6860	10555	10550	10550
PO_4-P μg dm^{-3}	10–250		140	60	<26			300		5	180	<5	14	18	500	100–1200	3500	39	(m) >1000	9			2	12		12
Vegetation (References)	v	h	jkb	r	d	h	f	p	no	h	n	rh	h	nl	p	v	r	n	g	h	h	h	h	r	h	h
Floating-leaved species																										
Potamogeton thunbergii			×		×		×																			
Nymphaea lotus	×		×		×			×			×	×					×									
Nymphaea caerulea		×	×		×		×			×		×					×									
Nymphoides indica		×	×										×													
Nymphaea capensis		×				×			×	×	×	×														
Submerged species																										
Myriophyllum spicatum		×		×		×			×	×																
Vallisneria aethiopica	×		×	×					×		×			×												
Potamogeton schweinfurthii	×	×	×	×	×	×	×		×	×	×	×	×	×												
Najas pectinata	×					×	×	×	×	×		×	×	×	×											
Ceratophyllum demersum	×	×	×	×	×	×	×	×	×	×	×	×	×	×	×	×	×	×								
Potamogeton pectinatus					×	×	×	×	×	×	×	×	×	×				×	×	×	×	×	×			
Najas marina								×	×		×			×	×			×		×	×	×				
Lamprothamnium papulosum																				×	?	×	×			
Ruppia maritima																				×		×				×
Ruppia spiralis																						×		×		
Zostera capensis																						×		×	×	

Legend to Table 2.2. The relationship between water salinity and euhydrophyte distribution in African lakes. A variety of athalassic and coastal lakes is arranged along a gradient of increasing salinity (expressed as increasing conductivity, K_{25}). Selected species of euhydrophytes are displayed on the vertical axis and the presence of a species in a lake is recorded. Pertinent chemical data of the lakes are included.
Chemical and floristic data have been obtained from various sources as indicated. Where necessary, conductivities have been converted from K_{20} to K_{25} by the factor $\times 1.1$ (Golterman, Clymo and Ohnstad, 1978). Where a range of chemical data for a particular lake is available, an average reading has been chosen for this table. The Lake Chilwa data are for March 1970. The chemical data for Lake Chad are only approximations as there are large seasonal changes. They were calculated from Carmouze *et al.* for the year 1970.

Reference	Waterbodies (in increasing order of salinity)
a = Allanson (1979)	Chad, south = Chad, south basin
b = Carter (1955)	6 = Bangazi-north
c = Denny (1972)	Vic = Victoria
d = Denny (1973)	Mal = Malawi
e = Denny, Harman, Abrahamsson and Bryceson (1978)	Bun = Bunyonyi
f = Gaudet (1977)	7 = Mgobezeleni, Maputaland
g = Harbott (1982)	Naiv = Naivasha, Kenya
h = Howard-Williams (1980b)	NYM NE = Nyumba ya Mungu NE, Tanzania
i = Howard-Williams and Davies (1979)	Tang = Tanganyika
j = Lind and Morrison (1974)	8 = Sibaya, Maputaland
k = Lind and Visser (1962)	Alb = Albert
l = Lock (1977)	4 = Teza, Maputaland
m = Talling and Talling (1965)	10 = Kosi: Amanzimnyama, Maputaland
n = Van der Ben (1959)	Edw = Edward
o = Van Meel (1952)	NYM NW = Nyumba ya Mungu NW, Tanzania
p = Welsh and Denny (1978)	Chad, north = Chad, north basin
r = Howard-Williams (unpublished)	Chil = Chilwa
s = Ferguson and Harbott (1982)	Kivu = Kivu
t = McLachlan (1979)	Turk = Turkana
u = Carmouze *et al.* (1972)	12 = Kosi: Mpungwini, Maputaland
v = Iltis and Lemoalle (1983)	16 = Poelela, Maputaland
w = Bruton, Wilson and Musil (1978)	3 = St Lucia (main lake), Maputaland
x = Begg (1978)	Swart lake = Swartvlei lake, Cape Province
	Swart est. = Swartvlei estuary, Cape Province
+ = estimated value	1 = Umshlatuzi, Maputaland
	13 = Lower Kosi system, Maputaland

μS cm^{-1}. The two species of *Najas* have fairly broad and overlapping salinity ranges, but, whereas *N. pectinata* occurs in less saline locations, *N. marina* is mainly a brackish and saline water species. The sample sites at Nyumba ya Mungu lie within the overlapping range; it is noticeable that, in the north-east sector (540 μS cm^{-1}), *N. pectinata* predominates over *N. marina*, whereas, in the north-west (1000 μS cm^{-1}), the dominance is reversed (Welsh and Denny, 1978). *Ceratophyllum demersum* has a wide salinity tolerance and occurs in waters from <100 to 1360 μS cm^{-1}. Nevertheless, *Potamogeton pectinatus* is the overall winner with an incredibly wide distribution. It is abundant in Lake Bunyonyi (260 μS cm^{-1}) (Denny, 1973), it is the dominant submerged plant in Lake Turkana (3620 μS cm^{-1}), and commonly forms very dense beds of vegetation in brackish coastal lagoons with conductivities of 10 000 μS cm^{-1}, or more. It can even tolerate occasional temporary inundation by seawater (Howard-Williams, 1980*a*). There are indications that it is a poor competitor in unpolluted freshwaters, and in those environments favourable to the establishment of competitor species it declines. This can be seen by examination of vegetation in the Rift Valley Lakes (see section 2.3). In Lakes Kivu and Turkana the salinity may be optimal for *P. pectinatus*; the conditions are too saline for *P. schweinfurthii*, and *P. pectinatus* forms large monospecific stands of vegetation. In Lake Albert it is in competition with *P. schweinfurthii*; it occurs rarely in Lake Edward and is unrecorded from Lake Victoria.

Salinity *per se* may not be responsible for floristic changes (Sculthorpe, 1967), but there are conflicting views as to the importance of the various components which make up the salinity (see Denny, 1980). Swindale and Curtis (1957) and Seddon (1972) favoured conductivity, Spence (1967) alkalinity, and Jones and Cullimore (1973) phosphate concentrations. These last authors found that, in a group of Canadian lakes, *Potamogeton pectinatus* became increasingly abundant with a rise in total phosphorus in the water. Data from Africa may appear to support this, but there are some inconsistencies – *e.g.* the comparatively low recorded concentrations of soluble phosphorus in Lakes Edward and Kivu where *P. pectinatus* abounds. Work on the brackish Wilderness Lakes in Cape Province (Howard-Williams, 1980*a*), and Lake Swartvlei (Howard-Williams, 1978; Howard-Williams and Allanson, 1981) goes some way towards an explanation. The Wilderness Lakes are described as relatively nutrient-rich whilst Lake Swartvlei is oligotrophic. Both have dense stands of *P. pectinatus*, which suggests that the plants are largely unaffected by nutrient status. In Lake Swartvlei, however, the principal source of nutrient supply is from the sediment, and the bulk of the phosphorus is either locked up in the sediment or vegetation. In the richer water of the Wilderness Lakes, the plants probably obtain their phosphorus from the water, but in Lake Swartvlei it is mainly taken from the sediment and cycled through the vegetation. In both cases the overall phosphorus level in the ecosystem is sufficient for *P. pectinatus* growth (see Chapter 5). The same explanation may be applied to Lakes Edward and Kivu. Therefore, levels of phosphorus alone (and indeed of other nutrients) in the water are unlikely to account for the floras in the lakes of varying salinity under discussion. This does not mean to say that the overall levels of nutrients in aquatic ecosystems are unimportant; there is a close correlation between sodium ion concentration and salinity, but insufficient information on the influence of major cations on euhydrophyte plant growth precludes any conclusions.

Is chloride ion concentration, perhaps, the main component of the salinity which affects euhydrophyte distribution? Comparative data from brackish waters and alkaline lakes of similar conductivity do not suggest it, at least in the middle range salinities. (Lakes Kivu and Bunyonyi have similar chloride concentrations (29–35 mg dm^{-3}), but the floristic composition of the two lakes if very different.) Likewise, Talling and Talling (1965) found no evidence that chloride radically affected the distribution of athalassic phytoplankton species. In estuaries and coastal lagoons high chloride concentrations may, indeed, limit freshwater plants, but overall salinity and its effect on salt uptake mechanisms may be catastrophic. Marine species, such as *Zostera capensis* (and the other plants grouped in the bottom right-hand corner of Table 2.2) on the other hand, are intrinsically different, being adapted to this particular environment.

This leaves pH and alkalinity to be considered. Chemically, they are interdependent, and so it is not easy to distinguish independent biological responses. In section 2.2.4 it was suggested that submerged plants could be divided into bicarbonate and non-bicarbonate users. The former frequently have a wide range of tolerance to carbon dioxide supply, but the latter are restricted to waters of low pH and conductivity where dissolved CO_2 is non-limiting. pH, therefore, can have both a direct effect at very low (say below about 4.0) and very high values (>10.5) where it may damage tissues, and an indirect effect. Very high pH values (or the accompanying high alkalinities) account for the dearth of species in the alkaline soda lakes. All the lakes under discussion, however, fall into a middle range of pH values (7–10), and, accepting that there may be wide seasonal or diel variation, Table 2.2 shows that pH has no obvious effect on floral composition. Comparison of the floral composition of Lake Bunyonyi with that of Nyumba ya Mungu north-west is an example. The pH values of the waters are similar, but whereas Lake Bunyonyi has a rich flora that of Nyumba ya Mungu is indicative

Plate 2.2. A highly saline crater lake in southern Uganda from which euhydrophyte vegetation is absent. During dry periods a distinct crust of salt develop over the littoral and on the shore. (Photo: P. Denny)

of more saline waters. What about the effect of alkalinity? Studies on algal distribution may provide a guide. Talling and Talling (1965) noted a reduction in desmid species at alkalinities above about 2.5 meq dm^{-3}, and a large change in the species composition of plankton in highly alkaline waters (above 80 meq dm^{-3}) in excess of, say, 7000$\mu S\ cm^{-1}$. The diatom, *Melosira granulata*, occurred mainly in waters below about 8 meq dm^{-3} but could tolerate values of up to 10.5 meq dm^{-3} if the pH was less than 9.0. The authors suspected that pH values >9.2 largely limited *Melosira* abundance. (They found it practically impossible to separate pH and alkalinity effects in more highly buffered waters). Hutchinson *et al.*, 1932, (in Talling and Talling, 1965) reported a profusion of desmids (which are normally associated with oligotrophic, neutral to acid waters) in a South African pan where a low alkalinity was accompanied by a pH value of 9.0, and they concluded that alkalinity was more significant than pH. In East Africa, Richardson (1968) also found a high correlation between diatom composition and alkalinity but could not isolate its influence from other components. Kilham (1971), for example, identified a reasonable correlation with silica concentrations.

Correlations between alkalinity and euhydrophyte distribution can also be found, Lake Chad providing one such example. Chad is an endorheic lake with the main inflow in the south, and apparent subterranean seepage to the north (see Beadle, 1981, for a fuller account). The lake is effectively divided into a south and north basin in which salinity increases steeply from the southern inflow to the north. Chloride concentrations are very low, the predominant anion being bicarbonate (Table 2.2). The extended Sahelian drought of the early 1970s caused a diminution of the lake size from 1972; O.R.S.T.O.M. refer to the period before this date as 'Normal Chad', and post-1972 as 'Lesser Chad' (Iltis and Lemoalle, 1983). The data in Table 2.2 and the current discussion refer to 'Normal Chad'. The vegetation in the southern basin of Normal Chad was rich, with submerged plants, including *Najas sp.*, *Ceratophyllum demersum*, *Vallisneria sp.* and *Potamogeton schweinfurthii*, growing profusely. This was accompanied by an abundance of swamp plants – especially *Vossia cuspidata*, *Cyperus papyrus* and *Phragmites australis* – though *Typha australis* (*T. domingensis*?) was rare. To the north, *Cyperus papyrus* was replaced by *Typha*; *Vossia cuspidata* became scarce, and there was a progressive disappearance of submerged species except *Ceratophyllum demersum* which was much reduced in abundance.

The exact cause of this change in vegetation is difficult to evaluate. It is certainly not due to chloride ion concentrations, nor can the increase in salinity *per se* be the main reason. Other lakes in Africa, such as Lake Edward, have similarly high conductivity values yet a wide range of euhydrophytes. The increase in alkalinity in Lake Chad is dramatic and may be sufficient to limit species distribution; but the alkalinity in Lake Edward reaches nearly the same value.

Table 2.2 suggests a correlation between species distribution and alkalinity with values above about 10 meq dm^{-3} being critical. Whilst the values for Lake Edward range *up to* 9.9 meq dm^{-3}, those of the north basin of Lake Chad are mainly >10.0. In the absence of data to the contrary, I believe that alkalinity values >10.0 are around the limit of athalassic euhydrophyte tolerance, and some submerged plants may be substantially more sensitive. Perhaps it can be concluded that the geographical distribution of submerged plants in African athalassic lakes is correlated with the salinity, and, further, the available evidence suggests that the alkalinity of the water exerts a strong influence. However, it would be wrong to assume that alkalinity is the only component of the salinity which influences events; rather, it is a major component of a complexity of factors which together govern the ultimate range of a species.

2.2.6 *Climate*

Although the primary conditions of light, temperature and nutrition need to be fulfilled, an intricate combination of factors will actually determine the floral composition of a particular body of water. For example, in Sierra Leone during the monsoon, rainfall is in excess of 500 mm $month^{-1}$ but, for the rest of the year, high pressure over the Sahara provides the hot, dry Harmattan. Rivers become deep and fast-flowing with heavy silt loads in the rains, but in the dry season water depths decrease by as much as 8 m resulting in shallow, meandering watercourses. Such conditions are unfavourable for the establishment of euhydrophytes, and, indeed, in the main rivers plants are notable for their absence. In Central and East Africa, there is a tendency towards two rainy seasons which ameliorate some of the more severe effects of a single, high, rainfall. Changes in water levels are generally less marked and, in lakes, are sufficiently gentle to allow the vegetation to adjust. Even during exceptional periods of prolonged heavy rain, such as in 1962/1963 in East Africa, when Lake Victoria rose to and remained more than 2 m higher, the vegetation zones soon adapted to the new shoreline.

Temporary pools have a specialised flora. Plants either complete their life cycle during the rains with the production of resistant fruits or seeds (*e.g. Najas pectinata*), or form drought resistant propagules. Alternatively, the pools become re-colonised from outside sources.

2.2.7 *Competition*

In tropical and equatorial regions, euhydrophytes are under considerable competitive pressures from surface-floating vegetation and emergent plants which grow out from shores and form rafts. These rafts may remain attached to the bottom, extending to the limits of the euphotic zone (*e.g. Phragmites autralis* in Lake Bunyonyi, Uganda, Denny, 1973), or float out on the surface over considerable depths of water. *Cyperus papyrus* has this growth habit (Lind and Morrison, 1974; also see Chapter 3) in the Nile and in Central and East Africa, and occupies large areas of water which otherwise could be colonised by beds of submerged and floating-leaved vegetation (Denny, 1984). The outer fringes of swamps become detached to form the floating vegetation typical of the Sudd. Floating islands, and surface-floating plants, especially *Pistia stratiotes, Eichhornia crassipes* and *Salvinia molesta*, get swept down rivers or onto lee-shores and form dense mats, underneath which the establishment of submerged plants depends upon the mat thickness and its transient nature. In quiet backwaters, vegetation dominated by members of the Lemnaceae may be sufficient to inhibit growth of a submerged flora.

2.3 Biogeography

When considering the biogeography of African plants it is convenient to divide the continent into zones. The broadest division would be the Tropical, and the North and South Temperate zones. However, this division is unsatisfactory as it does not take account of the diversity of climate or the biogeographical similarities of floras. Van der Ben (1959, after Lebrun, 1947) demarked six phytogeographical regions: he separated the Mediterranean and Sahara regions from a large Soudano-Zambezi region extending from about latitude 20°N through East Africa to southern Africa. The Zaire basin and the west African coastal area comprised a separate region, and the Cape and Mascarene Islands completed the regional divisions. The Herbarium, Kew, recognises even more specific floristic areas. As with most artificial classifications, anomalies occur. Nevertheless, the regional floras are generally distinctive although, in the aquatic habitat, they may become somewhat blurred.

In accounts of African floras (*e.g.* Van der Ben, 1953; Walter, 1971), present day plant distribution and the changing geographical and climatic conditions which have brought this about are discussed, but information on aquatic plants is very limited. Let us consider the geographical distribution of species within one genus, *Potamogeton*, which is reasonably well documented, rather than compare genera from incomplete floral lists of different regions. This genus is well-represented in aquatic vegetation and has comparatively few African species. Because the identity of closely related potamogetons is confused, some discussion of taxonomy has to be included.

2.3.1 *The potamogetons*

Perhaps the two most wide-ranging species are *P. pectinatus* and *P. panormitanus* (= *P. pusillus* L.).

Plate 2.3. Potamogeton sp. from Lake Kariba which produces 'terrestrial-like' growth when the water level recedes. When submerged, the same taxon develops *Potamogeton schweinfurthii*-like submerged leaves. (Photo: P Denny)

There are few areas of the world where *P. pectinatus* cannot be found. In Africa, it occurs in a great variety of lakes such as Lakes Edward and Turkana, and in the salt flats of northern Sudan, its distribution extending from Egypt through the equatorial region to the Cape. *Potamogeton panormitanus* is also reported throughout Africa, the Middle East, Europe and America, and across Asia to Siberia and Japan. The full extent of its distribution depends upon the nomenclature used. Whilst the *Flore d'Afrique Centrale* (Lisowski *et al.*, 1978) includes *P. pusillus* L. and *P. pusillus* L. var. *africanus* (A. Benn.) in with *P. panormitanus*, Tackholm (1974) considers *P. panormitanus* equivalent to *P. pusillus* non L., and both Dandy (1971) and Wickens (1976) include all under *P. pusillus* L., a procedure adopted also in the *Flora of the British Isles* (Clapham, Tutin and Warburg, 1962).

Non-tropical taxa, more normally distributed in the Northern or Southern Temperate regions, make sporadic appearances in Africa. *Potamogeton polygonifolius*, common in Europe and North America, is reported from Morocco and the Niger Province. *Potamogeton nodosus* and *P. lucens* are widespread throughout North and Central America, Europe, the Mediterranean, Asia and

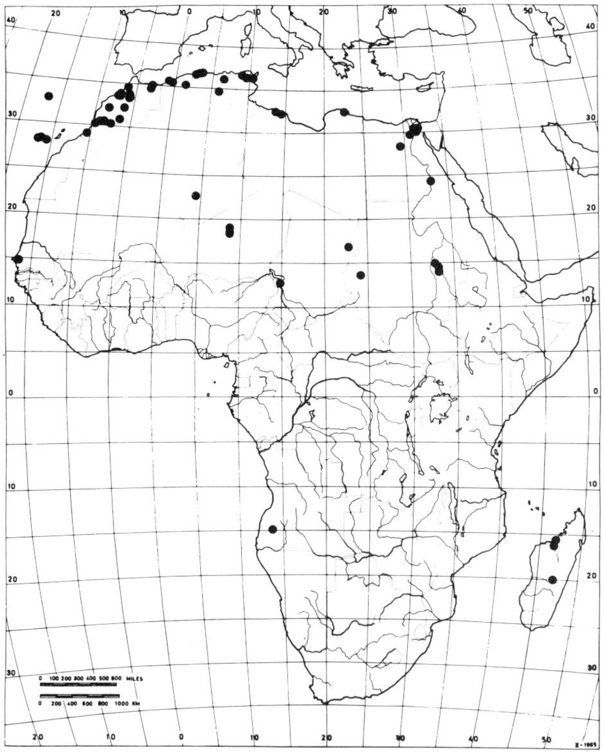

Fig. 2.2. The distribution of *Potamogeton nodosus* Poir. in Lam. in Africa. (From Symoens, Van de Velden and Buscher, 1979, Plate 2.)

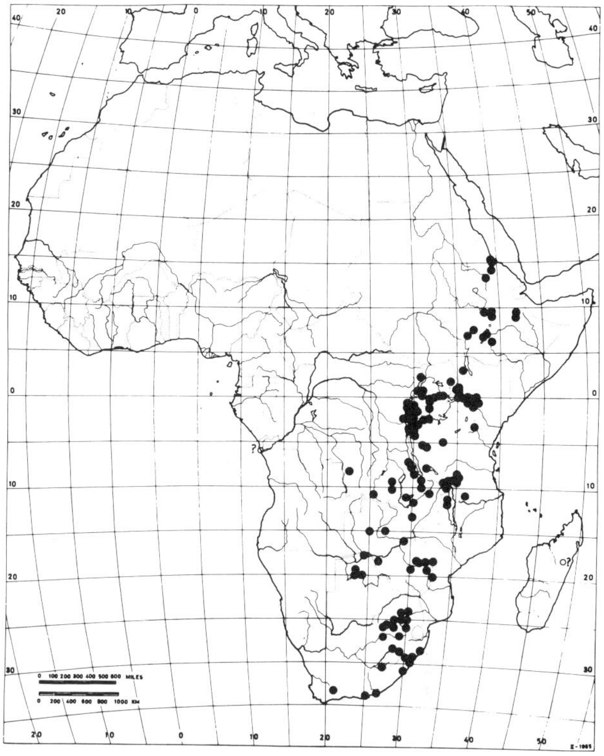

Fig. 2.3. The distribution of *Potamogeton thunbergii* Cham. & Schlechtend. in Africa. (From Symoens, Van de Velden and Buscher, 1979, Plate 3.)

the Middle East, and extend into North Africa and Egypt. *Potamogeton nodosus* is sometimes mistaken for *P. thunbergii*, but Symoens, Van de Velden and Buscher (1979) have now identified clear anatomical characters which separate these two species. *Potamogeton nodosus* is largely found to the north and west of the Sahara with occasional records down to latitude 10°N (Symoens *et al.*, 1979), and see Fig. 2.2. Wickens (1976) found it in Jebel Marra, and it mysteriously appeared in Angola, the only African record south of the Equator, although it does occur in Madagascar. *Potamogeton lucens* L. *sensu stricta* has been collected from Lake Mutanda, a high altitude lake in Uganda, but never with fertile spikes, which suggests that this is outside its normal geographical range (Denny and Lye, 1973). Other temperate climate potamogetons include *Potamogeton crispus* and *P. perfoliatus*. *Potamogeton crispus* is normally found in the temperate parts of Africa but has been reported from the Sudan (Andrews, 1956), and has recently become an invasive weed of the Blue Nile near Khartoum (Tag El Seed, 1981). *Potamogeton perfoliatus* occurs in Niger and Chad (Peyre de Fabregues and Lebrun, 1976), and across North Africa to the Nile at Allaqui, Egypt, and, surprisingly, as far south as Khartoum. It has also been recorded from the higher altitude regions in West Africa (Air Mountains, Niger) and central Sahara. *Potamogeton trichoides* appears to have a scattered distribution in Africa: North Africa, the Nile in Egypt at Assuit, Manfalout and Dayrout (El-Hadidi, 1965), the higher altitude lakes in East Africa and, occasionally, in southern Africa (Obermeyer, 1966). This distribution pattern may be incomplete as it can be overlooked in plant samples, or confused with other delicate fine-leaved potamogetons.

The three remaining potamogetons – *Potamogeton octandrus, P. thunbergii* (= *P. richardii* Solms) and *P. schweinfurthii* – are predominantly African. *Potamogeton octandrus* is a tropical and southern hemisphere species and extends outside southern Africa through Madagascar and tropical Asia as far as Japan and Australia. Its distribution in Africa ranges from west tropical Africa (Hepper, 1968; Lebrun *et al.* 1972; Lebrun 1973) across the continent south of the Sahara to equatorial and southern Africa, but it is absent from North Africa and Egypt. It occurs in the Nile but is rarely found north of Malakal, Sudan, (Lat. 9° 30′ N). *Potamogeton thunbergii* and *P. schweinfurthii* are strictly African. *Potamogeton thunbergii* is found in the Afro-montane region of Ethiopia, throughout the Soudano-Zambezi region and south to the Cape. It is absent from North Africa (where it is replaced by *P. nodosus*), from the Zaire basin and from West Africa (Symoens *et al.*, 1979), and see Fig 2.3. *Potamogeton schweinfurthii* is the most widespread African species. In its various forms it occurs in all the African regions specified for *P. octandrus* and *P. thunbergii* and has a scattered distribution in North Africa and Egypt. In Central and East Africa, a long, narrow, submerged-leaf form of *P. schweinfurthii* (with, or without occasional coriaceous floating leaves) predominates. Further north and south more of the populations have attributes akin to *P. lucens* (broader, shorter submerged leaves) and are often named *P. lucens* Aut. *non* L. This form is pronounced in the southern African flora and Madagascar, but not to the exclusion of other forms. In the presence of *P. thunbergii* it readily hybridises to produce *P.* × *bunyonyiensis*.

The pattern of distribution of *Potamogeton* species may be used as a guide to the expected floral composition of vegetation from different regions of Africa. The central position of Egypt in linking the Orient and North Africa, and the relatively narrow geographical barrier between Europe and North Africa, may provide the means for a degree of similarity between the African floras north of the Tropic of Cancer and those of southern Europe and the Middle East. The Sahara Desert, on the other hand, appears to be a major barrier restricting and separating species from the northern and southern hemispheres, although the Nile acts as an important link for population flow. Restriction on dispersal is unlikely to be the major cause of this biogeographical distribution as migratory birds could probably act as agents (see section 2.1.3). Climate, on the other hand, enforces restrictions on plant distribution such that some species are excluded from the hot, tropical zones whilst others are predominantly equatorial. Within the potamogetons only one species can be considered truly African with pan-African distribution, that is *Potamogeton schweinfurthii*.

2.4 Floristic compositions and vegetation types

Aquatic floras are sometimes conveniently separated into those of standing waters and those of running waters (*e.g.* Mitchell, 1978). However, as water flow is only one of many factors which may determine a particular flora, and as some large rivers such as the Zaire and Nile have stretches so sluggish that the riverine and lacustrine characters are barely distinguishable, this distinction will not be made here.

Let us consider the Rift Valley lakes, especially Lakes Kivu, Edward, Albert and Tanganyika, which were surveyed by l'Institut Royal des Sciences Naturelles de Belgique (1952-54), and on which Van der Ben (1959) carried out vegetation studies. In general, the flora of the first three lakes is species-poor, none reaching the richness of Lake Tanganyika (and the waters of the Zaire basin, Van der Ben, 1959). There is a gradation from Lake Kivu, the poorest, to Lakes Edward and Albert, where aquatic plant associations are more complete and better developed. The reasons for the poverty of species in Lake Kivu were obscure to Van der Ben, although he considered water turbulence, the steepness of the shore and calcareous deposition to be significant. However, the rise in salinity and the accompanying alkalinity of lakes, in the sequence Victoria → Tanganyika → Albert → Edward → Kivu (Table 2.2) cannot escape one's attention; this is, most likely, the controlling factor.

The submerged flora of Lake Kivu is dominated by monospecific stands of *Potamogeton pectinatus* from 0.5 to 3.0 m, with *Najas marina* subsp. *armata* in more sandy areas. In Lake Edward *Potamogeton pectinatus* again predominates, in association with *Najas marina* and *Vallisneria aethiopica*. Lock (1977) recorded occasional *P. schweinfurthii* and *Najas pectinata* from the eastern shores. Van der Ben observed that *P. pectinatus* tended to occur in moderately agitated water, especially in large shallow bays where there were fine sediments. *Vallisneria* and *Najas*, on the other hand, preferred highly oxygenated water and a slightly inclined substratum of sandy texture, conditions which were found near rivers. In the richer flora of Lake Albert, *Potamogeton schweinfurthii* replaced *P. pectinatus* which occurred only rarely.

In the deltas of rivers such as the Semliki, the submerged lacustrine associations are usually incomplete as there is a luxurious development of swamp vegetation. On the lakeward side of fringing swamps (predominantly, *Cyperus papyrus, Vossia cuspidata*, and *Phragmites mauritianus*), Van der Ben described two floating phytosociological associations: the free-floating *Lemneto-Pistietum* and the rooted, floating-leaved *Nymphaeetum*. The *Nymphaeetum* was dominated by *Nymphaea* spp. (particularly *N. lotus* and *N. calliantha*) which were sensitive to depth and mostly occurred in a zone between one and two metres deep. The submerged plants that occurred amongst the *Nymphaea* included the free-floating *Ceratophyllum demersum* and *Utricularia thonningii*, and also *Potamogeton* spp., *Vallisneria aethiopica* and *Najas marina* ssp. *armata* which were rooted in the substratum. Floating species included *Pistia stratiotes*, members of the Lemnaceae and *Azolla nilotica*.

The flora of Lake Tanganyika is less well described. Van Meel (1952) found that, in the inflow rivers, *Ceratophyllum* was particularly abundant and in the deltas *Azolla pinnata* sometimes formed immense floating mats of green or red-brown vegetation, whilst *Ottelia ulvifolia* 'with various species of *Potamogeton, Utricularia* and *Najas*' formed extensive beds. In contrast to Van der Ben's assertions, Van Meel considered the vegetation of the lake to be poorly developed. He mentioned observing diverse species of *Potamogeton, Najas* and *Vallisneria* with a predominance of *Potamogeton filiformis* (= *P. pectinatus*?) and *Najas horrida* (= *N. pectinata*?). However, De Wildman (1912, in Van Meel, 1952) describing the flora of Albertville-M'Toa, recorded from the inflow rivers and lakes *Ceratophyllum demersum, Myriophyllum spicatum, Utricularia thonningii, U. exoleta, Najas marina, N. horrida, Potamogeton schweinfurthii, Ottelia lancifolia* and *Vallisneria spiralis*.

The flora of Lake Victoria is richer. An early vegetation survey of Namanve Swamp, adjacent to Murchison Bay on the north shore of the lake, recorded a water-lily zone fringing papyrus (Eggeling, 1935). The zone was dominated by *Nymphaea lotus*, and *N. heudolotii* + *N. zanzibarensis* (= *N. caerulea*) with *Trapa bispinosa* (= *T. natans*?), *Brasenia peltata, Nymphoides* (*Limnanthemum*) *nilotica* with an understorey of *Ottelia ulvifolia, Ceratophyllum demersum, Utricularia thonningii* and *Potamogeton richardii* (= *P. thunbergii*). Lakeward of Namiro Swamp, Entebbe, Lind and Visser (1962) found a well-developed zone of *Nymphaea caerulea* and *Trapa natans* and, in deeper water, a submerged community of *Utricularia foliosa, Ceratophyllum demersum* and *Vallisneria spiralis* was well-established. I have observed extensive beds of *Potamogeton schweinfurthii* in sheltered bays and *Ottelia ulvifolia* in the drainage channels. This type of vegetation, which is also found in swamp pools, extends down the White Nile into the Sudd (Denny, 1984; also Chapter 5), and can be considered typical of the equatorial Nile region.

A vegetation study of Lake Naivasha, one of the Kenyan Rift lakes, has been made by Gaudet (1976; 1977) and his collaborators. His account (1977) of a zonation and succession through a papyrus swamp and water-lily zone to open water during natural drawdown is one of the few instances in which a comprehensive checklist of euhydrophyte species is provided. Beds of *Nymphaea caerulea* fringed the papyrus, and *Ceratophyllum demersum* was the most common submerged plant in the swamp. In open water, *Potamogeton* spp. were well-represented together with members of the Characeae and Lentibulariaceae. Species uncommon in Kenya included *Potamogeton octandrus, Drepanocladus sparsus* and *Wolffiopsis welwitschii*.

Lake George, a small lake attached to Lake Edward, and Lake Chilwa in Malawi, are notable for the absence of euhydrophytes in their open waters. Turbulence and turbidity, due to algae in Lake George (Lock, 1973) and to fine-grained clay sediments in Lake Chilwa (Howard-Williams and Lenton, 1975), cause rapid light attenuation which submerged macrophytes cannot survive. In the clear water of the inflow river of Lake George *Ceratophyllum demersum* and *Utricularia inflexa* were found in substantial beds and as an understorey to scattered plants of *Nymphaea caerulea*. A similar community was found in disturbed clearings amongst the *Typha domingensis* swamps of Lake Chilwa where turbulence is much reduced (Howard-Williams and Walker, 1974). This contrasts with Lake Bunyonyi, a high altitude lake in south-western Uganda, where a rich zonation has been recorded (Denny, 1971; 1972; 1973). This can be seen particularly well fringing a small island, Habukara Island (Fig. 2.4), and is described fully in Chapter 5. The shape of the basin and the cooler conditions at higher altitudes probably reduce swamp development so typical of the Rift Valley lakes, allowing the full zonation to develop.

The pre-impoundment survey of the River Niger at Kainji is one of the few surveys of euhydrophytes from West Africa (Cook, 1968). Within the phytosociological order, Nymphaeetalia loti Lebrun (1947) (see Chapter 1), Cook recognised eight associations. Three associations were surface-floating communities in which the *Lemno-Pistietum* was dominant. Submerged plants grew under the floating mats if the water was clear (as it was at the headwaters of the River Swashi), but usually the water was too turbid for colonisation. In unshaded areas of very shallow water (< 50 cm), *Eichhornia natans* and *Ranalisma humile* formed the *Eichhorneo-Ranalismetum* association, regarded by Léonard (1951) as a pioneer community. Under shade and in *Raphia sudanica* swamp forest, *Nymphaea maculata* co-existed with *Ottelia ulvifolia* and *Nitella furcata*. Water-lily beds (*Nymphaeetum loti* association) occurred in the shallows (50 – 150 cm) of the river where flow was negligible or very gentle. These were composed of mixed communities of *Nymphaea lotus* and *N. micrantha* together with swamp plants such as *Polygonum salicifolium*, and various submerged species. Submerged plants, characteristically *Najas affine* and *Nitella furcata* of the *Naiado-Nitelletum furcatae* association, thrived in clear water deeper than 100 cm. In the dry season extensive entanglements of *Utricularia* spp. developed, but these were usually washed away by the rains. Typically, in the backwaters and in the mouths of streams, *Ceratophyllum demersum* could be found in quantity. It is particularly interesting to see that, after the inundation of the River Niger by the Kainji Dam, rooted euhydrophyte development was restricted largely by the fluctuating water level, which could amount to 9 metres (Hall, 1975). However, the non-rooting *Ceratophyllum demersum* and *Utricularia inflexa* formed troublesome floating mats, especially in the early years.

2.5 Man-made lakes

Euhydrophytes appear to be of minor importance compared with the mats of floating and emergent water-weeds associated with man-made lakes (see Chapters 4 and 7). For example, *Utricularia inflexa*, an early coloniser of Lake Kariba, was soon displaced by *Salvinia molesta*. *Ceratophyllum demersum* was one of the first plants recorded in the lake, and although occasionally very troublesome and abundant (especially in river inlets) it never reached epidemic proportions. Masses of *Ceratophyllum* and *Utricularia* were reported from Lakes Kainji and Volta in their early years, and, in the latter, formed substrata for 'sudd'-forming species such as *Ludwigia leptocarpa*.

Rooted euhydrophytes are slow to become established in man-made lakes. In Lake Kariba, *Potamogeton pusillus* gradually occupied suitable areas to a depth of 4 metres, and *Lagarosiphon ilicifolius*, *Potamogeton schweinfurthii* and *Vallisneria aethiopica* were observed later. Mitchell (1969) believed that the rooted flora of Kariba will develop slowly and may be restricted by the fluctuating water level. Similarly slow patterns of development were reported for Lake Kainji (Hall, 1975) and Nyumba ya Mungu Reservoir, Tanzania, (Welsh and Denny, 1978).

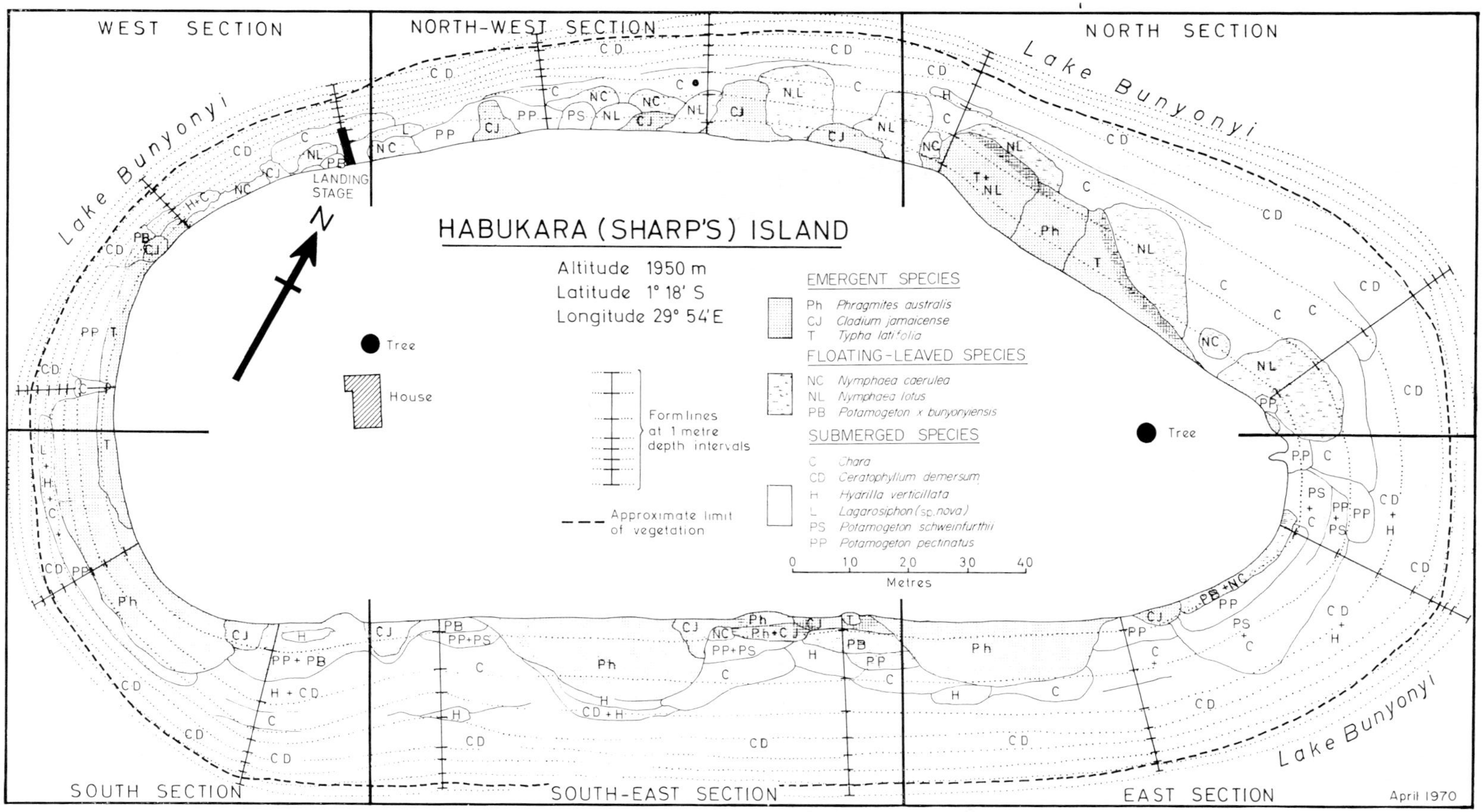

Fig. 2.4. Map of littoral vegetation around Habukara Island, Lake Bunyonyi, Uganda. (From Denny, 1973, Fig. 2.)

Plate 2.4. A bay filled with *Lagarosiphon ilicifolius* in Lake Kariba at low water level. The effect of the burrowing may-fly larvae, *Povilla adusta* on the normally submerged trees is also revealed. (Photo: P. Denny)

The occurrence of *Ceratophyllum* in association with *Utricularia* would appear to be a feature of most man-made lakes during their stabilisation. These plant associations are also prevalent in natural water bodies, particularly in swamps and at the mouths of inflow rivers. *Ceratophyllum* is the only submerged plant which has been reported as a nuisance. It can form large mats which are a focus for sudd formation; it restricts navigation; and it is a major refuge and feeding ground for bilharzia vectors.

2.6 Discussion and conclusion

This chapter may have demonstrated the lack of attention given to the euhydrophyte flora of African lakes and rivers. In biological studies of the aquatic environment it is not unnatural to emphasise applied aspects relating to fisheries, waterborne diseases and navigation, especially where full utilisation of the inland waters is essential; but neglect of floating-leaved and submerged floras is unacceptable. It is no longer sufficient to report the presence of *Ceratophyllum* or large beds of *Potamogeton*; a complete record of the flora (with carefully collected and correctly named herbarium specimens) should be a basic requirement in any survey. In this way, the biogeographical distribution of species can be obtained, and the information utilised in anticipating floras of man-made lakes. Euhydrophytes provide a variety of habitats for both beneficial and troublesome aquatic fauna, and occasionally grow to excess. *A fortiori*, studies on vegetation in relation to water depth, water chemistry and substratum, and the phytosociologi-

Plate 2.5. A decomposing fringe of *Potamogeton* and *Lagarosiphon* left behind as the water level falls in Lake Kariba. (Photo: P. Denny)

cal association of species are also needed.

Three genera which warrant special attention are *Potamogeton, Ceratophyllum* and *Utricularia*. The abundance of *Potamogeton pectinatus* in many of the African lakes is an example: it provides food; it can be heavily grazed by hippopotami in the dry season (Harbott, *pers. comm.*); and it has a very high dry matter production (Howard-Williams, 1978; 1979; also Chapter 5). *Ceratophyllum* and species of *Utricularia* which do not have roots are unaffected by fluctuations in water levels and are, at present, the only submerged plants of consequence in man-made lakes. Both are found at the mouths of inflow rivers and seem to thrive in clear, gently flowing water. However, as they can float just below the surface of the water, their photosynthesis may not be so much affected by water turbidity, whether from phytoplankton or abiotic suspensions. *Ceratophyllum* has a wide range of habitats, and it may be an excellent indicator of water quality. Clearly, all three genera provide extremely important habitats for aquatic fauna.

2.7 References

Allanson, B.R., 1979. The physico-chemical limnology of Lake Sibaya. In: B.R. Allanson (ed.) Lake Sibaya. Monographiae Biologicae 36, Dr W. Junk, The Hague: 42–74.

Andrews, F.W., 1956. The flowering plants of the Sudan I–III. Sudan Government Publication.

Arber, A., 1920. Water Plants. University Press Cambridge.

Beadle, L.C., 1981 (2nd. Edition). The inland waters of tropical Africa. Longman, London.

Begg, G.W., 1978. The estuaries of Natal. Natal Town and Regional Planning Commission, Pietermaritzburg, Report No. 41:1–657.

Bruton, M.N., Wilson, G. & Musil, C.F., 1978. Environmental

impact assessment: Proposed Umfolozi dams on Lake Tiza. Unpublished Report to Natal Parks Board.

Carmouze, J-P. *et al.*, 1972. Grandes zones ecologiques du Lac Tchad. Cahiers O.R.S.T.O.M., serie Hydrobiologie, 6:103–169.

Carter, G.S., 1955. The papyrus swamps of Uganda. Heffer, Cambridge.

Clapham, A.R., Tutin, T.G. & Warburg, E.F., 1962. Flora of the British Isles. Cambridge University Press.

Cook, C.D.K., 1968. The vegetation of the Kainji Reservoir site in Northern Nigeria. Vegetatio: 15:225–243.

Cook, C.D.K., Gut, B.J., Rix, E.M., Schneller, J. & Seitz, M., 1974. Waterplants of the world. Dr W. Junk, The Hague.

Cook, C.D.K. & Luond, R., 1982*a*. A revision of the genus Hydrilla (Hydrocharitaceae). Aquatic Botany, 13:485–504.

Cook, C.D.K. & Luond, R., 1982*b*. A revision of the genus Nechamandra (Hydrocharitaceae). Aquatic Botany, 13:505–514.

Cook, C.D.K. & Luond, R., 1982*c*. A revision of the genus Hydrocharis (Hydrocharitaceae). Aquatic Botany, 14:177–204.

Cook, C.D.K. & Luond, R., 1983. A revision of the genus Blyxa (Hydrocharitaceae). Aquatic Botany, 15:1–52.

Dandy, J.E., 1971. Potamogetonaceae. In: Flora Iranica. Akademische Druck – u. Verlagsanstalt Graz, Austria.

Denny, P., 1971. Zonation of aquatic macrophytes around Habukara Island, Lake Bunyonyi, S.W. Uganda. Hidrobiologia (Bucuresti) 12:249–257.

Denny, P., 1972. Lakes of south-western Uganda. I: Physical and chemical studies on Lake Bunyonyi. Freshwater Biology, 2:143–158.

Denny, P., 1973. Lakes of south-western Uganda. II: Vegetation studies on Lake Bunyonyi. Freshwater Biology, 3:123–135.

Denny, P., 1980. Solute movement in submerged angiosperms. Biological Review of the Cambridge Philosophical Society 50:65–92.

Denny, P., 1984. Permanent swamp vegetation of the Upper Nile. Hydrobiologia, 110:79–90.

Denny, P., Bowker, D.W. & Bailey, R.G., 1978. The importance of the littoral epiphyton as food for commercial fish in the recent African man-made lake, Nyumba ya Mungu reservoir, Tanzania. Biological Journal of the Linnean Society, 10 (1):139–150.

Denny, P., Harman, J., Abrahamsson, J. & Bryceson, I., 1978. Limnochemical and phytoplankton studies on Nyumba ya Mungu reservoir, Tanzania. Biological Journal of the Linnean Society, 10:24–48.

Denny, P. & Lye, K.A., 1973. The *Potamogeton schweinfurthii* complex in Uganda. Kew Bulletin 28 (1):117–120.

Denny, P. & Weeks, D.C., 1968. Electrochemical potential gradients of ions in an aquatic angiosperm, *Potamogeton schweinfurthii* (Benn.). New Phytologist, 67:875–882.

Denny, P. & Weeks, D.C., 1970. Effects of light and bicarbonate on membrane potential in *Potamogeton schweinfurthii* (Benn.) Annals of Botany, 34:483–496.

Durand, J.-R. & Lévêque, C., 1980. Flore et Faune aquatique de l'Afrique Sahelo-Soudanienne. Tome I. O.R.S.T.O.M., Paris.

Eggeling, W.J., 1935. The vegetation of Namanve swamp, Uganda. Journal of Ecology 23:422–435.

El-Hadidi, M.N., 1965. *Potamogeton trichoides* Cham. & Schlecht. in Egypt. Candollea 20:159–165.

Ferguson, A.J.D. & Harbott, B.J., 1982. Geographical, physical and chemical aspects of Lake Turkana. In: A.J. Hopson, (ed.) Lake Turkana. Overseas Development Administration, London. Published by Institute of Aquaculture, University of Stirling, Scotland. 6 volumes; 1:1–107.

Gaudet, J.J., 1976. *Salvinia* infestation on Lake Naivasha, East Africa. Proceedings of the UNESCO Regional Seminar on Noxious Aquatic Vegetation in the Tropics and Sub-Tropics. Junk, New Delhi, The Hague.

Gaudet, J.J., 1977. Natural drawdown on Lake Naivasha, Kenya, and the formation of papyrus swamps. Aquatic Botany, 3:1–47.

Gerrath, J.F. & Denny, P., 1979. Freshwater algae of Sierra Leone. I. Euglenophyta. Nova Hedwigia 31:525–565.

Gibbs Russell, G.E., 1975. Taxonomic bibliography of vascular aquatic plants in Southern Africa. Journal of the Limnological Society of Southern Africa, 1:53–65.

Gibbs Russell, G.E., 1977. Keys to vascular aquatic plants in Rhodesia. Kirkia 10:411–502.

Golterman, H.L., Clymo, R.S. & Ohnstad, M.A.M., 1978. Methods for physical and chemical analysis of freshwaters. Blackwell Scientific Publications, Oxford.

Good, R., 1974. The geography of the flowering plants (4th edit.). Longman, London.

Hall, J.B., 1975. The vascular flora of Lake Kainji and its shores. In: A.M.A. Imevbore & O.S. Adegoke, (eds.), The ecology of Lake Kainji. University of Ife Press.

Hall., J.B., Laing, E., Hossain, M. & Lawson, G.W., 1969. Observations on aquatic weeds in the Volta Basin. In: L.E. Obeng (ed.), Man-made lakes. Ghana University Press, Accra: 331–336.

Harbott, B.J., 1982. Studies on algal dynamics and primary productivity in Lake Turkana. In: A.J. Hopson (ed.), Lake Turkana. Overseas Development Administration, London. Published by Institute of Aquaculture, University of Stirling, Scotland. 6 volumes: 1:109–161.

Hepper, F.N., 1968. Potamogetonaceae. In: F.N. Hepper (ed.), Flora of West Tropical Africa, 2nd Ed. Rev. 3(1):16–18.

Howard-Williams, C., 1978. Growth and production of aquatic macrophytes in a south temperate saline lake. Internationale Vereinigung für Theoretische und Angewandte Limnologie, 20:1153–1158.

Howard-Williams, C., 1979. Distribution, biomass and role of aquatic macrophytes in Lake Sibaya. In: B.R. Allanson (ed.), Lake Sibaya. Monographiae Biologicae 36. Junk, The Hague: 88–107.

Howard-Williams, C., 1980*a*. Aquatic macrophyte communities of the Wilderness Lakes: community structure and associated environmental conditions. Journal of the Limnological Society of South Africa 6(2):85–92.

Howard-Williams, C., 1980*b*. Aquatic macrophytes of the coastal wetlands of Maputaland. In: M.N. Bruton & K.H. Cooper (eds.), Studies on the ecology of Maputaland. Rhodes University: 42–51.

Howard-Williams, C. & Allanson, B.R., 1981. Phosphorus cycling in a dense *Potamogeton pectinatus* L. bed. *Oecologia (Berl)* 49:56–66.

Howard-Williams, C. & Davies, B.R., 1979. The rates of dry matter and nutrient loss from decomposing *Potamogeton pectinatus* in a brackish south-temperate coastal lake. Freshwater Biology, 9:13–21.

Howard-Williams, C. & Lenton, G.M., 1975. The role of the littoral zone in the functioning of a shallow tropical lake ecosystem. Freshwater Biology, 5:445–459.

Howard-Williams, C. & Walker, B.H., 1974. The vegetation of a tropical African lake: Classification and ordination of the vegetation of Lake Chilwa (Malawi). Journal of Ecology, 62:831–854.

Hutchinson, G.E., Pickford, G.E. & Schuurman, J.F.M., 1932. A contribution to the hydrobiology of pans and other inland waters of South Africa. Archiv für Hydrobiologie, 24:1–136.

Iltis, A. & Lemoalle, J., 1983. II. The main types of communities and their evolution during a drought period. 5. The aquatic vegetation of Lake Chad. In: Lake Chad. Dr W. Junk, The Hague.

Johansson, D., 1976. Aquatic plants and man-made lakes in Africa. In: Ecological Studies of the Mtera Basin. Sweco. Stockholm: 101–113.

Jones, G. & Cullimore, D.R., 1973. Influence of macronutrients on the relative growth of waterplants in the Qu'Appelle Lakes, Canada. Environmental Pollution 4(4): 283–290.

Kilham, P., 1971. A hypothesis concerning silica and the freshwater planktonic diatoms. Limnology and Oceanography, 16:10–18.

Knipling, E.G., West, S.H. & Haller, W.T., 1970. Growth characteristics, yield potential, and nutritive content of water hyacinths. Proceedings: Soil and crop science society of Florida, 30:51–63.

Lebrun, J-P., 1947. La végétation de la plaine alluviale au sud du lac Edouard. Institut Parcs National Congo Belge. Bruxelles. 2 vols.

Lebrun, J-P., 1973. Enumeration des plantes vasculaires du Senegal. Inst. d'Elevage et de Medecine Veterinaire des Pays Tropicaux.

Lebrun, J-P., Audru, J., Gaston, A. & Mosnier, M., 1972. Catalogue des plantes vasculaires du Tchad Meridional. Instit. d'Elevage et de Medecine Veterinaire des Pays Tropicaux.

Léonard, J., 1951. Apercu preliminaire des groupements végétaux pionniers dans la region de Yangambi (Congo Belge). Vegetatio 3:279–297.

Lind, E.M. & Morrison, M.E.S., 1974. East African vegetation. Longman, London.

Lind, E.M. & Visser, S.A., 1962. A study of a swamp at the north end of Lake Victoria. Journal of Ecology, 50:599–613.

Lisowski, S., Malaisse, F., Symoens, J.J. & Van de Velden, J., 1978. Potamogetonaceae. In: Flora d'Afrique Centrale. (Flore du Congo, du Rwanda et du Burundi.)

Lock, J.M., 1973. The aquatic vegetation of Lake George, Uganda. Phytocoenologia 1(2):250–262.

Lock, J.M., 1977. The vegetation of Ruwenzori National Park, Uganda. Bot. Jahrb. Syst. 98(3):372–448.

Lowden, R.M., 1982. An approach to the taxonomy of *Vallisneria* L. (Hydrocharitaceae). Aquatic Botany, 13:269–298.

McLachlan, A.J., 1974. Development of some lake ecosystems in Tropical Africa with special reference to the invertebrates. Biological Review, 49:365–397.

McLachlan, A.J., 1979. The aquatic environment: I. chemical and physical characteristics of Lake Chilwa. In: M. Kalk, A.J. McLachlan & C. Howard-Williams (eds.), Lake Chilwa. Dr W. Junk, The Hague: 59–78.

Mitchell, D.S., 1969. The ecology of vascular hydrophytes on Lake Kariba. Hydrobiologia, 34:448–464.

Mitchell, D.S., 1978. Freshwater plants. In: M.J.A. Werger (ed.), Biogeography and ecology of Southern Africa. Dr W. Junk, The Hague:1115–1138.

Moreau, R.E., 1967. Water-birds over the Sahara. Ibis, 109: 232–259.

Moreau, R.E., 1972. The Palaearctic-African bird migration systems. Academic Press, London and New York.

Musil, C.F., 1973. Water plants of Natal. Wildlife protection and conservation society of South Africa. Durban.

Obermeyer, A.A., 1966. Potamogetonaceae. In: L.E. Codd, B. De Winter & H.B. Rycroft (eds.), Flora of Southern Africa. 1:60–70.

Peyre de Fabregues, B. & Lebrun, J-P., 1976. Catalogue des plants vasculaires du Niger. Institut d'Elevage et de Medecine Veterinaire des Pays Tropicaux.

Richardson, J.L., 1968. Diatoms and lake typography in East and Central Africa. Internationale Revue der gesamten Hydrobiologie und Hydrographie, 53:299–338.

Ruttner, F., 1963. Fundamentals of Limnology. 3rd ed. University of Toronto Press, Toronto.

Sculthorpe, C.D., 1967. The Biology of Aquatic Vascular Plants. Edward Arnold, London.

Seddon, B., 1972. Aquatic macrophytes as limnological indicators. Freshwater Biology 2(2):107–130.

Spence, D.H.N., 1967. Factors controlling the distribution of freshwater macrophytes with particular reference to the Lochs of Scotland. Journal of Ecology 55:147–170.

Spence, D.H.N., 1976. Light and plant response in fresh water. In: G.C. Evans, R. Bainbridge & O. Rackham, (eds.), Light as an Ecological Factor II, Blackwell Scientific Publications, Oxford.

Spence, D.H.N. & Chrystal, J., 1970*a*. Photosynthesis and zonation of freshwater macrophytes I. Depth distribution and shade tolerance. New Phytologist, 69:205–215.

Spence, D.H.N. & Chrystal, J., 1970*b*. Photosynthesis and zonation of freshwater macrophytes. II: Adaptability of species of deep and shallow water. New Phytologist, 69:217–227.

Spence, D.H.N., Milburn, T.R., Ndawula-Senyimba, M. & Roberts, E., 1971. Fruit biology and germination of two tropical *Potamogeton* species. New Phytologist, 70:197–212.

Swindale, D.N. & Curtis, J.T., 1957. Phytosociology of the larger submerged plants in Wisconsin lakes. Ecology 38:397–407.

Symoens, J.J., Van de Velden, J. & Buscher, P., 1979. Contribution à l'étude de la taxonomie et de la distribution de *Potamogeton nodosus* Poir et *P. thunbergii* Cham. & Schlechtend. en Afrique. Bulletin de la Société Royale de Botanique de Belgique. 112(1):79–95.

Tackholm, V., 1974. Students' Flora of Egypt (2nd Edit.) Cairo University.

Tag El Seed, M., (ed.), 1981. The Blue Nile Invaded. Sudan Environment 1(3):13.

Talling, J.F. & Talling, I.B., 1965. The chemical composition of African Lake waters. Internationale Revue der gesamten Hydrobiologie und Hydrographie, 50:421–463.

Van der Ben, D., 1959. Exploration hydrobiologique des lacs Kivu, Edouard et Albert (1952–1954). iv: La vegetation des rives des lacs Kivu, Edouard et Albert. Institut Royal des Sciences Naturelles de Belgique, Bruxelles.

Van Meel, L., 1952. Le Milieu Vegetal. In: Exploration Hydrobiologique du Lac Tanganika (1946–1947). Institut Royal des Sciences Naturelles de Belgique, Bruxelles: 51–68.

Viner, A.B., 1977. The influence of sediments upon nutrient exchanges in tropical lakes. In: H.L. Golterman (ed.), Interactions between sediments and freshwaters. SIL-UNESCO Symposium, Amsterdam, 10–16 September, 1976.

Walter, H., 1971. Ecology of tropical and subtropical vegetation. Oliver and Boyd, Edinburgh.

Welsh, R.P.H. & Denny, P., 1978. The vegetation of Nyumba ya Mungu reservoir, Tanzania. Biological Journal of the Linnean Society 10:67–92.

Welsh, R.P.H. & Denny, P., 1979*a*. The translocation of lead and copper in two submerged aquatic angiosperm species. Journal of Experimental Botany, 30:339–345.

Welsh, R.P.H. & Denny, P., 1979*b*. The translocation of ^{32}P in two submerged aquatic angiosperm species. New Phytologist, 82:645–656.

Wickens, G.E., 1976. Jebel Marra. H.M.S.O., London.

Wild, H., 1961. Harmful aquatic plants in Africa and Madagascar. Kirkia, 2:1–68.

Author's address:
P. Denny
School of Biological Sciences
Queen Mary College, The University of London
London E1 4NS, England

CHAPTER 3

Emergent plants of permanent and seasonally-flooded wetlands

KEITH THOMPSON

Detailed contents

3.1 Introduction

Emergents typically emerge from water (such as a lake margin); floodplain plants may also do so, but only on a seasonal basis. When drainage is impeded, swamps develop, and these may be either permanent or seasonal (see Chapter 6). All aquatic and floodplain plants, therefore, grow on 'wetlands', and the I.U.C.N. definition given in Chapter 1 applies. However, because some floodplains

Patrick Denny (ed.), The ecology and management of African wetland vegetation.
 ISBN 90 6193 509 1

are only inundated for perhaps two or three weeks in the year, it may be helpful to add a further definition of 'wetland' to those already offered:

> 'Wetlands' is a collective term for permanently or temporarily wet areas, shallow-water and land-water interfaces. The water may be fresh, brackish or saline, and wetlands are characterised, in their natural state, by plants and animals which are adapted to living in wet conditions (Environmental Council, 1984).

Thus, *Themeda triandra*, a common grass of the 'better' soils, some of which are seasonally flooded, is not especially adapted to flooding; *Leersia hexandra*, on the other hand, clearly is so, because its distribution is limited to the lower reaches of floodplains. All floodplain plants are emergents at one time or another (*i.e.* they are 'seasonal emergents') and, in order to be typical of floodplains, they must also be adapted in some way to being emergents, or to at least being very tolerant of waterlogging.

Unfortunately, although the African wetland and aquatic plant literature is quite extensive (see Chapter 10), it is almost exclusively regional and descriptive by nature. This chapter attempts to identify patterns on a continental basis and also to explain them in terms of autecology and synecology.

Although this book does not deal specifically with coastal or inland saline wetlands, such as coastal lagoons, saltmarshes, mangroves, dune lakes and slacks, seagrasses, salt pans, *etc.*, it is neither possible nor appropriate to ignore coastal and saline ecosystems altogether. The present chapter aims, wherever possible, not only to describe but also to explain the distribution of freshwater aquatic and semi-aquatic macrophytes. No such explanation can be valid unless the ecological extremes have been studied as well as the ecological norms. Much attention is paid in the following pages to seasonally flooded communities – the extreme case of plants tolerant of both flooding and of seasonal drought. It is equally important to examine the interface between saline and non-saline habitats, because the addition of osmotic stress to flooding/waterlogging tolerance imposes a very powerful selection pressure on species occurrence. It is as important to understand the selection pressures introduced by salinity as it is to appreciate that some semi-aquatic plants are restricted to sandy soils and others to base-rich substrates. Figure 3.13 shows how 'wetness' and salinity interact.

In contrast with temperate regions, there is no clear distinction in Africa between saline and non-saline soils. It is simply a matter of degree; tropical floodplains and seasonally wet soils are often saline, particularly on sand, because of the accumulation of evaporites. This chapter, therefore, ranges from the most salt-tolerant species, such as *Suada articulata, Diplachne fusca* and *Scirpus maritimus*, to aquatics such as *Miscanthidium* and *Scirpus cubensis* which are characteristic of high quality waters. It also covers the ground between halophytic shrubs such as *Tamarix, Pluchea* and *Pandanus* to large freshwater swamp emergents such as *Mitragyna* and *Syzygium*.

3.2 Distribution and diversity of species and communities

The diversity of wetland emergents is vastly greater than all euhydrophytes combined, and no attempt will be made here to reproduce a full list for the 30 million km^2 of the African continent. Rather, it is intended merely to characterise the different wetland types by discussing typical plant species and communities. The only literature of continent-wide value is Huntley and Walker (1982), Knapp (1973), Menaut (1983), Thompson and Hamilton (1983) and White (1983). The following regional references were also of particular value in the preparation of this Chapter: *Eastern and Central Africa* – Lind and Morrison (1974), Thompson (1976), Vesey-Fitzgerald (1963b, 1970); *Southern Africa* – Howard-Williams (1979), Mitchell (1978), Smith (1976), Thompson (1974); *Sudanian Zone* – Iltis and Lemoalle (1983), Lock (personal communication); *Guinea/Congo Zone* – Boughey (1957), Koechlin (1961), Lebrun and Gilbert (1954), Robyns (1947–1955), Sita (1970), Thompson (unpublished data), Van der Ben (1959). Additional literature is cross-indexed in Chapter 10.

It is quite instructive to look at some general statistics. Menaut (1983), for instance, notes that, in contrast with both the neotropics and the Asian tropics, the species richness of African savanna regions (*ca.* 1750 spp per 10 000 km^2) is not much less than that of tropical forest regions (*ca.* 2020 spp per 10 000 km^2). Although this is mainly due to species-poor forest (Hamilton, 1976, points out the role of altitude and Pleistocene climatic changes), there is also an especially diverse savanna zone flora, particularly south of the Sahel. There are other factors contributing towards the overall diversity of African vegetation, in particular the enormous altitudinal range. The highest diversity is, of course, to be found below 1000 m, but most of the African land surface falls between 800–1300 metres. A significant proportion (Hamilton, 1982; Thompson and Hamilton, 1983) is above 2000 metres. The African continent also spans some 65 degrees of latitude.

Species diversity rises with habitat diversity. Thus Robyns (1947–1955) recorded, for the Parc National Albert (which includes forest, swamp, savanna and afro-alpine zones), 2000 higher plant species per 8000 km^2. Lock's (1977) study of the Rwenzori National Park (immediately to the east of the Albert) listed at least 1400 species in its 2000 km^2 (which does *not* include alpine zones, but does have a particularly rich and varied flora on its largely nutrient-rich volcanic and alluvial soils). It is to be expected that the species diversity would be relatively low on the base-poor, sandy soils which characterise the northern and southern parts of Africa, but even the lowland forests of Zaire are species-poor when they are compared with their Far East and New World counterparts. White (1983), for instance, remarks upon the paucity of tree species in the swamp forests of the Guinea-Congo Zone, relative to the enormous size of the area. Of course, most of this zone has always been coherent, so that floristic similarities throughout are to be expected.

Moreover, waterlogging and flooding constitutes a powerful ecological stress factor which does itself restrict species diversity. Since Africa boasts over 25% of the world's readily exploitable hydroelectric potential, there must clearly be vast areas of wetlands on the continent. Although the major wetlands are mostly isolated from each other (*see* Table 8.1; Fig. 8.1; Fig. 11.3 of Thompson & Hamilton, 1983), and Africa has quite distinct vegetation zones (Fig. 3.1), Brenan (1978) observes that tropical African species frequently show especially wide geographical ranges. Even so the overall level of endemism in some regions (*e.g.* the Zaire Basin) is very high. The wide distribution of many wetland species (and even wider of genera) must be an indication of the antiquity of the African flora. The major disjunctions of Africa, north of the Sahara (see Morgan & Boy, 1982) and South Africa, are to be expected because of the occurrence of the Sahara and Kalahari Deserts. Even so, southern African wetlands retain many distinct floristic similarities with regions further north, and it will be noted that the Kalahari sands do not extend east of Zimbabwe and, moreover, the Okavango River was probably once a tributary of the Limpopo (see Thompson, 1974). There have at no time, in the recent past, been similar 'species migration corridors' to northern Africa.

Figure 3.1 delimits the main vegetation zones of Africa; it incorporates, but simplifies, features of Keay (1959), Menaut (1983), Monod (1957) and White (1971). Boughey (1957) applies the Yangambi classification scheme (Anonymous, 1956) to West African vegetation types, including wetlands categories, and his community types are identifiable within the African wetland classification developed for the present chapter (Table 3.1). Mavuti's (1982) wetland types are based more upon geomorphological criteria and are therefore probably more relevant to Chapter 8.

In order to simplify disussion, the phytogeographic regions of Africa used in this chapter are:

(a) North African zone wetlands: Mediterranean-type climate and European floristic affinities;

(b) Dry Savanna zone wetlands: (1) Sudanian region; (2) Zambezian region. Floristically and ecologically, these may be linked together, but, since the species migration corridor (eastern Kenya) has never been very wide, there are still distinct regional differences, which will become apparent later.

(c) Moist Savanna zone wetlands: (1) West African region; (2) East African region (=

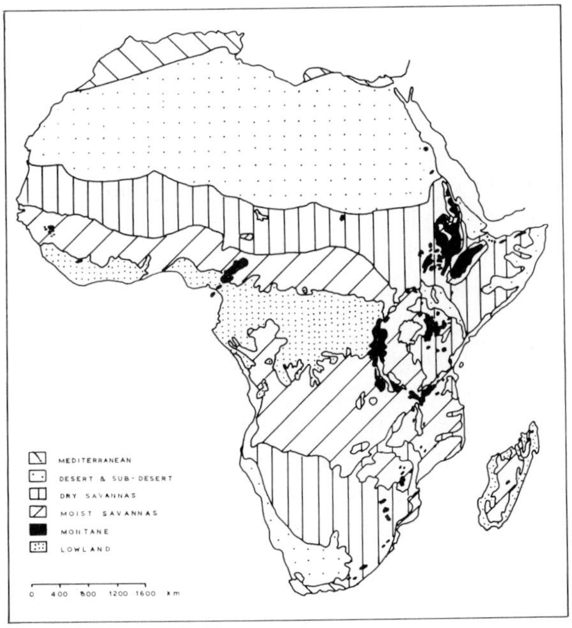

Fig. 3.1. The major vegetation zones of Africa. Zone boundaries are not usually abrupt, and 'transition zones' are therefore occasionally referred to in the text. Compare also with the maps of Anon (1956), Keay (1959), Menaut (1983), Monod (1957) and White (1971).

Lake Victoria region); (3) Central African region.

(d) Guinea/Congo Forest zone wetlands: (1) West African region; (2) Zaire Basin.

(e) Southern Africa zone: mostly confined to the Republic of South Africa, although there are distinct affinities with the Okavango Swamps in Botswana and with the coastal zone of Moçambique. The main species migration routes were undoubtedly coastal (to the east).

(f) Montane/Alpine zone wetlands: these are best developed in eastern and southeastern Africa and are only briefly referred to in the present chapter; they are covered by Thompson & Hamilton (1983), Killick (1979) and Hedberg (1964).

(g) Coastal zone wetlands: best developed in the Guinea-Congo zone and in Moçambique, although several South African estuaries have extensive, though less complex, wetlands without the forest component.

3.3 The major species

This section lists the most important genera of emergent freshwater macrophytes in African wetlands, together with most of the more important species. Taxonomic and ecological notes are provided as appropriate.

3.3.1 *Monocotyledons*

This section is subdivided into the grasses, the sedges/rushes/*etc.*, and the woody monocotyledons (mostly palms).

3.3.1.1 *Graminae (= Poaceae)*

As indicated earlier, plants need specific adaptations in order to flourish in water for prolonged periods of time. The selection pressures are very strong and the adaptations required are quite major; many of the genera are mono-specific, or nearly so. Thus, Vesey-Fitzgerald (1963*a*) collected 378 grass species from 102 genera in that part of the Zaire River catchment which falls within Zambia and Tanzania. Half of the genera were mono-specific, and at least nine of these are dominants of a major Central African wetland community.

3.3.1.1.1 *Small wetland genera.* The following list covers small genera with less than five ecologically important wetland species having wide distribution on the African continent.

Acroceras spp. *A. macrum* is the most characteristic, and often the dominant, component of the 'water-meadow grasslands' (*sensu* Vesey-Fitzgerald, 1963*b*) of floodplains. *A. amplectens* occurs in Nigeria, and throughout the Guinea-Congo zone (Streel, 1963), but not to the complete exclusion of the widely-distributed *A. macrum*. *A. zizanioides* is an East African species of restricted distribution, preferring shaded locations. *Acroceras* is usually closely associated with *Leersia* and, in backwater swamps and dambos, with *Echinochloa* species (see Figs. 3.6, 3.7 and 3.10).

Cynodon dactylon. A cosmopolitan, rhizomatous, mat-forming species characteristic of mildly-saline, shallow-flooded riparian and floodplain grasslands and floodplain margins (see Figs. 3.2, 3.9, 3.10 and 3.13). A good account of *Cynodon* grassland is provided by Heeg and Breen (1982). It is extensively grazed, frequently forms a mosaic with *Vetiveria* (a much more robust floodplain grass), and occasionally even with *Hyparrhenia rufa* (Fig. 3.2).

Diplachne fusca. A genus of only 20 species which is also ecologically important in Asia and Australia. *D. fusca* is usually the only macrophyte present in shallow, semi-permanent alkaline swamp and it persists as a dense sward even if the lake dries out. It is both an edaphic- and a fire-climax community. (See Fig. 3.13 and Plate 3.2.)

Echinochloa spp. Although there are about 20 species continent-wide, there are no more than 30 world-wide. Only four or five are at all common in African aquatic and semi-aquatic habitats. All of these are important food species for grazing animals. (See Plates 3.4, 5.4, 7.1.)

E. pyramidalis may exceed 2 metres in height, and it dominates on heavy alluvial clay soils in the medium/shallow-flooded zones of riverine and lacustrine floodplains throughout Africa (see Figs. 3.2, 3.6 and 3.15). In the Guinea/Congo zone it is firmly restricted to fertile soils near rivers, since much of its upper floodplain range is occupied either by swamp forest or by *Jardinea* grassland. On the toich soils of the Upper Nile Swamps, Sudan (Fig. 3.6), *E. pyramidalis* forms extensive swards considerable distances from open water.

E. scabra (formerly *E. stagnina*), known as 'burgu' or 'bourgou' in West Africa as, for instance, in the dune lakes of coastal Nigeria (Clayton, 1957), is a less robust species which may occur with *E. pyramidalis*, but is more characteristic of deeper-flooded areas lower on the floodplain. *E. scabra* is often mixed with *Vossia cuspidata* (*see* Fig. 3.11), with the latter dominating the deeper, faster flowing sites and the former the shallower, static or slower-flowing parts. *E. scabra* has an extensive distribution outside Africa (*e.g.* south-east Asia).

The perennial *E. haploclada* and the annual *E. colonum* are characteristic species of dambos (Fig. 3.8) and seasonally waterlogged (often mildly saline) ground away from rivers (*e.g.* Lock's, 1977, depressions in silt/clay soils of the Rwenzori National Park in Uganda).
The annual *E. crus-pavonis* may be a component of sudd (Table 3.3) communities (floating vegetation mats), but may also exhibit a tufted growth habit on firm swamp substrates.

Hemarthria natans. A genus of 10 species with a single important African member, *Hemarthria* is rhizomatous with floating stems. It may either be a minor sudd component or, more commonly, is associated (but not as a dominant) with *Panicum repens* on sandy soils (*e.g.* Lock, 1977; Smith, 1976).

Imperata cylindrica. This is a widespread species throughout the tropics from India to Malaysia (where *I. arundinacea* is the dreaded 'lalang weed' of rice paddies) and Australia. *Imperata*, with its massive, deep-rooted rhizomes, is a fire-climax grass, typical of old, heavy, easily-waterlogged, clay soils and marginal valley grasslands, and is particularly common in the forest zone. Few animals, other than hippo, will attempt to graze this plant.

Jardinea spp. There are only four species in this genus and all are restricted to the Guinea/Congo zone of Africa. *J. congoensis* and *J. gabonensis* (formerly *Pobeguinea gabonensis*) are by far the most important, together characterising the lower parts of seasonally inundated floodplains in the forest zone (Fig. 3.11). In this respect, *Jardinea* occupies part of the *Oryza longistaminata, Echinochloa pyramidalis* and *Vetiveria* zones so typical of this degree of inundation elsewhere in Africa.

Leersia spp. All 17 species of this genus are tropical marsh grasses. By far the most important species is *L. hexandra*, which is widely distributed both within Africa and outside the continent (Papua New Guinea, Amazonia, *etc.*). There are 9 African species, some of which are difficult to separate taxonomically from *Oryza* (Launert, 1965). The geographical ranges of all African species (*e.g. L. angustifolia* and *L. drepanothrix* in the Sudan) other than *L. hexandra* are very limited.

L. hexandra, Oryza, Acroceras and *Sacciolepis* characterise the water-meadow grasslands: medium/deep flooded, mid/lower floodplain soft-leaved grasses with very high grazing value (Fig. 3.2). *L. hexandra* is also a common component of sudd communities (Table 3.3).

Miscanthidium spp. Seven species are in the genus, but only two are widely distributed in African swamps. The distribution of the broad-leaved *Miscanthidium violaceum* (= *Miscanthus violaceus*) is centred on the Lake Victoria catchment (Fig. 3.5), although it extends as far as eastern Nigeria and southern Sudan. The terete-leaved *M. junceum* is strictly southern African, centred on the Okavango Delta (Plate 3.1). Both species are typical of acid, low-nutrient, permanently shallow-flooded sites, usually behind other types of reed-swamp. *M. capense* and *M. rectum* have very restricted distributions (from Zambia to Natal) in southern Africa.

Monocymbium ceresiiforme. There are only four species in this tropical genus, but a single one has ecological importance in African wetlands. It is rarely a dominant, but is a consistent and often major component of the widely distributed *Loudetia simplex* edaphic grasslands (Fig. 3.8).

Odyssea paucinervis. The genus contains only two or three species. The highly salt-tolerant and widely distributed *O. jaegeri* was recently transferred to *Psilolemma*, but *O. paucinervis* has a similar ecological range in south-western Africa. It is, for instance, a major component of the fringing vegetation of the world's largest salt pan: the Makgadigadi in Botswana. The species is not found outside this region.

Oryza spp. The genus contains only 25 species, all tropical and all with C_3 metabolism (see 3.6.3.4). All but a few (relatively) mesophytic forest species (such as *O. eichingeri* in central Africa) are aquatic with shallow/medium flooding adaptations. A few *Oryza* species are quite rare, such as *O. punctata*, which is found locally in perennial pools in high forest. *O. sativa*, of course, is the Asian cultivated rice, first introduced to Zanzibar about 2000 years ago. There has been much confusion regarding the taxonomy of the two common African *Oryza* species, and it is therefore appropriate to try to clarify the position. *O. barthii* is an annual, distributed from Tanzania to Senegal. *O. glaberrima* was selected from the *O. barthii* agg. and has been cultivated (in, for instance, the inland delta of the Niger in Mali) as 'African wild rice' for some 4000 years, although the genetic variability of its range of cultivars is considered to be rather less than that of the *O. sativa* which now dominates world rice markets (Buddenhagen & Persley, 1978). It is very likely that all material of *O. stapfii* and *O. breviligulata* should be referred to *O. barthii* agg. *O. longistaminata*, on the other hand, is a perennial, distributed mainly in southern and central Africa, but with a considerable *O. barthii* overlap and also having a considerable tropical distribution outside Africa. *O. longistaminata* is often wrongly referred to *O. barthii*, although the specific *O. perennis* is correct, and it is frequently used in the more recent literature.

In other words, only two *Oryza* species are *both* ecologically important *and* widespread in Africa: the annual *O. barthii* and the perennial *O. longistaminata*. Both are typical of water-meadow grasslands (see *Acroceras* and *Leersia*), although *O. longistaminata* tends towards the shallower-flooded end of this community and *O. barthii* is a common sudd component. In West Africa, *Oryza* often replaces *Vetiveria* on clay soils (Clayton, 1957), as it does also (Fig. 3.6) on the toich soils of the Sudan (Lock, personal communication; Tothill, 1948).

Paspalidium geminatum. There are twenty, mainly Old World, species in the genus, but only one is ecologically important in African wetlands. *P. geminatum* is widely distributed and locally common in sheltered, relatively deep water on sandy substrates (Fig. 3.9). It is tolerant of seasonal changes in water levels and, for example, dominates the fringe vegetation of Nyumba ya Mungu Reservoir in Tanzania (Welsh and Denny, 1978). Howard-Williams (1979) describes *Paspalidium* swards in Lake Chilwa (Malawi) growing in the sort of exposed situations typical of *Scirpus* and *Eleocharis*. *Paspalidium* clearly prefers sand, but it is probably also intolerant of competition. It may

Plate 3.1. Permanently shallow-flooded *Miscanthidium junceum* community with extensive coverage throughout the seasonal swamp region of the Okavango Delta, Botswana. Location: Moanachira distributary at Xaxanaxa. The shrub is *Ficus verruculosa* and a stand of *Scirpus inclinatus* can be seen on the water's edge to the left of centre. (Photo: K. Thompson.)

also be present in the water-meadow grasslands with *Acroceras, Oryza* and *Leersia*.

Phragmites spp. There are only three or four species worldwide, and all are aquatic reeds with adaptation to permanent flooding or waterlogging. *P. australis* (formerly *P. communis*) is cosmopolitan but, though well distributed in tropical regions, is more typically a temperate zone species. *P. mauritianus* (Fig. 3.15) may grow to 5 metres and has an exclusively tropical distribution in Africa. Where it overlaps with *P. australis, P. mauritianus* is more characteristic of riverbanks, with *P. australis* dominating the silted flats and backwater swamps. *P. karka* is also a much larger plant than *P. australis*, but its ecological range is apparently identical, so that it replaces the common reed in the Upper Nile Swamps of the Sudan, its only African locality (Fig. 3.6). The worldwide distribution of *P. karka* is quite disjunct, and the species may need some taxonomic attention; Campbell (1983), for instance, records it from tropical Australia.

Plate 3.2. Expanding swards of *Paspalidium geminatum* and *Diplachne fusca* in Lake Chilwa, Malawi. The shallow, saline water is a typical habitat of *Diplachne* and the *Paspalidium* is an indicator of a sandy substrate. (Photo: C. Howard-Williams.)

Psilolemna jaegeri. The single African species in this genus (formerly *Odyssea jaegeri*) is found throughout the continent, often forming a mosaic with the equally salt-tolerant *Cyperus laevigatus*. *P. jaegeri* is usually found adjacent to *Diplachne fusca* swards in saline lake basins such as those of the Eastern Rift Valley, Lake Chilwa or Lake Chad.

Sacciolepis spp. This genus has about 30 species, all flooding-tolerant. Several species exist in Africa, but only *S. africana* is widespread (water meadow grasslands). *S. auriculata* is also common around swamp margins in Uganda.

Suddia sagittifolia. This remarkable new monospecific genus in the Poaceae was discovered only in 1978 and described by Renvoize, Lock and Denny (1984). Its distribution is still imperfectly known, but it is apparently endemic to the Upper Nile Swamps (Plate 9.1). It is rhizomatous, with nodal rooting, attains a height of 2.5 metres and grows within papyrus swamp. Its enormous 12 cm wide leaves are well adapted to function in the shade of a reedswamp sward and gives it a rather similar habit to *Limnophyton* (Alismataceae).

Themeda triandra is one of less than 10 species in the genus. It is widespread on entirely mesic sites on the better soils (volcanic, alluvial), but is also tolerant of flooding and waterlogging because of the nature and distribution of its preferred substrates. It is a highly palatable species and is extensively grazed.

Vetiveria nigritana. About 10 species occur throughout the Old World tropics, but only two or three are African. *V. nigritana* is extremely important ecologically and frequently dominates the perimeter of the regularly-flooded zone of floodplains (Fig. 3.2). It prefers light, sandy soils (*e.g.* sandy silts), where it may be associated with *Acacia albida*. *Vetiveria* is replaced by *Oryza* on clay soils. *V. fulvibarbis* occupies a similar niche in the coastal grass savannas of the Guinea/Congo zone.

Vossia cuspidata is a monospecific genus found in Africa, India and south-eastern Asia. It characterises (and usually dominates) a major aquatic community type. There are two growth forms: the 'typical' very deep-flooded, floating-rhizome form which is able to colonise quite strongly flowing waters outside reed-swamp or along levée fringes (Fig. 3.6). Where flows are less, *Vossia* swards become mixed with *Echinochloa scabra* (Fig. 3.11), the latter species dominating in the pools and backwaters well behind the current-breaking fringe of *Vossia*. It is under these conditions that *Vossia* occurs in the 'erect' form. One form can change rapidly to the other if habitat conditions alter. With still lower flow rates *Vossia* becomes a 'sudd' component (see Table 3.2). (Plates 1.5, 3.3.)

Species of the above 18 genera are all exclusively aquatic or semi-aquatic. The small size of the genera is very typical of stress-adapted taxa. Put another way, the stress (flooding) tolerant grasses are usually taxonomically quite isolated.

3.3.1.1.2 *Large grass genera characteristic of wetlands.* The larger the genus, the less specialised are its species likely to be. Thus the more mesophytic genera usually have numerous species: *e.g. Eragrostis*, 300 spp world-wide, 35 African; *Hyparrhenia*, 90 spp world-wide, over 40 African.

There are no permanent swamp species in the large genera, and the following supplementary list, therefore, covers only seasonally flooded grasses. All of the larger genera also have non-wetland species.

Andropogon. This large genus contributes several seasonal swamp species, all of which are typical of clay soils. *A. gayanus* is continent-wide: in Nigeria it dominates a major floodplain community immediately above permanent swamp and is typically seral to forest (Fig. 3.12); in Botswana it characterises the 'red band' around the permanent swamps of the Okavango Delta, where it is replaced by *Imperata* swards on drier ground (Smith, 1976). *A. gabonensis* occurs with *Imperata* in the Congo Republic and *A. africanus* is seral to *Hyparrhenia* in Zaire (Fig. 3.11). All wetland species are, therefore, typical of heavy soils in the upper parts of floodplains.

Brachiaria. Most species are characteristic of soils with impeded drainage in the upper part of the seasonal floodplain (Fig. 3.8), but in the Guinea/Congo zone, *B. deflexa* may occur in the deeper-flooded parts (Fig. 3.11) as also does *B. mutica* in the Zambezian region (White, 1983). The most widespread wetland species in East and Central Africa is *B. brizantha*, but *B. obtusiflora* dominates parts of the Blue Nile floodplain in the Sudan (Bunting & Lea, 1962) and *B. rugulosa* and *B. deflexa* are the commonest species of this genus in the Okavango (Smith, 1976). Harker & Napper (1961) record *B. deflexa* in Uganda as 'a dry grassland species and weed of arable land': either Harker & Napper are not recognising the true seasonal wetland status of the Ugandan material, or *B. deflexa* is a species with exceptionally wide (for a grass) ecological tolerance, or there is a taxonomic problem.

Eragrostis. E. barteri, a component of permanent *Vossia* swamp in Lake Chad (Iltis & Lemoalle, 1983) is probably the only fully aquatic species of this genus in Africa. *E. mildbraedii* occurs in *Miscanthidium* swamps in Uganda (Table 3.4), but its flooding tolerance is more similar to that of the upper floodplain species (such as *E. stapfii* and *E. lappula*) listed by Smith (1976) for the Okavango Delta.

Hyparrhenia. Most species are found on dry savanna, but several are typical of the highest parts of floodplains and dambos. *H. rufa* (Figs. 3.2, 3.6, 3.11) is found at, or above, the upper limit of flooding in higher rainfall zones. It prefers heavy colluvial soils. Several other *Hyparrhenia* species, such as *H. filipendula* and *H. bracteata*, may also occur in this ecological zone, where flooding is rare but seasonal waterlogging is normal. *H. diplandra* (Fig. 3.11) belongs to a slightly lower flooding zone and may be found on a lighter soil type with *Panicum* as well as on the clays typical of *Andropogon* grasslands.

Loudetia. L. simplex is widespread on sandy soils

in the savanna zone, where it characterises the upper parts of floodplains. Its counterpart of the heavier soils is *Setaria* (Figs. 3.7, 3.8). *L. kagerensis* is important on shallow sandy soils around Lake Victoria. The giant *L. phragmitoides* is also centred in the Lake Victoria region, but extends to the southern Sudan and to Nigeria.

Panicum. This genus has over 500 species, mostly mesophytes and many with excellent grazing value. All African wetland species are highly palatable and typical of light, usually sandy, soils on the lower half of the shallow-flooded (seasonal) part of riverine floodplains and dambos. *P. repens* is the most widespread wetland species, often dominating regularly-flooded sandy substrates. This species is rhizomatous and low-growing; it may form a close turf with *Cynodon* (Fig. 3.9) or occur as a minor component in water-meadow grassland (Fig. 3.10). *P. maximum*, on the other hand, is a sward-former growing to 3 metres in height; it also prefers light sandy silts and is widely distributed throughout Africa (Figs. 3.8, 3.11). Smith (1976) records *P. hymeniochilum* as a permanent swamp species in the Okavango, but it is more likely to be a component of a seasonally-flooded part of the permanent swamp zone.

Paspalum. This is again a very large genus of over 250 species, mostly mesophytes. The common African wetland species occur rather below *Panicum* species on 'typical' floodplains: *P. polystachyum* is even recorded by Adejuwon (1971) as an associate of *Cyperus papyrus* (Fig. 3.12) in coastal swamps of Nigeria. *P. scrobiculatum* (which probably includes *P. commersonii*) has a wide tropical distribution, being cultivated in India (for fodder) as the 'Kodo millet'; in Africa it is a frequent component of water-meadow grasslands in savanna regions (Figs. 3.7, 3.10).

Pennisetum. This large genus contains only very few species tolerant of flooding. *P. glaucocladum* is an important component of the deeper flooded parts of the Bangweulu Floodplain (Zambia), but the giant (up to 5 metres) *P. purpureum*, though tolerant of waterlogging, is more typically a species of river banks and forest margins in the forest zones.

Setaria. Many wetland species exist in Africa, all typical of heavy black clay soils high up on the floodplain (above the water meadow grasses and below *Hyparrhenia*, Fig. 3.8). *S. sphacelata* typifies this distribution in savanna zones (Fig. 3.10, and *S. anceps* in the forest zone (Fig. 3.11). *S. avettae* may occur on lighter soils with *Vetiveria* (Fig. 3.7) and *S. angustifolia* is found on sandy silts in the Okavango.

Sorghum. This is, of course, a very important genus agriculturally, but several species are highly tolerant of flooding. *S. verticilliflorum* grows to 4 metres with *Echinochloa pyramidalis* in the Kafue Swamps (Sheppe & Osborne, 1971) and *S. lanceolatum* is widely distributed, though nowhere dominant, on the toich soils of the southern Sudan swamps (Tothill, 1948). *S. purpureo-sericeum* grows on heavy, seasonally-waterlogged clays in Uganda (Harker & Napper, 1961) and *S. rigidifolium* is a component of *Miscanthidium* swamps in Natal (Knapp, 1973). In fact, *S. rigidifolium* has now been transferred to the small (*ca.* 25 species) genus *Sorghastrum*, and the species is widely distributed in Africa, usually in the *Setaria* zone of floodplains.

Sporobolus. This is another very large genus (*ca.* 150 spp), containing many species, particularly *S. spicatus* and *S. pyramidalis*, tolerant of water-logged alkaline soils. Indeed, *Sporobolus* is one of the best indicator genera of saline or 'sour' soils on savanna (Fig. 3.9).

3.3.1.2 *Other herbaceous monocotyledons*

This list is not exhaustive, but it does cover the most important genera.

Alismataceae. *Limnophyton* is a common associate of *Cyperus papyrus* in the tropical zone (Fig. 3.5; Plate 1.5).

Amaryllidaceae. Several species of *Crinum* occur in Africa, in savanna and miombo woodland zones, where its underground storage organs are an ideal adaptation to enduring drought. However, *Crinum* is more characteristic of swamps and, in the Guinea/Congo zone, the enormous *C. giganteum* with leaves over 2 metres long, is a common com-

Plate 3.3. Massive development of papyrus swamp extending from the mouth of Nakivubo inlet on the northern shore of Lake Victoria. The city of Kampala lies beyond. In some places (for instance, just above centre), *Vossia* communities are expanding at the expense of papyrus; this is difficult to explain, but it may be related to the particularly high nutrient status of sewage works effluent discharged by the Nakivubo Channel. (Photo: K. Thompson.)

ponent of the ground flora in some types of swamp forest.

Araceae. There are several genera, but only three are important in wetlands: *Culcasia scandens* (not to be confused with the aroid *Colocasia esculenta*, the edible marsh taro, now introduced to West Africa from tropical Asia) is a climbing/scrambling aroid of West African swamp forests. *Cyrtosperma senegalense* is another giant (with petioles up to 2 metres long and peduncles up to 3 metres), which occurs in West African swamp forests, and may dominate the ground flora of some swamp forests in Zaire, particularly where canopy gaps occur. Indeed, White (1983) suggests that *Cyrtosperma* largely replaces *Cyperus papyrus* in the Zaire Basin. This may well appear to be the case, but it would be misleading to imply a straightforward instance of competitive exclusion, since there are other reasons for the scarcity of papyrus in the Zaire forest zone. *Typhonodorum lindleyanum* is an aroid restricted to swamp forests of coastal eastern Africa (Greenway, 1973).

Commelinaceae. Although *Commelina* may be better known as a weed of arable land and forest margins, some species (*e.g. C. diffusa, C. fluviatilis*) are common emergents in swamp margins and may also be sudd components.

Cyperaceae. This is by far the most important plant family in permanently herbaceous swamps. The major genera will be mentioned alphabetically: *Carex* is only of minor importance in tropical Africa, although it is often a sward-former at high altitude. Thus *C. runssoroensis* and *C. monostachya* dominate large areas of bog and moorland on the mountains of Central Africa. *Carex* also increases in importance north of the Sahara and south of the Kalahari. *C. clavata*, for instance, is a significant component of the flora of Groenvlei Fen on the coast of Cape Province (Martin, 1960).

Cladium is similarly distributed, although its single (African) species, *C. jamaicense*, becomes ecologically important at lower altitudes and latitudes than *Carex*. *Cladium* is not a common species in Africa and, although a sward-former, is of very restricted distribution in sites of rather better nutrient status. In fact, *Cladium* partly replaces papyrus in the more temperate climates, such as Lake Bunyoni, Uganda (Denny, 1971), the high altitude swamps of Rwanda (Deuse, 1966) and some coastal fens of South Africa (Martin, 1960). The few tropical occurrences of *Cladium* – *e.g.* very locally in the Okavango Swamps (Thompson, 1976*b*), small scattered stands around the shores of Lake Victoria (Lind and Visser, 1962), a small floating stand in the Rwenzori National Park of Uganda (Lock, 1977), etc. – are more difficult to explain in ecological terms.

Cyperus. A huge genus of over 600, largely tropical and sub-tropical, species which is widespread in Africa on (usually) poor, heavy soils or in deep water. *C. papyrus* is by far the most ecologically important aquatic emergent in this genus and, apart from one remaining occurrence (now reduced by agriculture to a 10 ha reserve) in the Huleh Swamps of Israel (latitude 32°N) has a natural geographical distribution which is entirely African and does not now extend above latitude 15°N (Thompson, 1976*a*). Papyrus is widely distributed in eastern and central Africa, but is relatively rare in the Guinea-Congo zone (*see* Fig. 3.14 for a typical wetland succession in the forest zone) and in any wetland with a seasonal amplitude greater than 2.5 metres. *C. papyrus* is the largest sedge in the world, attaining almost 9 metres in Lake Tana, Ethiopia, although 4–5 metres is an average for most locations. It is frequently a sudd dominant (Table 3.2; Fig. 3.5) and only remains in a healthy condition when the substrate is permanently flooded (Figs. 3.6, 3.7). *C. articulatus* is also a deep-flooded species, but does not typically form dense swards; it is a regular associate of *Scirpus inclinatus* in the lower parts of seasonal floodplains in the savanna zone (*e.g.* the Okavango). *C. latifolius*, on the other hand, is a shallow-flooded sedge of heavy clay soils near permanent water, or of peats and peaty gleys at high altitude (Deuse, 1966). Lock (1977) records permanently flooded *C. latifolius* swamp in the Western Rift Valley of Uganda.

Species such as *C. denudatus, C. longus, C. haspan* and *C. digitatus* all have wide geographical distributions and occur on the poorer clay soils behind permanent reed (particularly papyrus) swamps. Vesey-Fitzgerald (1970) records *C. longus* on heavy montmorillonite clays, in the Rift Valley of Tanzania, with *Scirpus maritimus* and *Cynodon dactylon*; this would make it as salt-tolerant as *C. procerus* (Fig. 3.13). *C. laevigatus,* however, is an 'obligate halophyte', being the most characteristic species of the more swampy margins of saline lakes (Fig. 3.13).

Eleocharis. This genus has some 200 species, with many in Africa. *E. acutangula* is probably the most widespread African species, either emergent from deep, permanent waters beyond floating sudd, or a minor component of *Vossia*/*Echinochloa* communities and water-meadow grasslands from the Sudan to South Africa. It is frequently associated with *Scirpus inclinatus*. *E. plantaginea* occupies the deep water niche in the Zaire Basin (Fig. 3.11), and *E. dulcis* occurs with *E. acutangula* in Bangweulu (Mitchell, 1978) and Okavango (Smith, 1976). Several species, such as *E. atropurpurea, E. intricata* and *E. nigrescens*, are more typical of shallow-flooded, but always waterlogged, *Cyperus (denudatus, longus*, etc.)/*Fimbristylis*/*Fuirena*/*Pycreus* communities on seasonal floodplains and lake margins.

Pycreus. By far the most common species is *P. mundtii* (sometimes recorded as *Cyperus mundtii*). Throughout Africa it is a component of sudd (Table 3.2) and of shallow-flooded, permanently waterlogged substrates; it is also moderately tolerant of alkaline soils (Fig. 3.13). *P. nitidus* is found in southern Africa. *Pycreus* is often a peat-former at high altitudes (over 2000 metres) in East Africa, as it is in some backwater swamps of the Okavango Delta.

Scirpus. This genus has a very similar ecology to that of *Eleocharis*. It therefore includes the widespread deep-water species *S. inclinatus* and *S. cubensis*, as well as the shallow-flooded species of permanently waterlogged ground, such as *S. isolepis* and *S. microcephalus*. *S. inclinatus* is typically associated with *Cyperus articulatus* as a sparse, deep-water community, but it may also be a sward-former on seasonally shallow-flooded terrain (as it frequently is in the Okavango Swamps).

S. cubensis is often a sudd component (Table 3.2). *S. setaceus* is a temperate zone species of the afro-alpine wetlands of eastern Africa, and there are several salt-tolerant species. *S. maritimus* is mainly coastal, but also occurs in several inland basins where alkaline soils develop (such as the Eastern Rift Valley saline lake basins, the Makgadigadi Salt pan *etc.*). *S. littoralis* (plate 1.8) is also a predominantly coastal species, but it occurs inland too, where it is tolerant of moderately saline conditions (*e.g.* parts of the Lake Chilwa swamps). Whereas *S. littoralis* is a mainly southern species, *S. holoschoenus* has an exclusively North African distribution in saline depressions and oases of the Sahara region.

Lastly within the Cyperaceae, there is a group of genera which is best considered collectively. *Ascolepis, Bulbostylis, Fimbristylis, Fuirena, Kyllinga, Scleria* and *Websteria* are characteristic of permanently or seasonally waterlogged clay soils with generally poor nutrient status, such as the wetter parts of dambos. Several *Cyperus* species, and a few *Juncus*, are also associated ecologically with this group of genera, and it is typical to find several members of this group present in the same community, often dominating it as a 'mixed cyperaceous sward'.

This mixed cyperaceous sward is often found with water-meadow grasses.

Juncaceae. Not a common family in the tropical zone, although *Juncus maritimus* is a characteristic component of many saline pans and lake margins. *J. effusus* is found in almost all zones, but is never a prominent member of any community. *J. bufonius* is North African only.

Liliaceae. *Kniphofia* (the 'red hot poker') is worth a mention as the most obvious, even if not the dominant (which is usually *Pycreus*), component of many montane (over 2000 m) fens in the Kigezi region of Uganda (Fig. 11.23 in Thompson and Hamilton, 1983). *K. thomsonii* is also a distinctive feature of (tussock grassland) moorland and *Alchemilla* mires on Mts. Kenya, Kilimanjaro and Elgon.

Marantaceae. *Marantochloa leucantha* and *M. purpurea* are large-leaved herbs, up to 3 metres in height, found in swamp forests of the African tropical region. *Thalia geniculata* is a 2 metre high understorey component of West African swamp forests, but it also forms a common swamp association in the upper reaches of the Zaire River drainage basin, where it is co-dominant with *Oryza longistaminata* and associated with *Vossia, Phragmites* and *Pennisetum glaucocladum*. *T. welwitschii* is a component of shallow-flooded 2–3 metre high wooded *Echinochloa*/*Sorghum* grassland in the Chad Basin.

Restionaceae. The 'jointed rushes' are close taxonomically to *Juncus*, and the 150 African species are all South African. Almost all are characteristic of sandy soils which are seasonally water-logged due to hard-pan formation (see Fig. 3.4). The family also occurs in Australia, and there are two peat-forming high-bog species in New Zealand.

Typhaceae. There are only two African species: *T. capensis* (formerly *T. latifolia*) and *T. domingensis* (formerly *T. australis*), of which the latter is widespread in the tropical and sub-tropical zones. *T. capensis* is more or less restricted to northern Africa, South Africa (including the Okavango Swamps) and some high altitude sites in the equatorial zone. Although *Typha* may persist for several years in the absence of a flooding regime, it is still always a reliable indicator of prolonged waterlogging in the root zone. It is both more drought tolerant (Fig. 3.6) and more salt tolerant (Fig. 3.13; Table 3.3) than papyrus, *Cladium* or *Phragmites*.

Xyridaceae. Almost all of the 250 species worldwide are marsh plants. In Africa, the genus is only ecologically important at higher altitudes, where it may even be co-dominant with *Miscanthidium*, *Erica* and *Sphagnum* (Deuse, 1966; Hamilton, 1982) as a peat-former. *Xyris* is, however, widely distributed at low altitudes, ranging from a *Loudetia/Jardinea/Cyrtosperma/Sphagnum* wetland community near Brazzaville, Congo Republic (Sita, 1970) to floating sudd communities in the Okavango Swamps.

3.3.1.3 *Woody monocotyledons*
Most woody monocots are palms, and roughly half of the 217 genera and 2700 species occur in the eastern tropics. With only 117 species in 17 genera, they are particularly poorly represented on the African continent, where their widespread distribution is attributable to only a handful of very common genera (Moore, 1973). Because even Madagascar has 132 species, it must be assumed that most palms are found in forests (Madagascar was once over 90% forested).

Curiously, there are no palms above 1000 metres altitude in Africa, even though the Palmae contains several species tolerant of quite temperate climates (*e.g. Rhopalostylis* of New Zealand). Even at 2000 metres altitude, the mean annual temperature is nowhere in Africa less than 15°C.

Several palms are wetland species, with *Hyphaene* and *Borassus* being indicators of high water-tables in the savanna zone and *Phoenix*, *Raphia* and *Elaeis* being the major swamp forest genera.

Calamus. This is a lepidocaryid palm and one of Africa's three genera of rattans. The lepidocaryids (which also include *Raphia*) are especially prominent in the Guinea/Congo forest zone, where *C. deeratus* is a swamp forest climber.

Elaeis. *E. guineensis*, the oil palm, has a natural distribution centred in West Africa, although it does extend to some swamp forests in East Africa. Its ecology is similar to that of *Raphia*, but is more restricted to riverine situations and forest fringes.

Hyphaene and *Borassus*. These 'savanna palms' are both widely distributed in grasslands which are regularly waterlogged or seasonally flooded to shallow depth. *Borassus* is the more flooding-tolerant genus, and it will persist when the duration of flooding has eliminated other woody plants of much less stature. *Borassus* is also a common feature of river levées. Corner (1966) considers elephants to be important in the dispersal of both *Hyphaene* and *Borassus* seed, but baboons are much more likely agents as they certainly do disperse the gigantic fruits of the moist savanna and riverine forest tree *Kigelia africana*. *Hyphaene* is unusual among palms in that its stem is typically branched. The genus extends to Arabia, and there are about 30 African species, the most widespread being: *H. ventricosa* (now incorporating both *H. benguellensis* and *H. petersiana*), which is distributed throughout the continent; *H. thebaica*, which extends from West Africa to the Sudan and East Africa; and *H. natalensis*, which is restricted to southern Africa. *Borassus* has only a single African species (*B. aethiopum*), but the genus is also found in India and south-eastern Asia.

Phoenix. *P. dactylifera* is the date palm, a familiar indicator of locally high water-tables in the deserts of Arabia. It was introduced to coastal East Africa by the Arabs, but it has no natural distribution south of the Sahara. *P. reclinata*, on the other hand, is widely distributed as a component of several types of permanent swamp throughout West, East and north-central Africa. It is restricted to the high rainfall zones and may form an edaphic climax with *Cyperus papyrus* in Uganda.

P. reclinata also dominates some West African coastal strand communities.

Raphia. There are about 15 species of this palm in Africa. The leaves are massive, exceeded in length only by those of the estuarine *Nypa* palm in tropical Asia. The commonest African species are *R. australis* (southern African swamp forests), *R. farinifera* (East African swamp forests), *R. hookeri* (West African swamp forests), and *R. sudanica* (West African savanna swamps). *Raphia* is typically scattered throughout a swamp forest, but it may occasionally be dense enough to be a co-dominant.

Two non-palm woody monocotyledons are also

Plate 3.4. Hibiscus diversifolius occupying a characteristic niche on a shallow levée in the forest zone of the Zaire River at Yangambi. *Echinochloa* (bottom left) would dominate this niche in the savanna zone. (Photo: K. Thompson.)

important in certain wetlands. *Pandanus* (Pandanaceae), the screw-pines, are typically co-dominants of the transition zone from mangroves to fresh-water swamp forest (Fig. 3.12), and also occur occasionally inland. Thus, *P. chiliocarpus* is prominant in the understorey of the Mongiro Swamp forest (*Mitragyna/Macaranga/Erythrina/Elaeis*) on the Semliki Plain in western Uganda (Lind and Morrison, 1974). *Dracaena laxissima* occupies a similar lower canopy position in the Lake Victoria zone swamp forests. *Dracaena* is also widely distributed in afro-montane forests and in gallery forests.

3.3.2 *Dicotyledons*

The monocots dominate virtually every herbaceous wetland plant community, particularly those which are only seasonally flooded. The dicots only assume a dominant role in swamp forest.

3.3.2.1 *Herbaceous dicotyledons*

Dicots are probably most prominent in lake fringe and sudd communities, where *Ludwigia* (Onagraceae) and *Polygonum* (Polygonaceae) may even be occasional co-dominants with *Vossia* or *Echinochloa scabra*. There are several African species of both of these dicot genera, with *Polygonum pulchrum* (Fig. 3.11), *P. senegalense* (Fig. 3.2) and *Ludwigia stolonifera* (Fig. 3.2) being the most widespread. There are very few additional non-climbing dicot genera and only *Hydrocotyle, Drosera* and *Alternanthera* can be listed as being at all common. However, the dicots redeem themselves by providing a large number of climbers and twiners, such as *Cassytha* (a *Cuscuta*-like parasite in the Lauraceae), *Cayratia* (Vitidaceae), *Cissampelos* (Menispermaceae), *Ipomoea* (Convolvulaceae), *Melanthera* (Compositae), *Mikania* (Compositae), *Vigna* (Leguminosae) and *Zehneria* (Cucurbitaceae). Climbers and twiners are especially prominent in papyrus swamps, although their contribution to community biomass rarely exceeds 2–3% (Plate 5.1).

3.3.2.2 *Woody dicotyledons*

About 20 dicot families contribute to African wetland communities, of which only about five contain important swamp forest dominants or co-dominants. Unfortunately, the literature on African forest species is almost entirely descriptive by nature and only regional in scope, so that, within the larger 'seasonally inundated' group in particular, very few existing accounts of an ecological nature can be regarded as being in any way definitive. As in the case of the grasses, a species may be wetland adapted, even though the substrate is 'dry' for 6–9 months in the year. A further important consideration is that, although some genera are more wetland-specific than others, some species of a particular genus may be more flooding-adapted than others.

Anacardiaceae. All wetland genera are seasonally inundated only. *Pseudospondias microcarpa* and *Lannea* species are often classified with forests described as 'dry types', and yet both are frequent components of seasonally waterlogged ground in forest regions. They clearly come within the 'waterlogging tolerant' group, rather than the 'flooding adapted' and in this respect are more typical of swamp margins than of swamps themselves. However, *Pseudospondias preussi* is recorded by Eggeling (1947) as being 'confined to swamps (in East Africa) and only very exceptionally found on dry sites'. *Rhus* is more characteristic of gallery forests in savanna regions, but has a similar ecological distribution (seasonal inundation).

Annonaceae. This family too is largely mesophytic, although *Cleistopholis patens* is a characteristic, though uncommon, component of some seasonal swamp and gallery forests in Uganda, (Eggeling, 1947; Hamilton, 1981; Lock, 1977) and in Nigeria (Clayton, 1957; Nielsen, 1964). *Xylopia aethiopica* and *X. rubescens* are large trees (up to 45 m) and are among the five or six most common swamp forest trees in the Zambezi Basin (White, 1983), although *X. rubescens* is also common in West Africa and just extends to East Africa (Nielsen, 1965). *X. holtzii* is southern African (Chapman and White, 1970) and *X. eminii* East African (Langdale-Brown *et al.*, 1964). *X. acutiflora* (Fig. 3.14), *X. parvifolium* and *X. staudtii* are all Guinea/Congo zone species. In freshwater swamps, *Xylopia* is usually associated with *Mitragyna*, but *X. staudtii* is recorded by Crawford (1983) as a mangrove. In fact, the latter species has wide ecological tolerance, since it is more common in entirely freshwater swamps (*e.g.* Bwindi Forest, Uganda: Hamilton, 1981) and also frequently grows in mesic sites (Longman and Jenik, 1974).

Apocynaceae. *Voacanga thouarsii* is one of the more important true swamp forest trees, often a co-dominant, throughout the Guinea-Congo forest zone, western East Africa and coastal Moçambique. In southern Africa, *Rauvolfia caffra* is sometimes an associate of *Voacanga* and *R. oxyphylla* dominates (with *Croton*) some seasonal swamp forests in the Western Rift Valley of Uganda (Langdale-Brown *et al.*, 1964). At higher altitudes, *Rauvolfia* species occur only on free-draining soils.

Aquifoliaceae. *Ilex mitis* has a distribution which is disjunct both altitudinally and ecologically; it occurs in seasonal swamp forests at 1200 m on the north-west shore of Lake Victoria (Lind and Morrison, 1974), as a stunted riparian forest with *Syzygium cordatum* at 1700m in Malawi (Chapman and White, 1970), on peat with *S. cordatum* at 1950m in Rwanda (Deuse, 1966; Hamilton, 1982) and on free-draining soils in the montane zone, with trees such as *Hagenia* (Rosaceae), from South Africa to the Sudan. *Podocarpus* and *Rapanea* share some features of this curious distribution pattern.

Bignoniaceae. *Kigelia africana* is widespread from the Sudan to southern Africa. It is recorded by Hamilton (1981) as a swamp forest tree, as too is the *K. moosa* of Eggeling (1951). However, *Kigelia* is much more typical of riverine forest and it also has a considerable distribution in open moist savanna woodland (White, 1983; Palgrave, 1977).

Compositae. The small shrubby *Pluchea* is probably the only composite woody wetland plant. It is tolerant of permanent water-logging and also of saline conditions (Fig. 3.13).

Ebenaceae. *Diospyros* is a large genus, with many African species. Several are characteristic of seasonal swamp forests throughout the continent, *e.g. D. abyssinica* in East Africa, *D. alboflavescens* 3.14) and *D. longiflora* in Zaire. The genus is most typical of the shallow end of seasonal flooding and of heavy soils which can be seasonally waterlogged.

Ericaceae. This family is only important at higher altitudes in Africa where, under high rainfall regimes, the definition of wetland becomes rather more vague. However, *E. kingaensis* grows on acid peat in the highlands of Central Africa (Deuse, 1966; Hamilton, 1982).

Euphorbiaceae. This is one of the most important families of woody wetland plants in Africa. About 8 genera are ecologically significant as aquatic emergents:
Alchornea cordifolia is almost restricted to the Guinea/Congo zone, where it typically forms a large hedge-like fringe on very shallow levées between water-courses and swamp forest (Fig. 3.14). It is particularly common in the upper reaches of the Zaire River (Lejoly, 1982). *Bridelia micrantha* is a small tree with a wide ecological range: it occurs in secondary, fringing and riverine forests throughout West (Nielsen, 1965), East (Hamilton, 1981), and southern (Palgrave, 1977) Africa, but is also a common component of seasonal swamp savanna in West Africa (Nielsen, 1965) and may even be found within papyrus swamps in Uganda (Lind & Morrison, 1974), although usually on slightly higher ground. *B. ripicola* is a component of riverine forest in Zaire (Fig. 3.14) and Lejoly (1982) records it on levées as a constant associate of *Alchornea*. In the wetter sites, both *Bridelia* species may develop stilts (see Fig. 3.16).

Cleistanthus polystachyus is recorded by Eggeling (1947) as a swamp forest species in Uganda, although others appear to regard it (elsewhere) as only mildly flooding tolerant, at most. *C. ripicola*, on the other hand, cannot be disputed as a seasonal aquatic from its location on the section given in Fig. 3.14.

Croton is a large genus in Africa, but has several species which are found regularly in seasonal swamp forest (*e.g. C. megalocarpus* in the Lake Victoria region).

Macaranga also contains a large number of species, most being colonising species and many being found regularly in swamps. A selection of some of the more important species includes *M. staudtii* of West African colonising forest (Fig. 3.12), the widely distributed Guinea/Congo zone seasonal swamp and colonising forest species *M. monandra*, and the Afro-montane *M. kilimandscharica*, which occurs on both swamp and free-draining soils. *Phyllanthus discoideus* is widely distributed throughout the Guinea/Congo and Lake Victoria regions in similar habitats to *Macaranga* (Fig. 3.12). The same genus, however, has the remarkable free-floating South American *P. fluitans*, with the habit of *Salvinia*.

Sapium ellipticum is recorded as a swamp forest canopy tree in South Africa (Palgrave, 1977), and occasionally as understorey in Lake Victoria swamp forests (Lind & Morrison, 1974), but it occurs widely in colonising and fringing forests.

Spondianthus preussii is a large swamp forest tree found throughout the Guinea-Congo zone and the Lake Victoria region.

Uapaca guineensis is an important medium-sized tree confined to swamp forests throughout the high rainfall areas of Africa. It is not recorded by Palgrave (1977) for southern Africa, although White (1983) lists it for the Zambezian moist savanna zone. The genus also contains several mesophytic species.

Guttiferae. *Symphonia globulifera* is one of the characteristic trees of seasonal swamp forest throughout the Guinea-Congo zone, although it also grows on free-draining soils above about 1000 metres. *Hypericum*, on the other hand, is a large temperate zone genus which is also strongly represented on tropical African mountains above 3000 m. *H. revolutum* is a widespread component of *Rapanea/Hagenia* montane forest up to treeline. Substrates are highly organic and often waterlogged.

Leguminosae. Probably the most important dicot wetland family of all, with wetland genera in all three sub-families and an ecological range from shrubs of permanent swamp to trees of seasonally waterlogged ground. The most important wetland genera are:

Acacia. Several species, such as *A. seyal* and *A. drepanolobium*, are characteristic of the seasonally waterlogged 'black cotton soils' (montmorillonite) in savanna regions. Indeed they are usually the only woody plants which can survive in this habitat. *A. albida* is more characteristic of lighter soils and frequently associates with *Vetiveria* high up on floodplains, or characterises gallery forest in savanna regions (Fig. 3.15). *A. albida* has considerable flooding tolerance. The yellow-barked *A. xanthophloea*, the 'fever tree', is also a gallery forest species, but more typically fringes lakes throughout the moist savanna region; it cannot endure prolonged flooding, but does require a very high water-table (Plate 3.7).

Aeschynomene is a genus of short-lived shrubs and small trees, widely distributed from Chad to South Africa. Some species, such as the southern African *A. trigonocarpa* and *A. nodulosa*, are mesophytes, but the most widespread species, *A. elaphroxylon* (known as 'ambatch' in the Sudan) and *A. pfundii* are both swamp species tolerant of prolonged flooding to about one metre depth. *A. elaphroxylon* is often found in papyrus swamps and may even be a sudd component (Drar, 1951). *A. uniflora* can be a sudd component in the Guinea/Congo zone (Koechlin, 1961). *A. elaphroxylon* is tolerant of quite saline conditions in Lake Chad (Iltis & Lemoalle, 1983). *A. pfundii* also occurs in Lake Chad, and Howard-Williams (1975) has shown that this species is indifferent to alkaline soils during the drying phase of Lake Chilwa, Malawi. There are several other fully aquatic species of restricted geographical distribution, such as *A. fluitans* (Okavango), and *A. afraspera* (Chad).

Baikiaea (Caesalpinoideae). Although *B. plurijuga* is a dominant of dry savanna in the Zambezi Basin, the Guinea-Congo zone species *B. insignis* is a dominant or co-dominant of some seasonal swamp forests on the northern and north-western shores of Lake Victoria.

Cassia contains mostly mesophytes, but there are several species which are characteristic of riverine forests. *C. mimosoides* is a regular associate of *Vetiveria* grasslands on the Accra plains (Ghana) and its counterpart in Uganda is *C. kirkii* (Table 3.4) in *Miscanthidium* swamps.

Dalbergia. Most species are mesophytes, but *D. melanoxylon* occurs throughout the Sahel, often on seasonally waterlogged black cotton soils. *D. laxifolia* is a species of riverine fringes in the Zaire basin (Fig. 3.14).

Erythrina (Papilionoideae). *E. abyssinica* is a mesophyte, typically of the forest/savanna transition in East Africa, but *D. excelsa* is a regular and characteristic component of seasonal swamp forest in the Lake Victoria region.

Gilbertiodendron. *G. dewevrei* is widely distributed as a type-dominant of a mesophytic forest community on moist, but well-drained, sandy soils in the Guinea/Congo zone. White (1983) describes this type near Kisangani, Zaire. However, it is also (locally) a swamp forest species (Fig. 3.14) on sandy soils at the top of the floodplain. IUCN (1979) refers to swamp forest in the Ivory Coast and Cameroon characterised by *G. splendidum*.

Guibourtia demeusii occurs locally in the Zaire Basin as a co-dominant, with *Oubanguia*, of seasonal swamp forest well down on the floodplain (Fig. 3.14).

Kotschya africana is a small tree often associated with *Aeschynomene, Sesbania, etc.* between reedswamp and swamp forest in the Lake Victoria zone.

Mimosa. This is one of the very few woody plant genera found in permanent, or deep-flooded seasonal, swamps. *M. pigra* is the commonest species (Fig. 3.14), but *M. asperata* occurs in the Zaire basin (Fig. 3.11). Mimosa is also small enough to be an occasional sudd component (Table 3.2).

Parkia (Mimosoideae). *P. filicoidea* is sometimes recorded as a swamp forest tree, (as by Eggeling, 1947, and Lind and Morrison, 1974, for the Lake Victoria region) and sometimes as a component of gallery forest (as by White, 1983), for the Zambezian region. It is clearly tolerant, at least, of prolonged waterlogging.

Sesbania. A leguminous shrub of deep-flooded swamps and lake margins, where *S. sesban* often occurs with *Mimosa pigra* and, where the hydrosere runs to swamp forest, also with the small leguminous tree *Kotschya africana*.

Malvaceae. By far the most important species, found almost throughout the continent is *Hibiscus diversifolius*. It is often a fringing riverine shrub, frequently associated with *Alchornea*, and is also a common component of papyrus swamp fringes. It tolerates permanent waterlogging and deep seasonal flooding. *H. asper* is found in wooded grassland in the Chad basin and *H. tileaceus* is a regular component of swamp forest in southern Africa.

Melastomataceae. *M. segregatum* is a shrub of deep-flooded swamps, often associated with the legumes *Mimosa pigra* and *M. asperata* (Figs. 3.11, 3.14). *Dissotis* is a melastomataceous shrub found in permanent swamps in both East and West Africa. *D. rotundifolia* is a deep-flooded species but *D. brazzei* is probably not so flooding tolerant (Table 3.4). Lind & Morrison (1974) record *Tristemma incompletum* throughout the hydrosere of the Lake Victoria zone herbaceous swamps. *Memecylon* (sometimes placed in its own family, the Memecylaceae), on the other hand, is largely a 'dry' ground tree, but *M. zenkeri* has shallow-flooding tolerance in the Zaire basin (Fig. 3.14).

Meliaceae. Both *Entandrophragma* and *Carapa* are best known for their non-swamp species (*e.g. E. angolense*, an important timber tree, and *C. grandiflora*). However, both *E. palustre* and *C. procera* are Guinea/Congo zone species found in seasonal swamps at low altitudes, the latter commonly but the former only locally. *C. procera* is also widespread above 1000 m on free-draining soils.

Moraceae. *Ficus* is often found in swamp forest, but two species in particular are confined to wetlands: *F. congensis* is typically a seasonal swamp species of West and East Africa, but it often floats over 3–5 metres of water in the permanently flooded parts of seasonal herbaceous swamps. *F. verruculosa* occupies the same niche in southern Africa, but it does overlap in the Lake Victoria region with *F. congensis*. This niche may require further explanation: in the Okavango, for instance, *F. verruculosa* clumps regularly grow into main river channels from the *Miscanthidium* fringe. Most of the *Miscanthidium* is only seasonally inundated, but the river channels are permanent. *Ficus* does not usually occur in the reed swamps of permanent swamp zones. (Plates 1.6, 3.1.)

Myricaceae. *Myrica* is not common in the tropical zone. *M. kandtiana* is a swamp species at medium (*ca.* 2000 m) altitudes in East and Central Africa. *Myrica/Rhus* communities dominate the fringes of some coastal South African fens (*e.g.* Groenvlei: Martin, 1960).

Myrsinaceae. *Rapanea rhododendroides* (in common with *Ilex mitis*: *see* above) grows on peat at 2000 m in Rwanda but on free-draining slopes at 3000 m in the afro-montane zone. This ecological disjunction needs some detailed study, although it may be observed that, at the treeline, most substrates are usually moist, and often waterlogged.

Myrtaceae. *Syzygium cordatum* is a small tree of permanent (or almost permanent) swamps in eastern Africa. As such, it is one of the very few truly adapted swamp trees in Africa (as opposed to the flooding-tolerant species). *S. cordatum* also occurs to over 2000 m in central Africa where (in contrast with some species, already mentioned, which show ecological disjunctions over their altitudinal ranges) it retains its swamp characteristics and dominates certain swamp forests on peat (Deuse,

1966). In southern Africa, too, *S. cordatum* may form pure stands in swamp forest in Natal and Moçambique (Palgrave, 1977). *S. owariense* occupies a similar niche in freshwater swamps in the Guinea/Congo zone from the West African coastal zone to inland and upland Zaire. In fact, *S. owariense* is also one of the main swamp forest dominants in the moist savanna zone of southern Africa (Fig. 3.1). *S. guineense*, with its numerous subspecies, is distributed throughout Africa and, although it is often a characteristic species of gallery forest, it is a tree only of free-draining soils and ridge sites at higher altitude.

Potaliaceae. *Anthocleista* is sometimes included in the Loganiaceae and sometimes allied with the Rubiaceae. *A. zambesiaca* is certainly tolerant of waterlogging, although its ecology has not been properly studied. It occurs frequently at higher altitudes in East and Central Africa and is recorded by Bouxin (1976) as occuring, with *Syzygium*, on the wettest sites in the Rugege Valley at 2000 m in Rwanda. *Anthocleista* pollen is considered by Hamilton (1982) to be a good indicator of swamp conditions at this altitude. The same species is a component of riparian/gallery forest in the Zambezi Basin but, although it clearly needs a high water-table, it does not often experience flooding there. *A. nobilis*, on the other hand, has peg-roots (Jenik, 1971) and is therefore a true swamp forest species; it is widely distributed in such sites throughout Sierra Leone and Ghana (Nielsen, 1964). Crawford (1983) notes *A. nobilis* only as a salt-tolerant mangrove, but it is actually more widely distributed in freshwater swamps.

Rhamnaceae. *Lasiodiscus mildbraedii* is tolerant only of brief shallow flooding but is rapidly replacing much of the natural seasonal swamp forests of the north-western shore of Lake Victoria after heavy exploitation of the podocarps. *Ziziphus mucronata* is a widely distributed shrub of gallery forest in savanna regions. *Maesopsis eminii* is a fast-growing tree typical of secondary forest regrowth in the Guinea/Congo and Lake Victoria zone. It has similar waterlogging tolerance to *Lasiodiscus*.

Rubiaceae. *Mitragyna* has several species, all of them confined to swamps. The genus frequently dominates swamp forest throughout the continent (except southern Africa) and is undoubtedly therefore the most important swamp forest tree in Africa. *M. stipulosa* (Fig. 3.12) is the most widespread species (Guinea-Congo, Lake Victoria and the moist savanna zone); *M. ciliata* has a predominantly West African (forest zone) distribution; *M. inermis* is found in woodland remnants on seasonally waterlogged clays or sands in the drier Sudanian region; *M. rubrostipulata* is predominantly Afro-montane, with a southern limit in Malawi. *Nauclea pobeguinii* is one of the most characteristic trees of Congo-Guinea swamp forest and *Nauclea* too has its Sudanian species, *N. latifolia*, which is typical of the dry woodland (usually on heavy, seasonally-waterlogged soils) transitional between riparian forest and true savanna.

Salicaceae. *Salix* is predominantly a northern hemisphere genus, but it does have a significant African distribution: North Africa and southern Africa, with two or three small montane outliers in the tropical zone (Meusel, Jager and Weinert, 1965). The North African species are mostly European (*S. cinerea, S. purpurea, etc.*), whereas the southern species (*S. woodii, S. capensis, S. hirsuta, etc.*) are all of local origin. All species are typical of sites which experience inundation or waterlogging, although one of the commonest *Salix* community in southern Africa, the *Rhus/Myrica/Salix* (Fig. 3.15 indicates a variant of this) riparian/gallery woodland, is rarely inundated.

Scytopetalaceae. *Oubanguia africana* will endure extended periods of flooding and almost permanently waterlogged soils. Together with *Guibourtia demeusei*, it characterises a fairly local swamp forest community in the Zaire Basin. Both species are endemic to the region.

Urticaceae. *Musanga cecropioides* is a fast-growing light-demanding species found throughout the Guinea/Congo zone and the western Lake Victoria region. It frequently occurs with *Macaranga* and *Mitragyna* and thrives in secondary forests on seasonally waterlogged soils. In swamp forest, *Mus-*

anga develops stilt roots, as too does the other common urticaceous tree of seasonal swamps, *Myrianthus arboreus*.

Although the above list of 23 dicotyledon families containing woody swamp species appears very diverse, it can be shortened by grouping related families. Thus the Melastomataceae is close to the Myrtaceae and the Annonaceae is related to the Ebenaceae; the Scytopetalaceae is probably quite close to the Malvaceae and the Urticaceae to the Moraceae; the Apocynaceae, Rubiaceae and Potaliaceae are fairly closely allied. It is interesting to note that related families often appear to be represented by species with remarkably similar ecological ranges, but this observation may well be spurious.

However, after completing this 'contraction' exercise, six family-groups emerge as the most important in African wetlands:
Rubiaceae/Apocynaceae/Potaliaceae, Urticaceae/Moraceae, Euphorbiaceae, Leguminosae, Melastomataceae/Myrtaceae, Malvaceae/Scytopetalaceae. The Rubiaceae/Apocynaceae/Potaliaceae (*Mitragyna, Nauclea, Voacanga, Anthocleista*), the Myrtaceae *Syzygium* and the Euphorbiaceae (*Uapaca, Spondianthus*) are probably the families most frequently represented in African permanent, and deep-flooded seasonal, swamp forests. However, the Leguminosae also provides a large number of swamp species. Many of the shrubs of permanent swamps (*Aeschynomene, Hibiscus, Melastomastrum, Mimosa, Pluchea, Sesbania, etc.*) come from the Leguminosae, which also provides such a high proportion of the trees of many seasonal swamp forests.

A further important conclusion to draw from this brief survey of woody swamp dicots is the considerable difficulty in distinguishing between permanent swamp species and those more characteristic of, or restricted to, seasonally-flooded sites. Much of the problem lies in a serious lack of good ecological data in the literature. Accounts are almost exclusively descriptive and attempts in the present chapter to interpret regional species lists in terms of actual habitat preferences and degree of waterlogging adaptation have, therefore, been based upon very little published autecological data (See also Section 3.8).

Even so, there is probably another, more fundamental, reason for the poor delimitation between permanent and seasonal swamp species of woody plants. This will be dealt with later under plant adaptations.

3.3.3 *Gymnosperms*

The only wetland gymnosperm in Africa is *Podocarpus*, and it is not exclusively so. Moreover, the species involved have the 'ecological disjunction' referred to earlier under *Ilex mitis*. *P. falcatus* (formerly *P. usambarensis*) and *P. latifolius* (formerly *P. milanjianus*) dominate, together with *Baikiaea insignis*, a seasonal swamp forest on the north-western shore of Lake Victoria. At higher altitude, the podocarps are characteristic of free-draining soils.

3.3.4 *Pteridophytes*

Ferns are present in several types of wetland communities. *Thelypteris striata* (Plates 1.2, 3.5) is exclusively (emergent) aquatic and is widespread as a sudd component and as a subordinate species of papyrus swamps (Fig. 3.5), whereas *T. confluens* is more typical of *Miscanthidium* communities (Table 3.4). *Cyclosorus interruptus* is also common in reed swamps at low and medium altitudes. Both *T. squamigera* (Table 3.2) and *Osmunda regalis* are probably restricted to higher altitudes. The former may be limited to East and Central Africa, but the latter is almost continent-wide, occurring on fen peats with *Pycreus* at 2000 m in Uganda (Hamilton, 1982) and between riverine forest and open water at 1700 m in Malawi (Chapman and White, 1970).

Ferns are particularly common in *Miscanthidium* swamps, and *Ceratopteris thalictroides* is a prominent species of river channel fringes (clumps often protruding out into open water) in the Okavango Swamps (Plate 1.6). *C. thalictroides* is one of the few species of edible ferns, and it has a curious disjunct distribution in Africa. It is, for instance, absent from West Africa.

The diversity of fern species in swamp forests is

Plate 3.5. Exceptional development of *Thelypteris* on a papyrus floating island on Lake Upemba, Zaire. The *Polygonum* (scattered, foreground) and *Typha* (extensive, beyond) are probably indicators of unstable hydrological conditions at this point on the island. (Photo: K. Thompson.)

much greater, but since so many are epiphytic or scandent, it is difficult to establish which are actually aquatic. *Arthropteris, Asplenium, Microlepis, Phymatodes* and *Thelypteris* are all common swamp forest genera. *Nephrolepis, Phaylopsis* and *Pteris* are characteristic of Elaeis/Raphia thickets in West Africa. *Stenochlaena tenuifolia* is a climbing fern typical of South African swamp forests.

Because most ferns require humid conditions, it is to be expected that a great diversity would be found in swamp forests. However, only those which are ground-living and self-supporting can be considered to be true aquatic emergents. Even then, many ferns are highly tolerant of waterlogging without being especially adapted to flooding; this would be expected of relatively slow-growing, shade-tolerant plants. The 'aquatic status' of *Ophioglossum reticulatum*, common in *Sporobolus pyramidalis* grassland in Rwenzori National Park, (Lock, 1977) is equally difficult to establish, and probably irrelevant anyway.

3.3.5 *Bryophytes*

The complete list of African wetland bryophytes, particularly if it includes Afro-montane sites, would be long, and not very relevant to the major themes of this chapter. *Sphagnum*, however, deserves a separate mention. At low and medium altitudes, *Sphagna* – such as *S. madagassum, S. pappeanum, S. planifolium, S. subsecundum* – are present only in the more acid, nutrient-poor sites, such as *Miscanthidium* swamps, or the mixed cyperaceous swards (*Cyperus, Bulbostylis, Fuirena, Kyllinga, Scleria*) often associated with *Jardinea* grasslands or dambos.

There are no '*Sphagnum* bogs' in Africa, but *Sphagnum* is sometimes an important component of mire vegetation at higher altitudes. Thus a *Sphagnum*/*Xyris*/*Erica* community is, very locally, an active peat former at 2000 m altitude in Rwanda and, in the Afro-alpine zone, species such as *S. davidii, S. mildbraedii* and *S. rugegense* are prominent members of tussock-sedge mires (Thompson and Hamilton, 1983).

3.4 Distribution and diversity of wetlands

The major vegetation types of Africa follow quite closely the rainfall distribution on the continent, but Fig. 3.1 is not a good guide to the distribution of *wetlands* (Table 3.1), as will be shown in this section.

3.4.1. *The major wetland complexes*

These complexes all have one feature in common: very shallow hydrologic gradients. The Bangweulu floodplain, for instance, slopes only one metre in 40 km. There are also similarities in their origins, since some of the larger swamps and lakes of Africa are the result of tectonic activity, many owe their existence to massive deposition of sediment, but most have needed both processes. Thus, the 200 km long Lake Kamulondo once filled a major graben in upper Zaire, and subsequent sedimentation from four very large rivers (Lualaba, Luvua, Lufira and Luvoi) has now evolved the Upemba Swamps (Thompson, Shewry and Woolhouse, (1979); the extensive swamps of southern Uganda (Thompson, 1976*a*) derived from tectonic tilting which reversed drainage lines; the lakes of the rift valleys all derived from faulting; the 20,000 km^2 Bangweulu Basin in Zambia is a very ancient (> 2000 million years) cratonic platform which has been subjected to gradual subsidence over the last 20 million years (Grimsdell & Bell, 1976); the Okavango Swamps may trace their origin to an ancient rift valley system (Reeve, 1972), now filled with 150

Table 3.1. The major wetland types of Africa.

1. Freshwater herbaceous wetlands
   - 1.1 Permanent
     - 1.1.1 Reed swamps
     - 1.1.2 Bog, fen and moorland
   - 1.2 Seasonal
     - 1.2.1 Black clays, pans and dambos
     - 1.2.2 Floodplains and valley grasslands
     - 1.2.3 Alkaline grasslands
2. Freshwater swamp forests
   - 2.1 Permanent
   - 2.2 Seasonal
     - 2.2.1 Floodplain forests
     - 2.2.2 Riverine and gallery forests

Plate 3.6. Part of the seasonal swamp zone of the Okavango Delta, Botswana. The swamps have two flood peaks per year: one resulting from local rains and the other from the wet season in the headwaters of the Okavango River in Angola. The flood peak takes about 3 months to work through the Delta to the Thamalakane River outlet, and the photograph shows the rising floodwaters, black-stained after passing through the permanent swamps, spreading across the 'melapo' of the floodplain. The tsetse fly is endemic throughout these swamps (see Section 8.3.5). (Photo: K. Thompson.)

metres of sediments from the continent's fifth largest river; the Upper Nile Swamps of the Sudan are the product of at least a million years of sediment accumulation from the continent's third largest river (Berry, 1976), and the swamp forests which dominate a large part of the 900,000 km² 'cuvette centrale' (central basin) of Zaire stand on sediments which have now displaced a vast lake (Lac Tumba is a remnant) which once filled a (postulated) ancient 'graben central'. The coastal deltas of the Rivers Niger, Zambezi and Zaire support large and very complex wetland systems (see, for instance, Fig. 3.12).

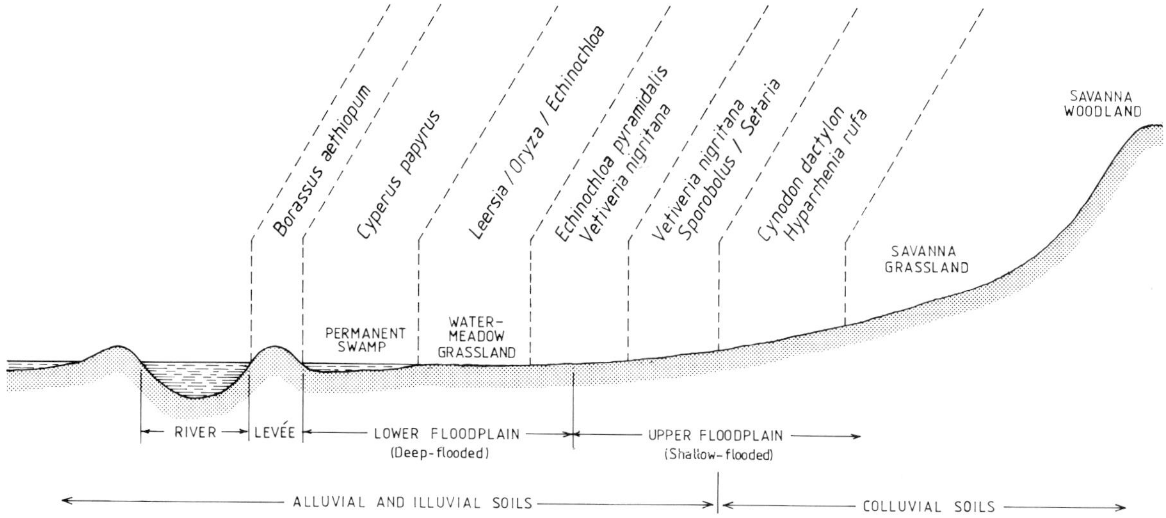

Fig. 3.2. A floodplain hydrosere in the upper Zaire River Basin, Shaba Province, Zaire. The hydrological regime has a 4–5 metre amplitude at this point and has therefore built up substantial levées. As the flood regime and the river flow rate decrease, the levée height falls.

3.4.2 *The seasonal wetlands*

All permanent inland water-bodies have floodplains. The lacustrine floodplain covers that area of land above low-water level which is usually inundated at least once per year. The riverine floodplain is bounded at its lower extremity by the 'levée'. Levées do not form around lakes, because the water is not flowing; levées are built up along the margins of river channels by (differential) sediment deposition. In flood, the coarser particles (sand, coarse alluvium) sediment out first, forming the levées, and the finer particles (silts) build up the backwater swamps. Levées are high where river flow is greater and where the flood regime develops rapidly; shallow levées are characteristic of slower flows and gradually-developing flood regimes.

Thus, as the Zaire (formerly Lualaba) River enters the Upemba Basin, it builds up substantial levées which rise several metres above normal river level. Figure 3.2 shows the floodplain structure, and its vegetation, at this point. As the river flows further into the Upemba Basin, the slope becomes less, the floodplain wider, the 'buffering capacity' towards the flood regime becomes greater, the grade of sediment carried by the river becomes finer (the coarse material has already been deposited where flows are higher), the levées (now silty rather than sandy) decrease in height and papyrus replaces *Borassus*. Sutcliffe (in Jonglei Investigation Team, 1954) gives a good account of levée formation and functioning in the Upper Nile Swamps. Thompson (1976b) describes levée formation in the Okavango Swamps, and explains how continued sediment build-up (particularly bed-load) leads to channel evolution in complex riverine swamp systems.

Figure 3.3 illustrates the seasonal development of a typical flood regime. Note that it takes only 6 weeks to attain its maximum coverage, but dissipation is spread over 6 months. Although the time scales are different for other riverine floodplains, the inequality between filling and emptying phases is always there. It is, of course, a critically important property of floodplain wetlands from the viewpoints of both wildlife management and water resource management. Chapter 8 deals with this property in more detail; here it is only appropriate to point out that the floodplain plant communities are a function more of the dissipation phase of flooding than they are of the expansion phase.

Although Table 8.1 lists most of the major African wetlands and associated (seasonally-inundated) floodplains it still under-estimates the total area (and greatly under-estimates the total num-

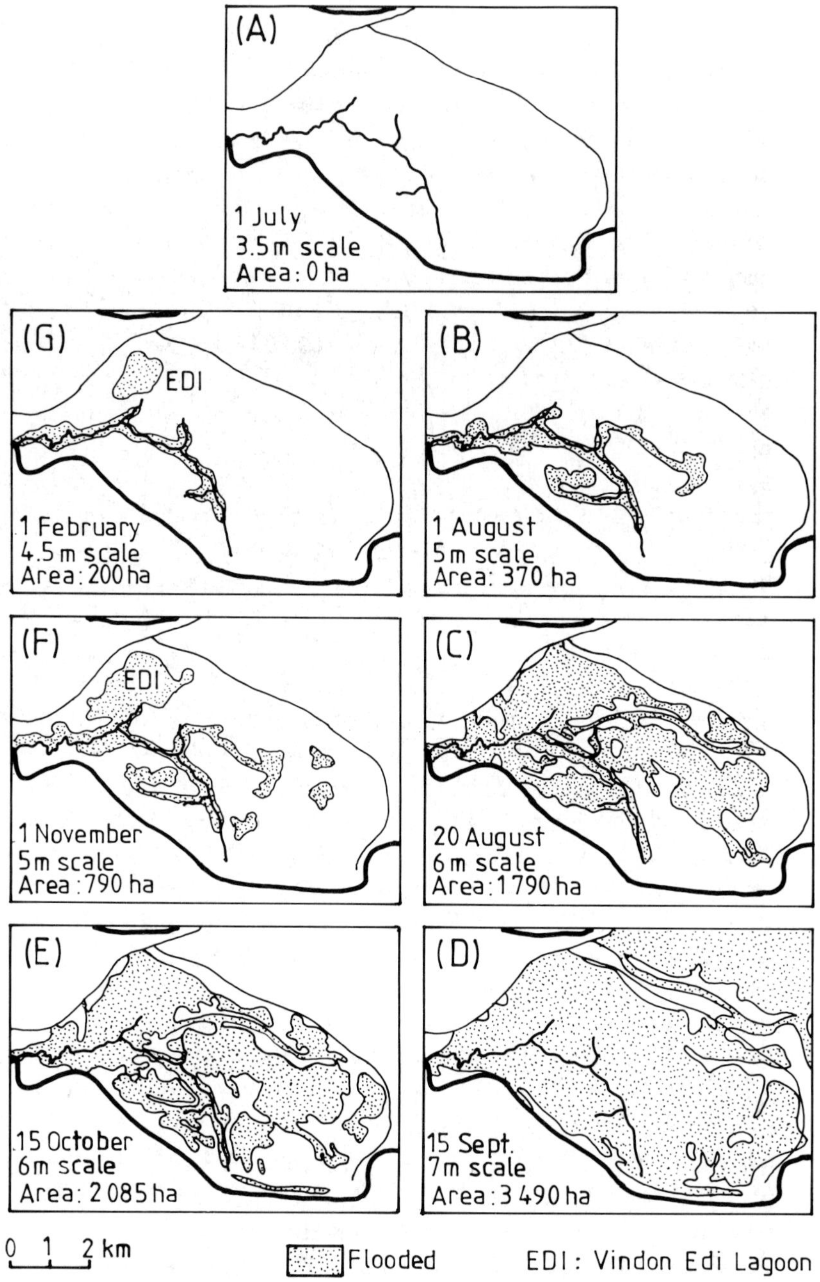

Fig. 3.3. An example of a typical seasonal flooding regime (after Reizer, 1974), showing the full extent of the floodplain. Note that the expansion phase takes 6 weeks, but the dissipation phase extends over more than 20. This asymmetrical cycle is typical of most floodplain systems.

ber) of wetlands on the continent. The reason for the considerable discrepancy is that it is exceedingly difficult to estimate the huge area of seasonally waterlogged land *remote* from the major watercourses and waterbodies. There are two types of wetland in this important category:

3.4.2.1 *The black clay wetlands*

Vertisols are heavy-textured, black or grey soils with very high (30–80%) montmorillonite/illite (2:1 expanding-lattice clay) content. The organic content rarely exceeds 1.5%, but it is well distributed as a thin film adsorbed over the whole clay lattice and, together with the high calcium content, imparts the dark colour. Vertisols include types called 'black cotton soils' (India, and also the Sudan, where these soils are considered ideal for cotton plantations), 'tropical chernozems' (because of their high calcium contents), 'black earths' (Australia), vlei soils (South Africa) and mbuga soils (Tanzania). Most vertisols derive from transported material and commonly dominate valley floors (vleis and mbugas). Most of Vesey-Fitzgerald's (1963 and 1970) valley grasslands have developed on black clays which originated in similar ways to Milne's (1947) Tabora catena in Tanzania: skeletal, acid grey loams on the granitic hills, through kaolinitic red earths and sands on the slopes, to seasonally waterlogged mbuga sandy clays in the valleys. However, some vertisols arise by direct weathering of basalts (usually volcanic soils). True vertisols are ancient, probably Early Holocene or Pleistocene (Buringh, 1970), and often cover quite extensive areas (such as most of the northern region of Uganda). Vertisol formation only takes place under conditions of alternating wet and dry seasons, in the presence of a montmorillonitic clay type. Although normal chemical weathering processes would degrade the expanding lattice clays to kaolinitic types, this does not occur under conditions of seasonal waterlogging.

Turning now to the behaviour of these black clays: they shrink when dry, leaving wide (a few centimetres) and deep (up to a metre) cracks, and swell when wet. Such is the pressure developed under expansion that, once it is saturated, the clay becomes almost impermeable to water and also frequently anaerobic. The widespread black clay wetlands of savanna regions are often waterlogged for as much as 4–5 months of the year. They also occasionally account for swamp forest communities, even on sloping ground.

3.4.2.2 *The dambos and pan wetlands*

Much of the African savanna is underlain by shallow, impermeable horizons, usually referred to (*e.g.* Mohr and Van Baren, 1954) as 'hardpans'. Tinley (1982), however, applies the collective term 'pan' to all subsurface impermeable soil horizons, whether indurated (cemented) or merely compacted. Vesey-Fitzgerald (1963) has an even wider definition of pan: 'depressions which are liable to flooding during the rains, but which are not necessarily part of the developed drainage system'. This definition thus also includes tectonic folds and obsolete drainage lines, neither of which need be underlain by impermeable soil horizons in order to generate flooding: the only requirement is high seasonal rainfall and lack of drainage outlets. Nevertheless, some type of impermeable horizon is usually involved – otherwise it is black clays. Tinley's cemented pan types include ferricrete, calcrete and silcrete, resulting from the leaching through the profile of iron compounds, calcium carbonate and alumino-silicates. Calcrete may also be formed *in situ* by precipitation of nodules from base-rich sands or black clays. The compacted ('claypan') types include illuvial gleys and also very fine-grained sands.

Not only can water accumulate above the pan horizon, causing waterlogging or flooding, but there may also be an aquifer immediately below the pan, where it is protected from evaporation losses. Where the cemented or compacted crust is broken, permanent water may occur. Most 'pan wetlands' are thus very small and, because the aquifer has been confined, the water supply may even be artesian. The Mzima Springs, in the Tsavo National Park, Kenya, is an example of a particularly large topographic depression in which a confined aquifer yields enough water to supply the town of Mombasa.

'Dambos' are depressions where the pan horizon comes to the surface and seasonal water accumulates. Occasionally, dambos are actually 'seasonal pans'. Vesey-Fitzgerald (1963) refers to dambos as 'headwater valley grasslands'.

Figure 3.4 illustrates the wide range of pan and dambo structure and shows how they develop. Pan horizons are also typical of the lower slopes (*i.e.* lower colluvial zone of Fig. 3.2) of catenas such as Milne's (1947) 'type' example at Tabora, Tanzania. Here they have the effect of enhancing run-off to the valleys and also producing seasonal waterlogging beyond the normal range of the floodplain. The rain-fed *Hyparrhenia rufa* grasslands on floodplain fringes often owe their existence to superficial pans. Dambos also frequently have fring-

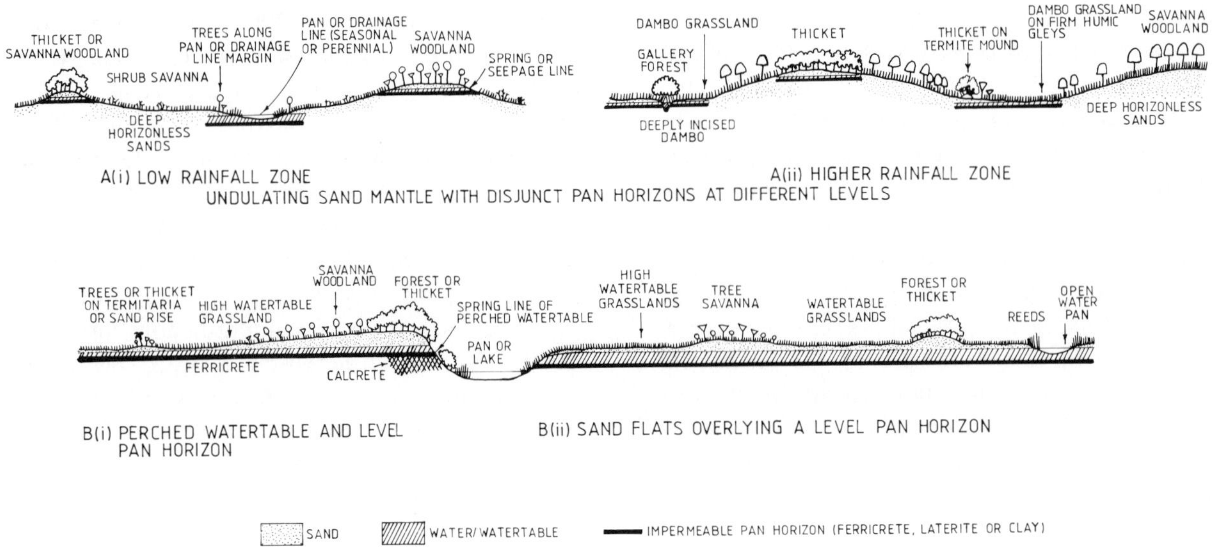

Fig. 3.4. Various types of dambos and pans, showing their relationships with subsurface impermeable horizons (after Tinley, 1982).

ing gallery forest or woodland (Fig. 3.15). Riparian forests are not usually pan-dependent communities, although in the higher rainfall savanna zones they may replace the proximal (to the river) part of the *Hyparrhenia* zone.

3.5 The major wetland plant communities

The wetland plant communities of Africa are discussed in this section within five broad categories: (a) floating sudd communities; (b) permanent reed swamps; (c) seasonal herbaceous swamps (edaphic grasslands); (d) swamp (edaphic) forests; (e) montane wetland communities.

3.5.1 *Sudd communities*

Sudd, or 'sadd', is an Arabic word meaning 'blockage' or 'barrier'. It first entered swamp terminology when it was applied to the massive papyrus blockages which made navigation of the waterways in the Upper Nile Swamps of the Sudan so difficult (Hope 1902; Drar, 1951). Indeed, this vast swamp complex between Lake No and Mongalla, on the White Nile, is often known as the 'Sudd Swamps'. More recently, the term 'sudd' has been borrowed extensively to describe *any* floating vegetation mat. The wider definition is used in the present chapter and, in order to demonstrate how wide the definition must be, floating plant communities from eight different accounts of African swamps have been compared in Table 3.2.

Many permanent swamps, particularly lacustrine types, have floating fringes, and the *Ludwigia*/*Pycreus*/*Leersia* type is most typical of this; these genera are also widespread as minor components of other sudd types, so that they appear in most sudd lists. Several other plants are commonly encountered as minor components, such as *Aeschynomene, Drosera, Hydrocotyle, Marsilea, Mimosa, Polygonum, Pistia, Sacciolepis, Thelypteris, Utricularia* and *Xyris*. An *E. scabra* dominated sudd is common in many West African lakes; in fact, this plant is under-represented in Table 3.2 in relation to its overall importance as a sudd plant in Africa. *Polygonum* and *Vossia* are also under-represented, the latter being the most typical fringe plant of riverine swamps. *Vossia* is, however, more usually rooted (see Fig. 3.6) than truly floating in riverine swamps. *Eichhornia crassipes*, too, is a very effective sudd-former which replaces *Vossia* fringes and *E. scabra* dominated lagoons in the middle reaches of the Zaire River: several hundred tonnes per hour of *Eichhornia* sudd now drift past Kinshasa, the Zaire capital (Plate 7.1).

It is relatively unusual to find a sudd type which does not have one or more *Cyperus* species associated with it, but *C. papyrus* is the only one which actually dominates a major sudd type: the fringes of almost all riverine and lacustrine papyrus swamps are floating. Drar's (1951) papyrus sudd community stands out as being somewhat different from the others cited in Table 3.2; the reason is that his sudd definition is based upon the true Arabic meaning of the word: a barrage. And that is precisely what the *Cyperus papyrus* community is in the Upper Nile Swamps and why Drar has listed all the main associates of papyrus in his sudd type. Because of its phenomenal growth rate, papyrus

Table 3.2. Sudd communities, as characterised by eight different workers. The species list is still not complete: there are several other types of floating vegetation mats. However, the inventory could be subdivided into two major classes of sudd plant: (a) the 'mat-formers' (e.g. papyrus, *Pycreus, Leersia, Vossia,* etc), and (b) the 'mat-users' (e.g. *Cissus, Melanthera, Thelypteris,* etc).

	Denny (1971)	Drar (1951)	El Din el Zarka (1973)	Koechlin (1961)	Lock (1973)	Sita (1970)	Smith (1976)	Iltis & Lemoalle (1983)
Aeschynomene elaphroxylon		★						
A. uniflora				★				
Cissus ibuensis		★						
Cladium jamaicense	★							
Cyperus auricomus						★		
C. cubensis						★	★	
C. imbricatus				★				
C. nudicaulis							★	
C. papyrus		★						
Drosera madagascariensis							★	
Echinochloa pyramidalis		★						
E. scabra			★					
Enhydra fluctuans					★	★		
Hydrocotyle ranunculoides					★			
Ipomoea aquatica		★						★
Leersia friesii							★	
L. hexandra				★		★		
Ludwigia stolonifera			★	★	★	★		★
Luffa cylindrica		★						
Marsilea minuta					★			
Melanthera scandens		★						
Mimosa asperata				★				
Neptunia oleracea								★
Phragmites australis		★						
Pistia stratiotes		★	★		★			
Polygonum senegalense			★					
Pycreus mundtii					★	★		★
P. nitidus							★	
Sacciolepis spp						★		
Scirpus cubensis						★	★	
Thelypteris squamigera	★							
T. striata		★				★		
Typha australis		★						
Utricularia inflexa			★				★	
Vigna nilotica		★						
Vossia cuspidata		★						★
Xyris spp							★	

quickly occludes channels and forms blockages as sections of swamp break off and drift with the flow until they jam in channel constrictions. Camus (1957) describes the hazards of drifting papyrus rafts in the Lualaba (now the Upper Zaire) River; Thompson (1976*b*) discusses the possible role of papyrus debris blockages in the evolution of major distributaries of the Okavango inland delta; Terry and Minto (1970) detail the annual operations in the port of Kisumu (Kenya), on Lake Victoria, to tow the large drifting islands out of the harbour; Iltis and Lemoalle (1983) describe the large papyrus islands ('kirtas') of Lake Chad, and Thompson and Hamilton (1983) refer to similar structures in the Upemba Swamps (Plate 6.3).

Sudd is bound together by rhizomes (papyrus, *Cladium*), or floating stems (e.g. *Vossia, Echinochloa, Ludwigia*), and the resulting raft provides a suitable habitat for non-mat-forming plants such as *Impatiens, Drosera, Aeschynomene, Scirpus*, etc. Sudd may thus comprise emergents, floating species (*e.g. Pistia, Marsilea, Eichhornia*) and submerged aquatics (e.g. *Utricularia*).

3.5.2 *Reed swamps*

All African reed swamps, as here defined, are found below 2500 m altitude, and they are almost always found as components of larger swamp complexes. Thus, reed swamps almost always grade into other wetland communities, such as edaphic grasslands or swamp forest. The most obvious exceptions to this general rule are the reed swamps (mostly papyrus) of relatively steep-sided valleys in Uganda and Rwanda, which are often the only wetland type present. Reed swamps are typically dominated by a single reed species: this usually means *Phragmites, Typha, Cladium, Cyperus* (particularly *C. papyrus*) or *Miscanthidium*. Although some investigators may be inclined to add a few of the larger grasses, such as *Loudetia phragmitoides, Echinochloa pyramidalis* or *Pennisetum purpureum*, these are considered in the present chapter under 'edaphic grasslands'. Because of their stand density, reed swamps are usually poor in associated species, although papyrus swamps are rich in climbers and twiners. Reed swamp *fringes* usually have high species diversity and species lists for wetlands do not usually make this distinction between swamp interior and swamp fringe.

It is difficult to compute the total area of reed swamp in Africa, but 70–80,000 km^2 would be a reasonable guess. Reed swamp is found in all biogeographical regions of Africa where conditions are suitable, but it is relatively uncommon in the Guinea/Congo region, where swamp forest types, and various edaphic grasslands seral to forest, predominate. As a rough guide, a large wetland outside the Guinea/Congo zone can be expected to comprise some 20–25% reed swamp (assuming that most 'permanent swamp' will be reed swamp). The 'big five' African wetlands support this rule-of-thumb, but also highlight the two extreme cases:

Wetland Complex	Seasonal Swamp (km^2)	Permanent Swamp (km^2)
Bangweulu	8,500	2,500
Okavango	12,000	3,000
Chad	13,800	2,400
Upemba	8,500	4,500
Upper Nile	80,000	8,000

Okavango, Chad and Bangweulu illustrate the 'normal' ratio of permanent to seasonal wetland, but the Upper Nile complex of the southern Sudan appears to have a disproportionately large seasonal floodplain. The explanation lies in the extreme seasonality of most of the rivers (Achwa, Sobat and Blue Nile) supplying the depression. The Achwa, for instance, dries up completely every year; the Blue Nile can rise 3 metres overnight. Most of the permanent reed swamps in the Upper Nile basin are associated with the fourth river supplying the basin, the White Nile. The White Nile's minimum flow is 'guaranteed' by the high rainfall of the Lake Victoria region, and at peak this flow is only twice the minimum because of the 'buffering capacity' of the central African lakes. The low-amplitude regime is clearly demonstrated again in the Upemba system, with its 4-river water supply (three of which originate in large lakes) and its consequent very high proportion of permanent swamp.

Reed swamps, therefore, are communities of stable hydrological regimes, and it is instructive to look at a relatively *un*stable system in order to appreciate the factors which control reed swamp development. Lake Chad is an ideal example to take in that the lake is particularly sensitive to variations in climate. Lake Chad (281 m O.D.) and its permanent wetlands normally fill 10% of a 250 000 km^2 basin in the dry Sahel zone. The mean depth of the lake is the only four metres and there is a rising salinity gradient from the fresh-water input of the Chari/Logone delta (dominated by *Vossia cuspidata*) and the southern basin (characterised by *Cyperus papyrus*), to the less stable and more saline northern basin (dominated by *Typha domingensis*). Iltis and Lemoalle (1983) describe the effects on the aquatic flora of the contraction, by 1974, of Lake Chad to a mere one third of its 25000 km^2 size some 10 years earlier. In fact, most of the regression occurred during the 1972–1975 drought. Within the space of only three years, the waters of the northern basin became even more saline as the shoreline receded, and permanent swamp virtually disappeared, leaving only remnants of *Typha domingensis* and stunted *Aeschynomene elaphroxylon*. As the lake receded in the southern basin, *Vossia* (and *Ipomoea aquatica*) expanded to replace the regressing *Typha, Phragmites*, and (particularly) the *Cyperus papyrus* communities. In 1974, with the lake area at its lowest, the most widespread plant communities in the southern basin were *Vossia* meadows and *Aeschynomene elaphroxylon* 'forests'.

After the end of the drought in 1974, the lake rapidly re-filled; during this phase, *Typha* and *Cyperus papyrus* did not, as might have been expected, re-establish, but in fact continued their decline for almost two years. *Vossia*, on the other hand, continued to expand its cover as the waters rose again, particularly around the fringes of the now-stabilised *Aeschynomene* communities.

The 'seasonal drawdown cycle' (Gaudet, 1977) is an important feature of tropical lakes: it regenerates nutrients through wetting/drying processes, it allows germination of seed on exposed mud, it determines the feeding cycles of fish and invertebrates, it controls macrophyte community boundaries. But it is the episodic, rather than the regular seasonal, events which provide the greatest insight into plant adaptation. Thus, in the Chad example, *Vossia* appears to adapt best to an unstable system, or at least to one with high amplitude; papyrus and *Typha* are more suited to stable hydrological regimes, although *Typha* is the more tolerant of drought and salinity; the deeper root system (and very effective seed dispersal mechanism) of *Aeschynomene* allows it to quickly colonise exposed mud, and it will tolerate up to a metre (but no more) of flooding when the waters return. Howard-Williams (1975), describing an episodic drawdown in Lake Chilwa, again demonstrates the *Aeschynomene* advance under drought conditions and the sensitivity of papyrus to large water-level changes and high salinity. The adaptations of the reeds will be covered in more detail in the following sections devoted to individual community types, but Table 3.3 provides a useful summary of the relative ecological tolerances of the major reed types (and *Vossia*) of Africa.

3.5.2.1 Vossia *swamps*

Vossia cuspidata is able to cope with unstable hydrological regimes, and it is therefore not typical of other reed swamps types. It is included here because it is even less appropriate to discuss it as an 'edaphic grassland'.

Table 3.3. Vegetation zones of the 'kirtas' (floating islands) of Lake Chad (after Iltis and Lemoalle, 1983). Each zone is concentric around the previous one. The northern basin of this lake is more saline, and less stable, than the southern and the middle basin is intermediate. *Pycreus, Vossia* and *Cyperus papyrus* communities do not occur where the hydrological regime is less reliable. Kirta-type islands are also well-developed in the Upemba Swamps of Zaire (Thompson and Hamilton, 1983), but there is no comparable salinity gradient there, the outer zone usually being *Vossia* (see Plate 6.3).

WETLAND COMMUNITY	LOCATION WITHIN LAKE CHAD		
	SOUTH BASIN	→	NORTH BASIN
Open water	~	~	~
Pycreus sudd	+	~	~
Vossia	+	~	~
Cyperus papyrus	+	+	~
Phragmites australis	+	+	+
Typha domingensis	+	+	+

SWAMP INTERIOR ↓ · INCREASING SALINITY ↓ · INCREASING SEASONALITY ↓

From the Lake Chad example, it is clear that the morphological feature (long, flexible stem) which adapts *Vossia* so well to growth in the flowing waters of the Chari/Logone delta also gives it exceptional tolerance of large water-level changes in lakes (note in Fig. 3.10 its tolerance of a 7.5 metre flood regime). Rapid growth then allows the younger parts of the floating stems to follow the receding waters, whilst the older parts root at the nodes on the newly-exposed alluvium. As the waters rise again, the buoyant stems grow fast enough to remain emergent at the surface, although still rooted below. This tremendous adaptability allows *Vossia* to dominate hydrologically unstable systems, although it cannot displace the compacted root mat of the 'typical' reeds which develop under hydrologically stable conditions. There is probably only one plant which can out-compete *Vossia* in its preferred habitats: *Eichhornia crassipes* does this by shading out the grass. The ability of water hyacinth to eliminate *Vossia* is giving rise to serious concern both in Zaire and in the southern Sudan, since it increases the possibility of river-bank or permanent swamp erosion (Plate 7.1).

The Chad example also explains why *Vossia*, often in association with *Echinochloa scabra*, is so much more common in permanent and semi-permanent herbaceous swamp in West Africa than any of the 'typical' reeds (see, for example, Fig. 3.11). West African rivers typically have substantial flood regimes and large seasonal amplitudes, whereas East African rivers, with their vast associated reed swamps, are hydrologically buffered. Most permanent swamps in Zaire are also characterised by *Vossia*, together with other associated grasses. Although *Phragmites* and *Cyperus papyrus* swamps are also much more common in southern Africa (than in West Africa), White (1983) records that most permanent swamps along the major rivers of the Zambezian zone are, in fact, characterised by *Vossia*, together with *Polygonum acuminatum, Brachiaria mutica, Echinochloa pyramidalis, E. scabra, Leersia hexandra, Panicum parvifolium* and *Cyrtosperma senegalense*. Once again, the abundance of *Vossia* can be used as an indication of hydrological instability or high seasonal amplitude. *Cyrtosperma* is an indicator of a natural succession to riparian forest.

Figure 3.7 compares the swamp fringe of a relatively stable hydrological regime (Kafue) with one (Chambeshi) of higher amplitude (greater seasonality). In the first, *Vossia* occupies its usual position in flowing water; in the second, it has entirely replaced the papyrus.

3.5.2.2 *Cyperus swamps*

Papyrus swamps are more typical of stable hydrological regimes than any other reedswamp type, and in this respect its adaptations are as different as they could be from those of *Vossia*. Although papyrus swamps are usually about pH 6.5, the plant will tolerate a wide pH range of 4–8, but not saline conditions. Papyrus swamps are found from sea level to over 2000 m altitude, and at the higher altitudes they can deposit peat, as they do in the Grand Marais, on the Rwanda/Burundi border (Deuse, 1966). Papyrus swamps cannot endure the large amplitudes of West African rivers, and the Chad example shows why: papyrus which survived the 1972–74 drawdown did so by rooting to the newly-exposed alluvium but, when the lake rose again, it was unable to detach from the substrate and was drowned by permanent deep-flooding. Papyrus is normally able to adjust its rhizome height to cope with water-level changes (Fig. 3.16), but it cannot do this as rapidly as *Vossia*.

It is generally believed that *Cyperus papyrus* became extinct in Egypt and the northern Sudan because of the increasing amplitude of the Lower Nile. On the other hand, the extreme rarity of papyrus in the forest zone of Zaire cannot be due to seasonal floods, as the Zaire River is very stable in the cuvette central. It is replaced mainly by *Vossia* along the river fringes (better adaptation to flowing water) and by *Cyrtosperma* in backwaters (better adaptation to forest shade).

It is convenient to classify papyrus swamps into three types, although there are, of course, intermediates.

(a) Lacustrine swamps are typically floating mats, although they are usually anchored on firm soil. They develop only under stable hydrological regimes. Figure 3.5 illustrates most features of lacustrine papyrus swamps in the Lake Victoria zone, and a fuller species list is given in Table 3.4.

Plate 3.7. A 5 metre culm of *Cyperus papyrus* from the interior of the well-developed swamp fringing Lake Naivasha, Kenya. The mature culm is flowering; the juvenile has not yet opened its umbel. The tree to the left of centre is *Acacia xanthophloea:* in its typical location as an indicator of high water-tables. (Photo: K. Thompson.)

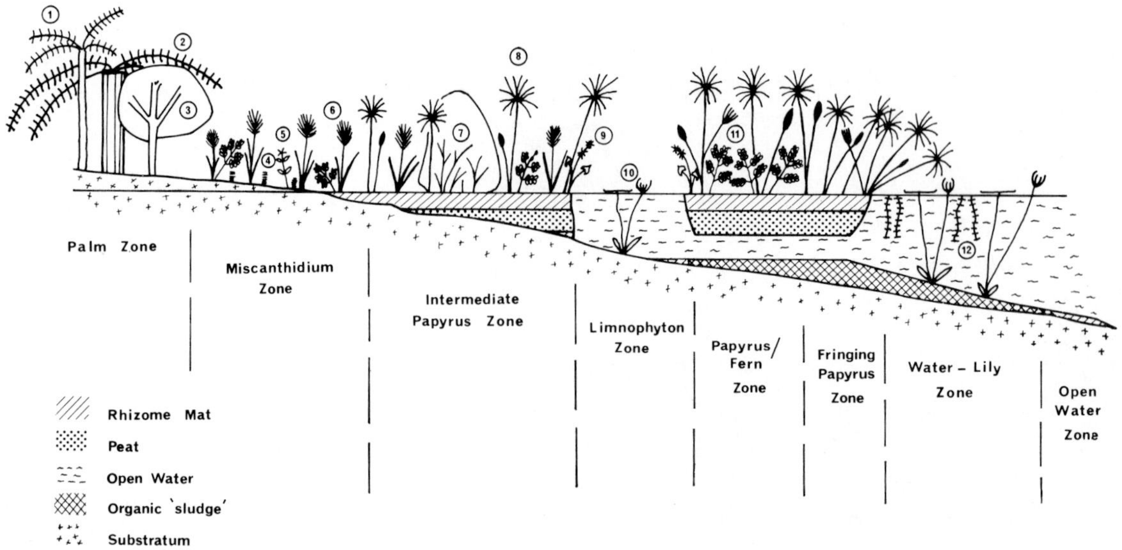

Fig. 3.5. Vegetation of a Lake Victoria swamp (after Thompson, 1976a). Vegetation zones are similar to those described by Eggeling (1935) for Namanve Swamp, Lake Victoria north shore, but certain features of Lind & Visser's (1962) Namiro Swamp are also incorporated. The sudd profile follows the terminology of Gaudet (see Chapter 6). The species represented are as follows (the list is not intended to be exhaustive, and only features some of the most typical plants): 1. *Phoenix reclinata, 2. Raphia farinifera,* 3. *Mitragyna stipulosa,* 4. *Sphagnum* spp., 5. *Dissotis brazzei* and *Leersia hexandra,* 6. *Miscanthidium violaceum,* 7. *Ficus verruculosa,* 8. *Cyperus papyrus,* 9. *Limnophyton obtusifolium,* 10. *Nymphaea caerulea* and *Trapa natans.* 11. *Thelypteris striata,* 12. *Ceratophyllum demersum, Utricularia* spp and *Potamogeton* spp. A zone of swamp shrubs and small trees, characterised by *Kotschya, Mimosa, Sesbania* and *Syzygium cordatum* is often present between the *Miscanthidium* and the *Mitragyna*/palm zones.

(b) Valley swamps occur in valleys with good hydraulic gradients and the papyrus is typically rooted in a firm mineral soil. This swamp type is particularly common in Uganda (see Plates 1.1, 6.6).

Around the northern shore of Lake Victoria, numerous 'drowned valleys' are completely filled with papyrus. Although they were originally valley swamps, tectonic activity and lake-level rises have flooded the valleys. These drowned valleys now support lacustrine swamps, floating over up to 10 m of water (Bishop, 1958).

Floristically, valley swamps do not differ significantly from lacustrine swamps, but the species list given in Table 3.4 would be greatly reduced due to loss of both the open water interface and the transition zone to dry land. Apart from the climbers, valley swamp interior is very species-poor.

(c) Riverine and floodplain swamps fringe rivers with small amplitudes and spread to permanently flooded or waterlogged parts of the floodplains. Figure 3.6 shows a typical hydrosere in the permanent swamp zone of the Upper Nile Swamps (Sudan); it differs from Fig. 3.5 mainly in the greater amplitude of the water body, which contracts the papyrus zone to a riverine fringe and leaves niches available for more drought-tolerant reeds (*Phragmites* and *Typha*). Papyrus will progressively exclude other reeds as hydrological conditions become more stable.

Figure 3.7 shows a hydrosere (Kafue), again from the savanna zone, from which *Phragmites* and *Typha* have been excluded by *Echinochloa* development, and permanent swamp is eliminated altogether under the even more highly seasonal regime of the Chambeshi River. A particularly interesting case is that of Lake Nabugabo, on the northern shore of Lake Victoria. This lake is one of Africa's most oligotrophic water bodies, and it supports only a minute amount of papyrus swamp, even though hydrological conditions are ideal for *Cyperus papyrus* development. Here, papyrus is replaced as a pioneer of open water habitats by *Vossia*. This is a very unusual occurrence for a small, shallow lake with stable water levels, and it presumably demonstrates that *Vossia* tolerates

Table 3.4. A comparison (retabulated after Lind and Morrison, 1974) between the species composition of the *Miscanthidium* zone of permanent herbaceous lake-edge swamp in the Lake Victoria zone, with that of the papyrus zone. Papyrus swamps are exceedingly well-developed in this part of Africa and have very rich floras. Less than 20% of the species are common to both zones: a good indication that swamp conditions are very different.

Family *Species*	*Miscanthidium* zone	*Cyperus papyrus* zone	Swamp/open water interface
Alismataceae			
Limnophyton obtusifolium		★	
Balsaminaceae			
Impatiens irvingii		★	
Bryophyta			
Sphagnum spp.	★		
Compositae			
Melanthera scandens		★	
Mikania cordata	★	★	
Convolvulaceae			
Ipomoea rubens		★	
Cyperaceae			
Acriulus greigiffolia	★	★	
Cyperus denudatus	★	★	
C. haspan	★		
C. papyrus		★	★
Eleocharis acutangula			★
Fimbristylis pilosa	★		
Fuirena umbellata	★		
Pycreus nitidus	★	★	
P. polystachyos	★	★	
Rhynchospora subquadrata	★		
Scirpus inclinatus			★
Scleria nyassensis	★		
S. nutans	★		
Euphorbiaceae			
Bridelia micrantha		★	
Graminae			
Andropogon canaliculatus	★		
Digitaria scalarum	★		
Eragrostis mildbraedii	★		
Hyparrhenia diplandra		★	
Leersia hexandra	★	★	
Loudetia phragmitoides	★		
Miscanthidium violaceum	★		
Paspalidium geminatum			★
Panicum parvifolium	★	★	
P. subalbidum	★		
Paspalum commersonii	★		
Vossia cuspidata			★
Guttiferae			
Hypericum lalandii	★		
Labiatae			
Hyptis lanceolata		★	
Neohyptis paniculata		★	
Leguminosae			
Cassia kirkii	★		

Table 3.4. Continued.

Family *Species*	*Miscanthidium* zone	*Cyperus papyrus* zone	Swamp/open water interface
Eriosema glomeratum	★		
Mimosa pigra	★		
Smithia elliotii	★		
Vigna luteola		★	
Lentibulariaceae			
Utricularia gibba	★		
Malvaceae			
Hibiscus diversifolius		★	
Melastomataceae			
Dissotis brazzei	★		
D. rotundifolia		★	
Melastomastrum segregatum		★	
Tristemma incompletum	★	★	
Menispermaceae			
Cissampelos mucronata		★	
Ochnaceae			
Sauvagesia erecta	★		
Onagraceae			
Ludwigia leptocarpa		★	
Oxalidaceae			
Biophytum petersianum	★		
Moraceae			
Ficus verruculosa	★	★	
Polygonaceae			
Polygonum pulchrum	★	★	
P. strigosum	★	★	
P. salicifolium		★	
Pteridophyta			
Thelypteris confluens	★		
T. striata		★	
Rubiaceae			
Oldenlandia goreensis		★	
Pentodon pentander		★	
Scrophulariaceae			
Torenia thouarsii	★	★	
Sterculiaceae			
Melochia bracteosa		★	
Tiliaceae			
Triumfetta macrophylla		★	
Vitaceae			
Cyphostemma adenocaule		★	

lower nutrient concentrations than papyrus. That is why *Vossia*'s position above papyrus in Table 3.3 has less to do with nutrient availability than it has with other ecological factors. And yet *Vossia* does not always replace papyrus in flowing water either. Consider the large Okavango River as it enters the Delta: with an electrical conductivity of less than 10 µS it is chemically similar to Nabugabo, and yet papyrus, not *Vossia*, fringes the levées. The real nutrient status is higher than it appears, because the water is flowing.

There are probably about 20 000 km^2 of *Cyperus*

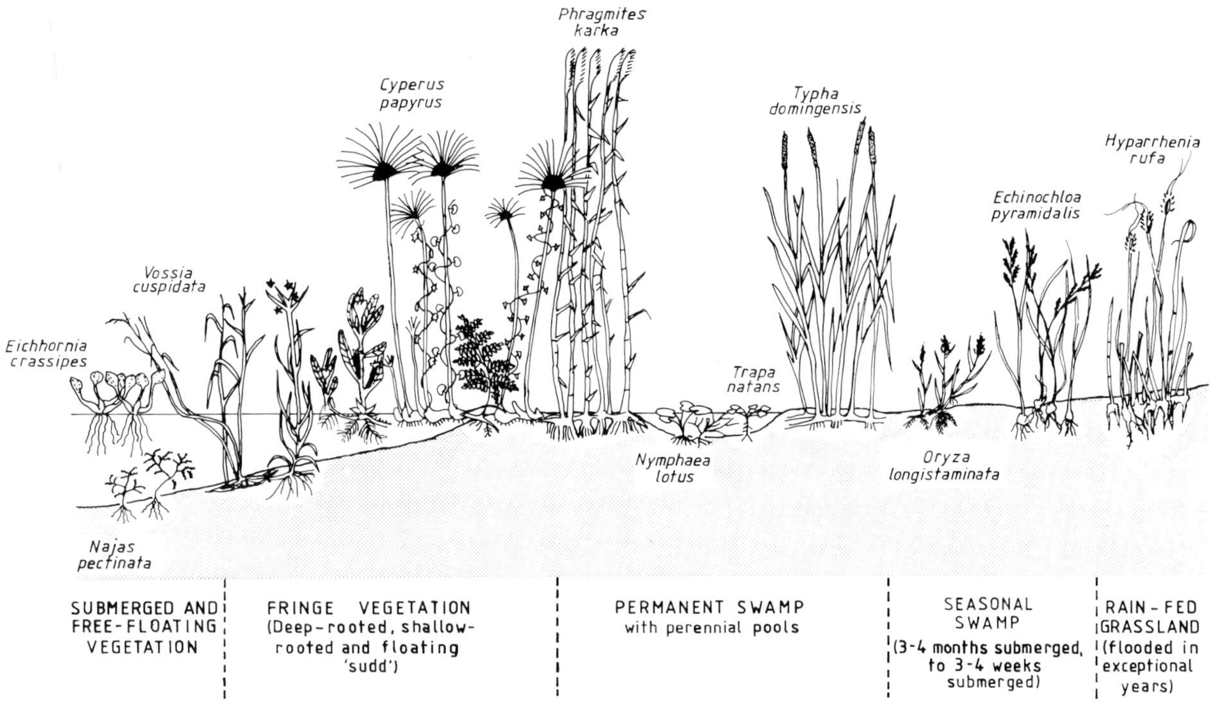

Fig. 3.6. A typical hydrosere in the Upper Nile Swamps of the southern Sudan (modified after Denny, 1984). The horizontal scale is much compressed, but the vertical scale is accurate with the *Phragmites* shoots at a height of 5.5 m. This sequence of grassland communities is found on most major floodplains in Africa.

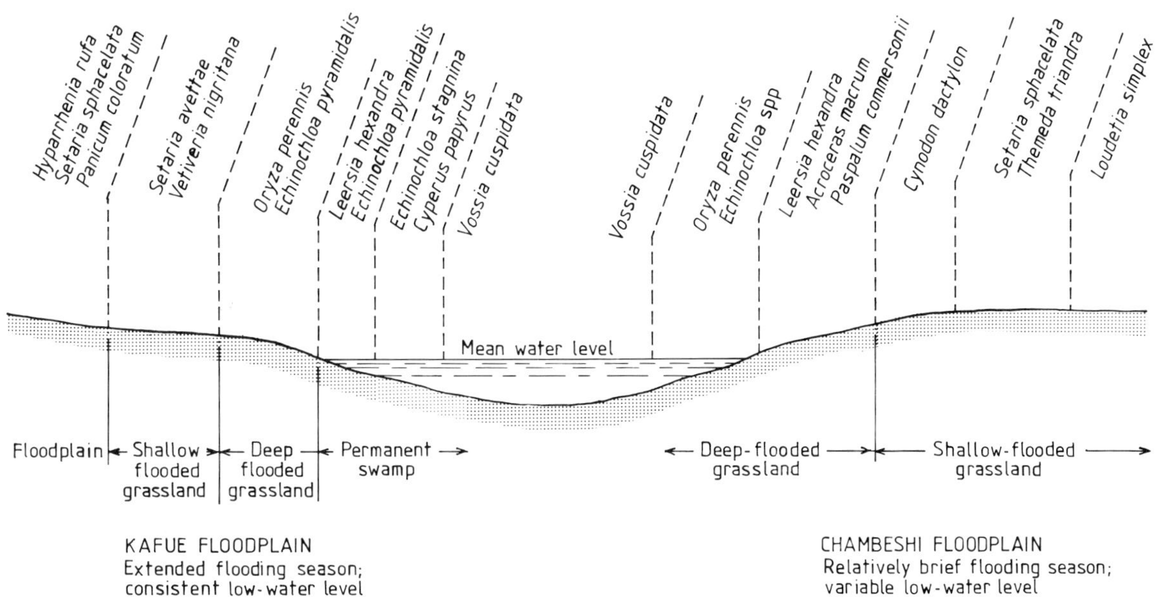

Fig. 3.7. Composite of two Zambian floodplains with different flooding regimes and soil types (after Astle, 1965 and Van Rensburg, 1968). The Kafue has the more stable hydrology.

papyrus swamps in Africa. Iltis and Lemoalle (1983) estimate some 240 km^2 for Lake Chad, which is about 10% of the total reedswamp cover. On the other hand, papyrus probably constitutes in excess of 50% of total reedswamp in Uganda, where river and lake regimes are very stable. Papyrus swamp is less than 20% of total reedswamp in both the Upper Nile and in the Okavango Swamps.

Several other *Cyperus* species can form reed stands, but two are particularly widespread and ecologically important. *Cyperus latifolius* is widely distributed as a reed. It dominates several peat bogs at almost 2000 m altitude in Burundi (Deuse, 1966; USAID, 1980), as well as seasonally flooded, perennially waterlogged, usually lacustrine clay soils at lower altitudes.

Cyperus articulatus and *Scirpus inclinatus* 'SICA' communities (Smith, 1976) form extensive swards throughout the seasonal swamps of the Okavango Delta, where they are usually associated with *Cyperus denudatus, C. longus* and *Paspalum repens*. In deeper water, *Eleocharis* or *Schoenoplectus* may be associated with the SICA community. Almost pure stands of SICA reeds occur where flow rates are low and flooding duration long (*e.g.* in the deep-water seasonal lagoons known locally as 'madiba'). *C. articulatus* may occur alone where flow rates are higher (Fig. 3.10), and *S. inclinatus* is sometimes more prominent in still waters. Although SICA reedswamps do not occur in permanent swamps, a 'SICA association' is widespread on the continent as a minor component of papyrus swamps (see Table 3.4).

3.5.2.3 *Miscanthidium swamps*

Where these reedswamps occur (*see* Section 3.3.1.1.1 and Fig. 3.5), they fill a niche which elsewhere is occupied partly by *Phragmites* or *Typha* and partly by *Echinochloa* or *Vetiveria* (in the Zaire basin, *Jardinea*). However, *Miscanthidium* is typical of low-nutrient sites which are too acid for the other reeds, and, since they are permanently waterlogged, also unsuitable for the floodplain grasses. *Miscanthidium* occurs with *Cyperus latifolius* on high altitude peats in Burundi (Deuse, 1966), although in the recent past its co-dominants on these bogs appear to have been *Xyris* and *Sphagnum* (Hamilton, 1982; Thompson and Hamilton, 1983).

Table 3.4 demonstrates a clear floristic distinction between the acid habitat of *Miscanthidium* communities and the relatively base-rich papyrus environment: only 15–20% of the total species list is common to both reed communities, and the *Miscanthidium* stand has twice the number of sedges and grasses found in the papyrus zone.

3.5.2.4 *Cladium, Phragmites and Typha reed swamps*

As with other reedswamp types, these are all typically mono-dominant stands, and they probably can be arranged in the following descending order of abundance on the African continent: *Phragmites australis – Phragmites mauritianus – Typha domingensis – Phragmites karka – Typha capensis – Cladium jamaicense*. Because *Phragmites* is *not* restricted to the most stable hydrological sites, it is probably more abundant even than *Cyperus papyrus*. As nutrient status declines, the genera would rearrange: *Cladium – Typha – Phragmites*, and *Miscanthidium* at the bottom of the sequence. Figure 3.6 and Table 3.3 show the typical spatial relationship between *Phragmites* and *Typha* stands on floodplains. All of the above species are sward-formers, but *P. mauritianus* is more typical of riverine fringes, particularly of narrow bands on levées. *T. domingensis* is the most salt-tolerant species, although its seedlings will not germinate on alkaline soils. *Cladium* is considered to be a 'calcicole' in Europe and North America, and there is similar evidence from its African distribution. Its sparse distribution may well be limited to wetlands of volcanic areas, coastal wetlands, montmorillonitic substrates, and a few other sites which are particularly base-rich.

3.5.3 *Edaphic grasslands*

Vesey-Fitzgerald (1963*b*) established this term firmly in the literature when he described the grasslands of the headwater catchment area of the Zaire River, which incorporates parts of south-western Tanzania and northern Zambia. He pointed out that valley grasslands, as defined by Gillman (1949) includes various edaphic-climax grasslands associated with the drainage lines. In fact, Michelmore (1939) had previously recognised the importance of relating grasslands to both climate and physiography, because he used the terms 'flood plain grassland' for those occurring in the lower parts of river systems (Vesey-Fitzgerald equates these with 'mbuga'), and 'small valley grasslands' (equivalent to 'dambo' or 'vlei') for those occurring in the upper reaches of the drainage lines. Vesey-Fitzgerald established further sub-divisions on the basis of characteristic genera, thus demonstrating that there is no single 'edaphic factor'.

Further synonyms for edaphic grasslands are 'seasonally-flooded (or -waterlogged) herbaceous wetlands' or 'swamp savanna', and, because of their great variety and complex inter-relationships, it is much more difficult to compile a definitive account of them than it is to describe the African continent's reedswamps. The species composition of edaphic grasslands is very variable, since factors such as rainfall, soil type and salinity, and flooding depth and duration all play important roles both individually and through interaction with each other and with each individual plant species. Nevertheless, it is still possible to generalise, since a number of quite distinct patterns recur throughout the continent.

Community boundaries are often indistinct because flood regimes vary from year to year, but there can be no doubt that the depth of flooding and the duration of the flood regime are the primary determinants of species composition in seasonal wetlands.

3.5.3.1 *Hydroseres*

The most common hydrosere in the savanna region is *Vossia – Oryza – Echinochloa – Hyparrhenia (rufa)*. This sequence is to be found, for instance, in the Sudan (Fig. 3.6), the Kafue Flats (Fig. 3.7) and the Bangweulu Floodplain (Bell & Grimsdell, 1973) and West Africa. However, the relative proportions of the above three floodplain emergents vary greatly from site to site: thus the deep-flooded *Oryza* dominates on the Kafue Flats, Zambia (Sheppe & Osborne, 1971), *Vossia* predominates in the fringing wetlands of the Zaire River throughout the cuvette central, *Echinochloa scabra* is the most important component of the inland delta of the Niger River in Mali (Bacalbasa-Dobrovici, 1971), and *E. pyramidalis* often dominates the small edaphic grassland community between reedswamp and gallery forest in the wetter dambos of the bush veldt in South Africa (Fig. 3.15), as well as the swamp communities of many Central African valley grasslands (Vesey Fitzgerald, 1970).

The above highly generalised sequence, though widely applicable, may be expanded to include other species and the following composite hydrosere can be applied, simply by deleting species (or zones) not present, almost anywhere in Africa: (Open water) – *Vossia – Cyperus papyrus – Phragmites – Typha – Miscanthidium – Echinochloa scabra – Oryza – Acroceras* etc.; (watermeadow grasses) – *Echinochloa pyramidalis – Jardinea – Panicum maximum/P. repens – Pennisetum glaucocladum – Vetiveria/Setaria/Cynodon/Loudetia – Hyparrhenia* – (rain-fed savanna).

Naturally, no single locality will have all of these types represented, so 'real' hydroseres are always contractions of the above sequence. *Jardinea*, for instance, only occurs in the Guinea/Congo zone, where true reedswamp, *Vetiveria, Cynodon* and *Loudetia* are typically very rare (Fig. 3.11). It should also be noted that quite subtle local edaphic conditions become more significant as determinants of sequencing towards the top of the floodplain, making it difficult to generalise concerning the 'order of species' on floodplains. For this reason, some genera in the above sequence have been bracketed together. In other words, although *Loudetia* and *Vetiveria* usually enter the hydrosere before *Cynodon*, because of local factors such as grazing pressure, burning frequency, etc., they do not always do so. To illustrate with examples, on the Chambeshi Floodplain all reeds are absent, so that *Vossia* merges into *Oryza* and *Echinochloa*, followed by the water-meadow grasses, then *Cynodon – Setaria – Loudetia* (in that order). Figure 3.8 shows the replacement of *Panicum* and *Loudetia* species, dominating the sandy soils, by *Setaria* and *Hyparrhenia/Brachiaria* on heavier soils with the same flooding regime. On the Bangweulu Floodplain, the 'normal' sequence of

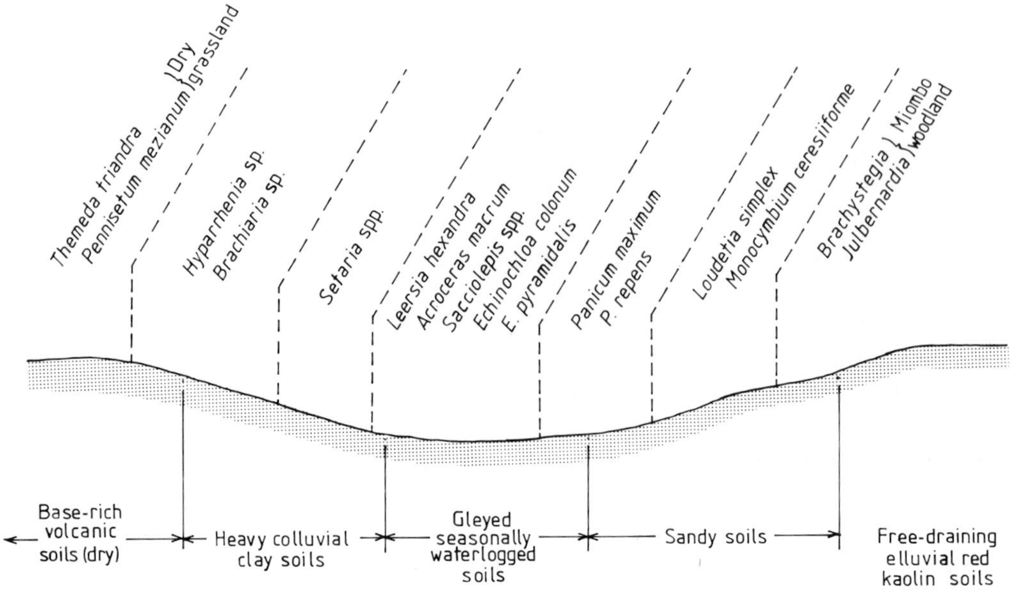

Fig. 3.8. A composite, illustrating some characteristic grasses of dambo hydroseres.

Oryza – Echinochloa – Hyparrhenia on heavy illuvial clays is interrupted on fine alluvium by one of *Oryza – Acroceras – Themeda*, which leads to *Loudetia* on oligotrophic sandy soils, or to *Hyparrhenia* again on the heavier materials (Grimsdell & Bell, 1976). Figure 3.9 represents a heavily grazed regime on sandy, mildly alkaline soils, and *Panicum repens* here occurs mixed with *Cynodon*. Figure 3.10 shows a vegetation section on a narrow riverine floodplain in the Okavango Delta: the sequence *Vossia – Oryza – Acroceras – Setaria – Cynodon* omits the 'swampy' *Echinochloa* species and the *Vetiveria* and *Loudetia* stands typical of lighter soils, but introduces *Imperata cylindrica* in a very characteristic location immediately prior to riverine forest. Thus, Streel (1963) records that, where gallery forest is present in Shaba Province (Upper Zaire), *Echinochloa* is largely replaced by *Hyparrhenia diplandra, Pennisetum purpureum* and *Imperata cylindrica* (*see* Fig. 3.11).

The *Cynodon/Chloris* community of the Okavango (Fig. 3.10) is widespread in Africa, and, where gallery forest is *not* present (as in much of the Sudan), *Brachiaria obtusifolia* replaces the *Imperata* component.

Smith (1976) gives a sequence for the Cubango River floodplain in Angola which runs from the riverine fringe of *Phragmites (mauritianus)* to *Miscanthidium (junceum)*, then *Pennisetum glaucocladum* on the steep river bank (*Pennisetum* is a typical replacement for the water-meadow grasses, which would have dominated this zone on a gently-sloping floodplain), leading to *Vetiveria* at the limit of flooding and finally a sward of *Cynodon*. *Cynodon dactylon* has a very wide ecological range: it may occur as an understorey in *Echinochloa pyramidalis*, as an associate of the water-meadow grasses, as a mosaic community in the floodplain margin *Vetiveria* and *Hyparrhenia* zones, or as a sward in its own right on regularly inundated pan margins (Heeg & Breen, 1982), lake margins (Fig. 3.9) or rarely-flooded flats (Smith, 1976). It is tolerant of moderate salinity (Fig. 3.13) and advantaged (when in competition with taller grasses) by moderate to heavy grazing regimes, as on the Pongolo Floodplain, South Africa (Furness, 1981), and the valley grasslands of East Africa (Vesey-Fitzgerald, 1970).

3.5.3.2 *Secondary grasslands*

Cynodon illustrates well the difficulty in deciding whether a plant is specifically adapted to flooding, or whether the flooding adaptation is but an incidental benefit of some other primary adaptation.

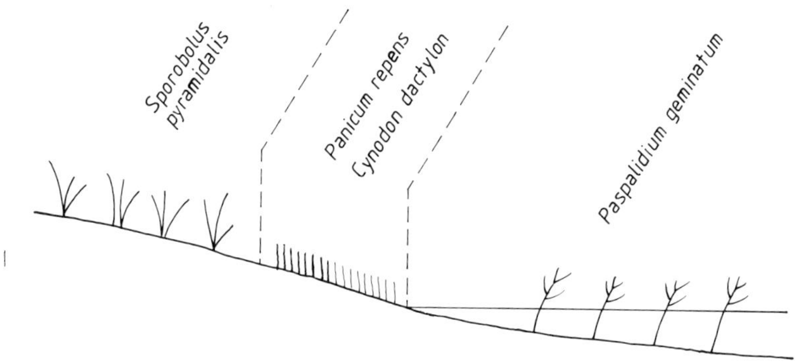

Fig. 3.9. A hydrosere on leached sands fringing Lake George, Uganda (after Lock, 1973). *Paspalidium* is typical of relatively sheltered locations. The *Panicum*/*Cynodon* zone is often enriched with *Hemarthria natans, Brachiaria radicans* and *Pycreus polystachyos*. On less alkaline substrates, *Sporobolus* is replaced by *Loudetia*/*Monocymbium*.

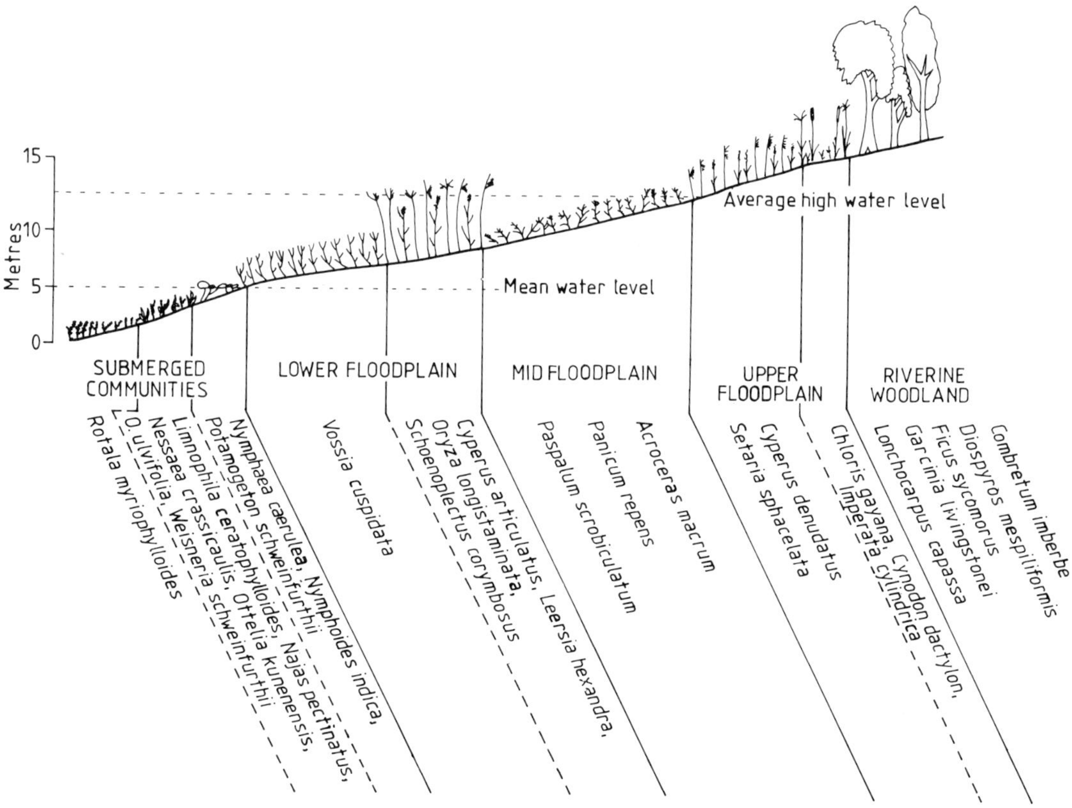

Fig. 3.10. Vegetation sequence on a section of the lower Boro River, Okavango Delta, Botswana (after Lubke, Raynham and Reavell, 1981).

This problem arises again in considering the components of secondary grasslands. Most of these are fire-induced, some being fire climax and others seral to other grassland types or to forest. The importance of fire as an ecological factor, a vegetation determinant, and a range management tool in Africa must not be underestimated (see Anonymous, 1971). In fact, 20 years ago, West (1965) estimated that about one third of African savanna was burned every year; it is unlikely that this statis-

Plate 3.8. Edaphic, fire-climax grassland in Ankole District, Uganda. The only woody plants able to withstand the harsh regime of seasonal waterlogging and fire are *Borassus* palms and the occasional *Acacia*. (Photo: K. Thompson.)

tic has changed. If anything, burning, particularly of wetlands, is a more widespread practice now even than it was 20 years ago. Thompson (1973) draws attention to the recent breakdown of traditional management practices, and Van Rensburg (1971) discusses the use of fire as a management tool in the Okavango Swamps. There can be no doubt that many savanna communities today are no longer 'natural'. This is particularly true of edaphic grasslands, which are always the first savanna types to be managed for increased productivity, since the seasonal flooding regime confers upon them greater potential for improved grazing than is the case for dry savanna. Edaphic grasslands (seasonal herbaceous swamps) in Africa are today almost always modified by fire in one way or another. In those rare cases where fire is uncommon, grazing pressure is still high enough to be a significant evolutionary force.

Schimper (1903) and Robyns (1936) classified certain African grasslands as 'climatic' when they were actually fire-climax. The confusion probably arose because many water-logging tolerant species are characteristic of fire-induced communities, particularly those of the forest transition regions of the Guinea/Congo zone. Forest has a higher water demand than grassland, so that, when trees are removed or suppressed, there is 'excess' water. If there is but a brief riverine seasonal flooding regime, or if the rainfall is high and there is a pan horizon in the soil, removal of a basically mesophytic forest can result in a herbaceous wetland.

Jardinea communities, for instance, are typical of floodplain grasslands on low-lying, gleyed clay substrates in Zaire and the Congo Republic, and also characterise the 'ésobe' forest/grassland mosaic wetlands of Robyns (1936) and Germain (1965) which are found around the western and southern perimeters of the forest zone of Zaire. The ésobes are seasonally waterlogged and appear to be fire-maintained: they are certainly very regularly burned today. Whether they would be seral to forest if fire could be excluded is uncertain.

Other species characteristic of these forest-zone edaphic grasslands are *Hyparrhenia mutica, H. diplandra, Anadelphia* spp., *Andropogon* spp., *Imperata cylindrica* and several Cyperaceae, such as *Bulbostylis, Fuirena, Rhynchospora* and *Scleria*; all of these are also common in fire-induced grasslands. Figure 3.11 demonstrates some of the interrelationships between forest-zone floodplain grassland communities.

3.5.3.3 *Communities of black clays, pans and dambos*

Most of the black clay wetlands, and the dambo

Plate 3.9. The fire-climax 'ésobes' of the forest/savanna transition zone in southern Zaire. A recently burned patch can be seen at the extreme upper right. Note the numerous termite mounds. Much of the surrounding forest is rich in flooding tolerant species and is best described as 'shallow-flooded seasonal swamp forest'. (Photo: K. Thompson.)

and pan wetlands, are also fire-climax systems. The wetter areas of the valley grasslands are characterised by *Echinochloa pyramidalis* and the water-meadow grasses (Fig. 3.8). The wetter areas of dambos are often dominated by the water-meadow grasses, or by Cyperaceae, such as *Ascolepis elata, Bulbostylis cinnamomea, Cyperus esculentus, C. margaritaceus, C. platycaulis, Fuirena pubescens, Kyllinga erecta, Mariscus deciduus, Pycreus aethiops, Scirpus microcephalus* and *Scleria bulbifera* (White, 1983).

Where waterlogging is less frequent, a variety of

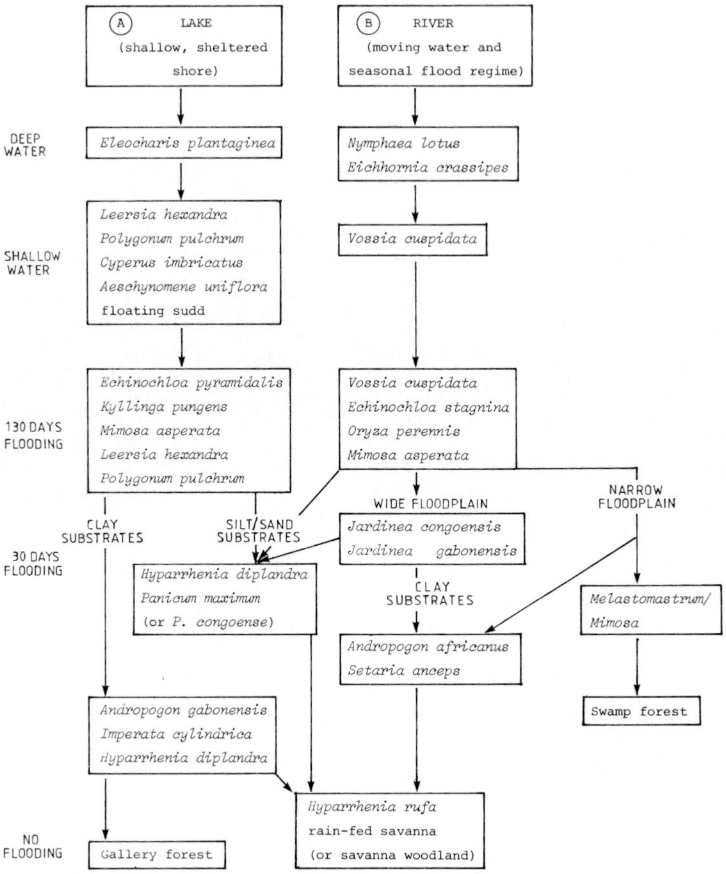

Fig. 3.11. Hydroseres in the Zaire (Congo) basin (Congo/Guinea forest zone).

Sequence A (based upon Koechlin, 1961) represents vegetation development from a sheltered shore with a relatively narrow swampy margin. *Eleocharis* emerges from deep water beyond the floating sudd and is an indicator of silty substrates. The sudd mat components also characterise the marginal swamp community, which is typically dominated by *Kyllinga* when the floating community is absent. *E. pyramidalis* becomes dominant on higher ground and indicates heavy clay soil of colluvial origin. This catena rises fairly quickly above the floodplain and, depending upon the nature of the colluvial material, the community is dominated either by *Andropogon gabonensis,* or by *Hyparrhenia diplandra* (with *Panicum maximum*). Both edaphic grassland communities lead to the climatic climax grassland of *Hyparrhenia rufa. Imperata,* though widely distributed in Africa, is very characteristic of heavy, seasonally-saturated clay soils in the forest zone. It may be replaced by *Pennisetum purpureum* along forest margins. Streel (1963) records, for the middle reaches of the River Lufira, Shaba Province (Haut-Congo, as opposed to the Bas-Congo of Koechlin (1961) and Sita (1970)), a community of *Echinochloa pyramidalis, Phragmites mauritianus, Cyperus thonningii, Mikania cordata, Acacia polyacantha* and *Polygonum salicifolium,* which is limited in the Congo/Guinea zone to fertile soils near rivers, although Streel's detailed study was actually conducted in the forest/savanna transition zone of Zaire. Streel also notes that this community is replaced by *Pennisetum purpureum, Hyparrhenia diplandra* and *Imperata cylindrica* where gallery forest is present.

Sequence B (after Sita, 1970) has a more pronounced flood regime which eliminates sudd communities. *Vossia* dominates the aquatic fringe and also, together with *E. stagnina,* the lower reaches of the floodplain. This is a widespread floodplain community type on low-lying, regularly flooded and permanently saturated silt substrates throughout the Congo/Guinea zone and the Niger inland delta. *Oryza* is almost always present in this 'burgu' community, and may dominate where *Jardinea* is absent. At this point, the catena may develop in any one of four directions, depending upon substrate type and floodplain width and seasonality: (a) to swamp forest where the substrate is permanently saturated (b) to *Jardinea* seasonal swamp grassland on low-lying, gleyed illuvial clay substrates where the floodplain is extensive, (c) directly to *Andropogon/Setaria* (colluvial clay substrates) or *Hyparrhenia/Panicum* (silt/sand substrates) where the ground rises more quickly. *Hyparrhenia rufa* is again the climatic climax grassland community, although it may either lead to, or be replaced by, forest.

communities predominates, depending upon substrate type, waterlogging duration, rainfall and fire frequency. *Loudetia simplex* and *Monocymbium ceresiiforme* are the most widespread and typical species of dambos. Although dambos are not described, as such, from West Africa, White (1983) notes a site in Ghana, close to the forest zone, which is characterised by *Andropogon, Loudetia simplex, Monocymbium, Setaria sphacelata* and *Sporobolus* and is clearly a dambo-type community. *Loudetia* and *Monocymbium* are replaced by *Setaria* spp., *Brachiaria* spp. and *Themeda triandra* on the heavier black clay soils of mbugas. Although all of the latter five genera are only rarely inundated, their soils are always gleyed, or at least mottled. They are, therefore, clearly tolerant of prolonged and regular waterlogging – well within the definition of 'seasonal emergent' used in this chapter.

Some *Acacia* species are also typical of vertisols. *A. seyal* is found on black clays where the annual rainfall lies between 600 and 750 mm; *A. drepanolobium*, together with *Setaria incrassata*, prevails on similar substrates receiving in excess of 750 mm. Where waterlogging exceeds 4 months' duration, *Acacia* spp. are usually replaced by *Borassus*, or even *Phoenix*.

The principal species of secondary (fire-induced) grasslands in the savanna zone are: *Andropogon gayanus, Brachiaria brizantha, Hyparrhenia diplandra, Hyparrhenia rufa, Imperata cylindrica, Loudetia phragmitoides, Loudetia simplex, Monocymbium ceresiiforme* and *Pennisetum purpureum*. *Setaria incrassata* is an edaphic climax replacing *H. rufa* on the heavier soils. *Andropogon, H. diplandra, Imperata, L. phragmitoides* and *Pennisetum purpureum* are also species typical of zones transitional to forest (*i.e.* they are disturbed communities seral to forest). Several of these species feature in Fig. 3.11, and the importance of *Andropogon gayanus* in the seasonally waterlogged park savanna community of Fig. 3.12 also suggests a fire-climax regime. The eutrophic *Cyperus papyrus/Paspalum polystachyum* community of Fig. 3.12 is regularly inundated and not burned; on the other hand, the *Cyperus distans/Brachiaria deflexa* grassland of the freshwater hydrosere is only seasonally flooded and is probably fire-controlled.

3.5.3.4 *Alkaline grassland and swamp*

Figure 3.12 also introduces the important question of alkaline (halophytic, saline) wetlands. African coastal wetlands are well developed only in Moçambique and the Guinea/Congo zone. Most of these wetlands are highly complex, grading from the entirely saline to the entirely freshwater, from the very alkaline to the very acid, from saltmarsh to swamp savanna, from mangrove to tropical high forest. One such wetland complex is illustrated by Fig. 3.12. The Moçambique counterparts are virtually undescribed, but Tinley *et al.* (1976) refers to large areas of edaphic grasslands in the Zambezi Delta which support considerable buffalo concentrations. Barradas (1951) briefly mentions the extensive coastal peatlands ('machongos'), many forested (such as the 2213 ha Marrangua machongos), on both modern and ancient dunes, and underlain by riverine alluvium and sometimes linked with mangrove systems.

Inland, seasonal alkaline swamps are widespread, from the Makgadigadi in Botswana (the largest salt pans in the world) and the Eastern Rift Valley lakes, to the enormous number of small dambos where accumulation of alkaline salts in the soil profile has simply resulted from evaporation of rainwater or fresh groundwater. Figure 3.13 shows the relative tolerances of the more important salt-tolerant species of the savanna zone. Vesey-Fitzgerald (1963*b*, 1970) and Greenway & Vesey-Fitzgerald (1969) give very lucid accounts of the plant communities of alkaline swamps and flats, and these are summarised by White (1983). Other useful material is provided by Howard-Williams, (1973, 1975, 1979) and Iltis & Lemoalle (1983). Dean (1967) gives the following sequence for Lake Rukwa: open (saline) water – *Sporobolus spicatus* – *Diplachne fusca* – *Sporobolus robustus* – *Cynodon dactylon* – *Echinochloa* – *Aeschynomene* – savanna woodland.

S. spicatus, though apparently more salt-tolerant than *D. fusca*, is not able to endure inundation. It is typically a 'beach-zone' species which surrounds alkaline lakes. For instance, it fringes (together with *Odyssea paucinervis* and chenopods such as

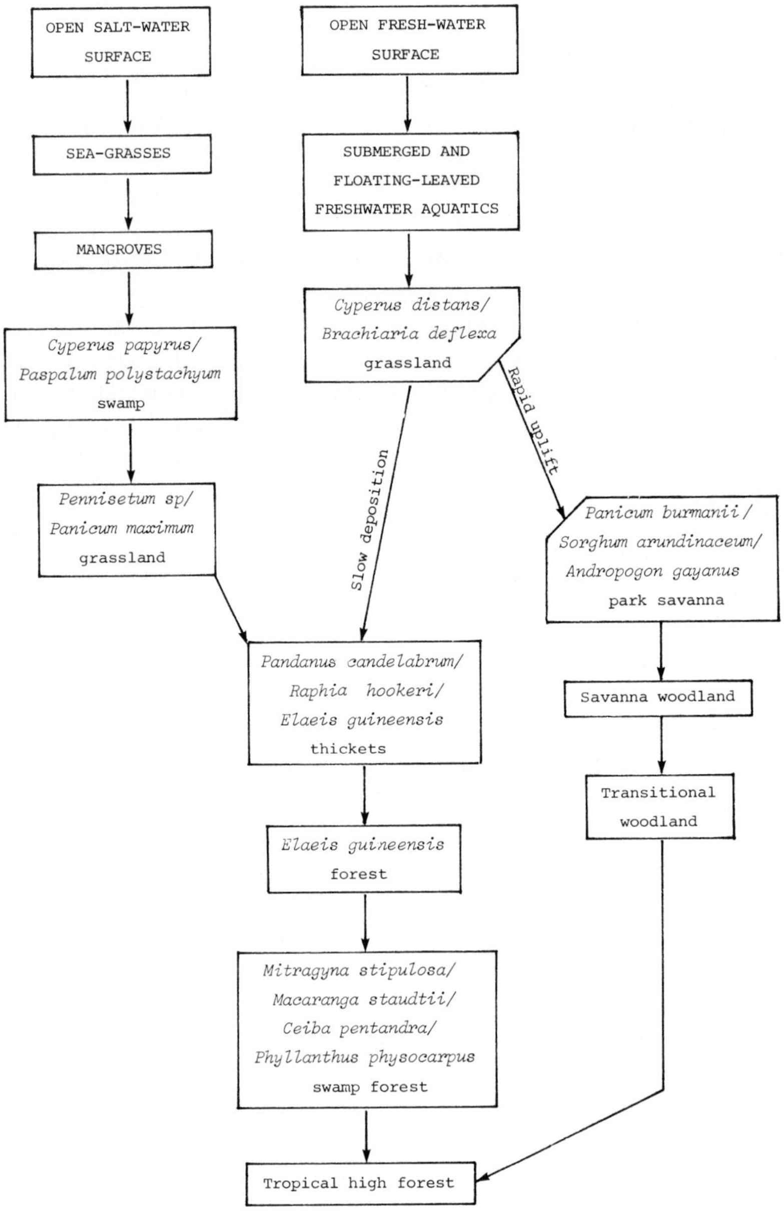

Figure 3.12. Plant succession on the floodplains of the River Ogun delta, Nigeria (after Adejuwon, 1971). Note the convergence of saltwater and freshwater sequences. The *Pennisetum* is probably *P. glaucoclaaum.* Since deltas are highly active systems and flow rates and sedimentation patterns are subject to change with time, so vegetation succession may proceed either 'forwards' or 'backwards'. Boughey (1957), for instance, regards afl *Raphia* dominated swamps as being retrogressive developments (either natural or man-induced) from broad-leaved swamp forest. This is almost certainly too narrow a view of the seral role of this ecologically important swamp palm.

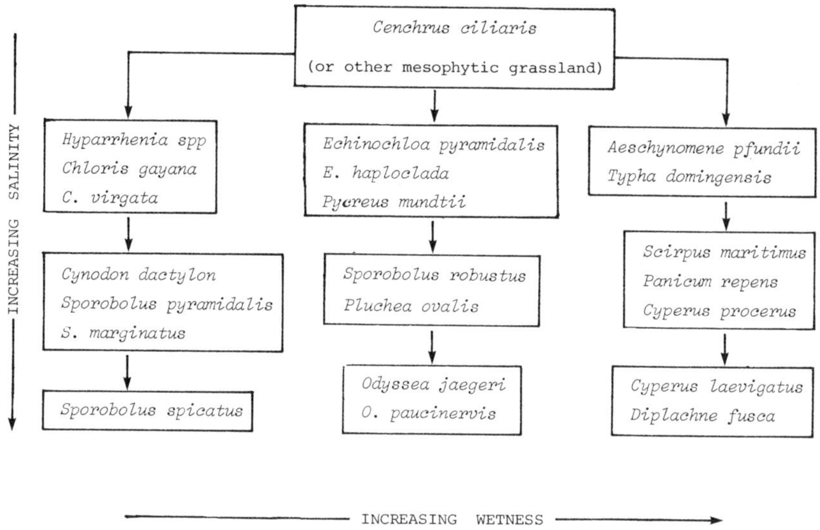

Fig. 3.13. Relative tolerances of some important wetland plants of the savanna zone to flooding and salinity. The arrows represent some common directions for plant succession.

Suada and *Atriplex*) both the Etosha and Makgadigadi Saltpans. *S. spicatus* always occurs with *D. fusca* or *Psilolemma jaegeri* (*O. paucinervis* in southern Africa), and usually all three are present. This is because they are each adapted to different ecological conditions. During dry periods, *S. spicatus* often expands across a dry lake bed, colonising the black mud under the soda crust, but as the lake rises again, *Psilolemma* replaces it and may persist indefinitely under shallow flooding, and in depressions which are flooded frequently. In East Africa, there are two rainy seasons each year, and this increases the flooding frequency of the Eastern Rift Valley lakes to the point at which *S. spicatus* is usually replaced by other, more flooding-tolerant *Sporobolus* species.

Diplachne fusca is more typical of semi-permanent (but shallow) saline swamp, but, since it is also capable of surviving long periods of drought as a monospecific, fire-climax sward, it is clearly a very adaptable species. Howard-Williams (1975, 1979) describes *Diplachne* as one of only three aquatic emergents (the others are *Cyperus laevigatus* and *Aeschynomene pfundii*) which survived a prolonged drying phase on highly alkaline soils exposed by the receding Lake Chilwa, Malawi. He also pointed out that *Diplachne* can persist for long periods in deep water, but will die after a year of continuous deep flooding. Permanent (shallow) saline swamp is usually dominated by *Cyperus laevigatus*.

Several predominantly coastal species may be associated with inland saline communities. Thus Howard-Williams (1973) recorded *Scirpus maritimus* in Lake Chilwa, and Symoens (1953) noted *Juncus maritimus* and *Sporobolus virginicus* in the Mwashya Salt Flats of the Lufira Valley. In fact, *J. maritimus* has a widespread inland distribution in Africa.

The most adaptable genus of African saline soils is probably *Sporobolus*: indeed, it is probably the most useful indicator of 'soil sourness' on rangeland. The range of adaptation extends from the highly salt-tolerant *S. spicatus*, through the moderately tolerant (but more flooding-tolerant) *S. robustus* to the mildly-tolerant *S. pyramidalis* of only rarely flooded soils.

3.5.4 *Edaphic forests*

The above account of edaphic grasslands, and the species characteristic of them, highlights the problem of distinguishing species which are specifically adapted to survive inundation from species which are merely tolerant of flooding or waterlogging. An even greater problem arises in considering the large number of trees and shrubs which experience seasonal waterlogging because so many are also normal components of apparently quite mesophytic forest types.

There are, of course, several different types of edaphic forest. Boughey (1957) suggested the following classification for West African communities: (a) mangroves; (b) swamp forest; (c) periodic swamp forest; (d) riparian and gallery forest.

This scheme recognises the importance of flooding duration and also acknowledges that there are forest types which only very rarely experience surface flooding but probably often endure waterlogging. However, the Boughey scheme uses unspecific terminology which means different things to different people, so Dawkins & Philip (1962), in their site type classification for the Ugandan Forest Department, were more precise with a scheme which recognises both flooding depth and duration: (a) raised dry; never flooded or waterlogged; (b) raised dry; water table only rarely (and briefly) within two metres of the surface; (c) raised wet; water table always, or frequently, within two metres of the surface; (d) swamp, seasonal; water table well below the surface for several months (= occasionally inundated); (e) swamp, permanent; water table only rarely below the surface (= regularly inundated).

Dawkins & Philip split the Boughey category (d) into two (b and c), which rightly draws attention to the wide range of adaptation among riparian and gallery forests. The Dawkins & Philip scheme also acknowledges that 'permanent swamp' is only 'an extreme case of seasonal'. It is, in fact, unlikely that more than a mere handful of African trees can endure persistent flooding for more than a year or so.

3.5.4.1 *Permanent swamp forests*

Swamp forests are best developed in the Guinea/Congo zone, where they are structurally little different from rain forest. They are, however, relatively species poor, the most common genera in this zone being *Carapa, Nauclea, Mitragyna, Pandanus, Phoenix, Raphia, Spondianthus, Symphonia, Uapaca* and *Voacanga*. The wettest areas are dominated by the palms and *Raphia* (*see* Fig. 3.12). *Carapa, Nauclea, Spondianthus* and *Symphonia* are strictly Guinea/Congo zone species. *Mitragyna* does not occur in South Africa but, even so, it is probably still the most important dicotyledon swamp forest tree in Africa. *Uapaca, Syzygium, Voacanga* and *Xylopia* are also widespread and important as swamp forest dominants. The *Oubanguia/Guibourtia* 'permanent' swamp forest (Fig. 3.14) is quite restricted in its distribution (parts of Zaire only). *Entandrophragma palustre* is not, strictly speaking, a member of this community, since it more typically dominates its own 'permanent' swamp forest type, with associates *Coelocaryon botryoides, Uapaca guineensis* and *Symphonia globulifera*.

3.5.4.2 *Seasonal swamp forests*

The seasonal swamp forests are more difficult to delimit, because they are so variable in their species composition. Most of the permanent swamp forest trees will also persist under the less rigorous, seasonally-flooded regimes, together with genera such as *Pseudospondias, Diospyros, Croton, Ficus, Rauvolfia* and *Anthocleista*. Thus White (1983) records the following swamp forest taxa for his 'Zambezian zone of regional emdemism' (more or less equivalent to the southern African sector of the moist savanna zone of Fig. 3.1): *Ficus, Mitragyna, Raphia, Rauvolfia, Syzygium, Uapaca, Voacanga, Xylopia. Mitragyna* and *Raphia* do not occur in the savanna zone proper, where flooding is much more markedly seasonal. 'Permanent' swamp forest is very rare outside the Guinea/Congo zone.

Africa's only podocarp swamp forest occurs in the Guinea/Congo forest transition zone, on the north-western shore of Lake Victoria. Although several forest zone trees are represented, the seasonal nature of the flooding regime (Dawkins & Philip category d) is also obvious from a list of the most prominent taxa: *Podocarpus/Baikiaea, Croton, Diospyros, Ficus, Garcinia, Lasiodiscus, Macaranga, Maesopsis, Mitragyna, Parkia, Sapium, Pseudospondias, Spondianthus, Symphonia, Syzygium.*

3.5.4.3 Riverine and gallery forests

Throughout the savanna zone of Africa, these high

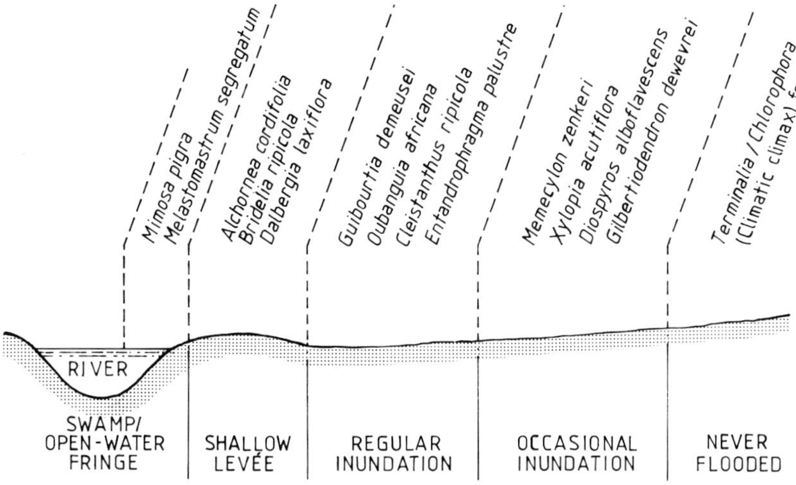

Fig. 3.14. A composite hydrosere for the forest zone of the lower Zaire basin (after Sita, 1970). The community subjected to 'regular inundation' corresponds, more or less, to Lebrun and Gilbert's (1954) 'periodic swamp forest', although *Entandrophragma palustre* is more typical of its own community type (see text). *Alchornia* and *Dalbergia* are regular components of the Amazonian igapo inundation forests (Gessner, 1968), where the Euphorbiaceae and the Leguminosae are also very prominent.

water-table, rarely-inundated forests (*see* Figs. 3.10, 3.15) are characterised by the following genera: *Acacia, Croton, Combretum, Diospyros, Ficus, Garcinia, Kigelia, Rauvolfia, Zizyphus.*

Most of the above genera show species differences between the northern and southern sectors of their ranges (*see* Section 3.3.2.2). *Acacia, Combretum* and *Zizyphus* are not characteristic of riverine and gallery forests in the Lake Victoria zone, but *Erythrina, Phoenix, Pseudospondias* and *Spathodea* are (together with *Croton, Diospyros, Ficus, Kigelia* and *Rauvolfia*). In the Guinea/Congo zone, the distinctive character of these fringing forests is blurred and they merge gradually with both swamp forest and rain forest (Fig. 3.14).

However, the river levée is usually a distinct feature in any vegetation zone. In the forest zone it is usually colonised by *Alchornea* and *Bridelia*, although higher levées in the Guinea/Congo zone will probably have *Pennisetum purpureum* stands. Outside the forest zone, where the hydrological regime is seasonally stable (*e.g.* the Upemba Swamps or the Okavango Swamps) shallow levées are colonised by *Cyperus papyrus*, but a more pronounced seasonal regime replaces the giant sedge with *Phragmites mauritianus.* As the flooding regime becomes more pronounced still, *Phragmites* is supplanted by *Borassus* groves on levée crowns, or by *Acacia albida/Zizyphus mucronata* communities. The *Ficus sycomorus/Rauvolfia caffra* community of southern African water courses (Fig. 3.15) is more typical of gallery forest, but it may also dominate levées as it does on the Pongolo floodplain (Heeg & Breen, 1983).

3.5.4.4 *Secondary forests*

One very noticeable feature of seasonal swamp forests is the abundance of tree species which are better described as 'secondary forest colonisers'. *Mitragyna* itself has many of the typical features of a coloniser: fast growth, open crown, high light requirement, *etc. Musanga cecropioides*, the 'parasol tree', is an even more typical coloniser, and yet it too has stilt roots and frequently occurs on swamp substrates. Other typical colonisers, which also occur widely in swamp forest, are *Croton, Macaranga, Maesopsis, Myrianthus, Lasiodiscus* and *Rauvolfia*. The *Podocarpus/Baikiaea* swamp forest described in Section 3.5.4.2 has many of these colonisers and the *Mitragyna/Macaranga/Ceiba/Phyllanthus* swamp forest of Fig. 3.12 is a classic example of the seral role of these trees on flooded or waterlogged ground. The seral role of certain swamp-tolerant grasses characteristic of the forest zone was referred to in Section 3.5.3.2.

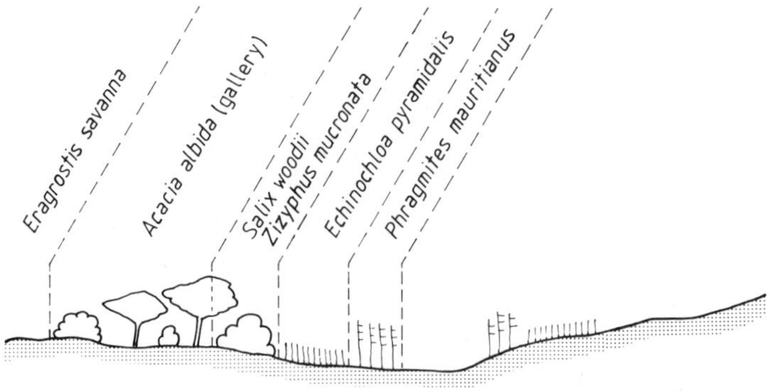

Fig. 3.15. Vegetation zones of a typical seasonal water-course in south African bushveldt (after Knapp, 1973). *Acacia albida* is a characteristic species of high water-table gallery forest in most sub-arid regions of Africa. In the Guinea/Congo zone it often occurs close to permanent rivers, commonly on levées (Streel, 1963). In East Africa it is often replaced by *Ficus* and *Acacia xanthophloea*. The lowest point on the profile will be underlain by a hardpan.

3.5.5 *Montane wetlands*

Some references have already been made in this chapter to certain high altitude wetland types in Africa, but it is outside the scope of this book to deal with them in detail, since an adequate account is already available in Thompson and Hamilton (1983).

3.6 Adaptation and tolerance in wetland plants

It was argued at the beginning of this chapter that definitions of the terms 'aquatic emergent' and 'aquatic habitat' cannot be complete unless they include the 'seasonal emergents' and the 'seasonal wetlands'. By the same token, no study of emergents or wetland communities can be complete unless it recognises the full ecological range from 'permanent aquatic' to 'waterlogging tolerant'. In this section, an attempt is made to analyse the nature of adaptations to the wetland environment.

Aquatic emergents fall into two broad ecological categories: the 'deep-flooded species' of the permanent swamps, and the 'shallow-flooded species' of the seasonal swamps and floodplains. The former are clearly specifically adapted to life in permanently water-logged conditions. Shallow-flooded species, on the other hand, can be sub-divided into those which are adapted to periodic flooding and waterlogging, and those which are merely tolerant of such conditions. Much of this present chapter has been devoted to demonstrating how difficult it is to distinguish between these two groups of seasonally-flooded emergents.

A 'tolerant' plant has a 'plastic' phenotype, which confers a high capacity for survival outside its optimal ecological range. The grass *Themeda triandra*, mentioned in the same context at the beginning of this chapter, is probably such a plant. It is a major ungulate food plant of grasslands dominating the richer volcanic and alluvial soils, but, because these soil types are sometimes subject to flooding or waterlogging, the plant must be able to *tolerate* these events.

Upper floodplain grasses, such as *Vetiveria* and *Setaria*, appear to have wide ecological tolerance, and yet their case is not the same as that of *Themeda*. A grass high on the floodplain may well have metabolic tolerance of brief flooding, but the major *adaptation* is to the drought which constitutes the majority of the year. The reverse is the case for deep-flooded species such as *Leersia* or *Oryza*; these plants tolerate the dry season and their major adaptations are to growth under aquatic conditions. Even so, they have their limits like all other species, and *Oryza* will die if it is not able to keep the tips of its leaves, and its inflorescence, above water.

Examples of trees with broad ecological tolerance are even easier to find. *Xylopia staudtii* will grow in saline swamp with mangroves, in freshwater swamp with *Mitragyna* or on mesic sites (where it fails to develop its peg roots) with a variety of 'dry' forest species. The mesic sites are never significantly drought-stressed, so it can be assumed that the tree is primarily adapted to endure waterlogged soils but is tolerant of a wide range of other conditions.

The case of some riverine and gallery forest species is again quite different. *Syzygium guineense* and *Rhus quartiniana* are the only woody species able to survive over 2 metres of seasonal flooding on the Barotse floodplain of the upper Zambezi (White, 1983). Neither species can be termed a 'swamp tree' (although *S. cordatum* certainly is), but they both have impressive tolerance of brief periods of very deep flooding – again likely to involve a metabolic, rather than structural, solution, since such a relatively short period of inundation is involved. *Borassus* is another extreme case. It is very rarely flooded, and yet it is often the only woody plant present on black clay soils in the savanna zone which are waterlogged for prolonged periods. *Acacia xanthophloea* is also an excellent indicator of high water-tables, but it is quite intolerant of either surface flooding or prolonged waterlogging. Reedswamp will not persist *unless* it is almost permanently waterlogged. And, yet again, there is the case of *Croton*, which tends to be a swamp forest type in the lowland Guinea/Congo zone, but predominantly a riverine/gallery forest component above 1000 metres' altitude (this curious feature is discussed at greater length in Section 3.7).

Clearly, flooding, and plant response to it, is far from being a simple ecological factor. Flooding depth, duration of flooding, water-table height during the dry period, soil type, temperature, and several other factors, are all involved to a greater or lesser extent.

Adaptations can be classified as morphological, anatomical or physiological (metabolic), but it is not always obvious which mechanisms are being employed, as for instance, with the well known substrate preferences of many grasses.

Fine illuvial silts and clays	*Echinochloa, Oryza*
Alluvium	*Acroceras, Themeda*
Sandy silts	*Vetiveria*
Sands	*Paspalum repens*
	Loudetia, restiads, Paspalidium
Heavy black clays	*Imperata, Setaria, Brachiaria, Hyparrhenia*
Alkaline soils	*Sporobolus, Diplachne, Psilolemma, Odyssea*

3.6.1 *Morphological adaptations*

3.6.1.1 *Stems*

Deep-flooded grasses are always rhizomatous, and usually also stoloniferous, with nodal rooting (*e.g. Leersia, Paspalidium, Panicum repens*). Most lower-floodplain grasses, therefore, have the basic adaptations of sudd components, for tolerance of water flow and rapid adjustment to water depth, and for sward-forming on swamp fringes. Among the dicots, *Ludwigia, Commelina* and *Polygonum* are obvious counterparts, with their prostrate stems and adventitious nodal roots.

The long, floating rhizomes of *Vossia* are described in Section 3.5.2.1. All the true reeds are rhizomatous and creeping, except *Miscanthidium*, which is basically tufted (rhizome internodes are very short). *Miscanthidium*, however, is only a shallow-flooded plant which occurs higher up the floodplain than any other reed; perhaps it should not be called a reed at all. Because the reeds do not have stolons or flexible rhizomes, their responses to prolonged water-level changes are slow. Figure 3.16B shows how papyrus re-sets its rhizome height under an altered hydrological regime.

The adaptations required of upper floodplain plants are quite different, and the tufted habit is characteristic, as in *Loudetia, Vetiveria* and *Setaria*. The tufted growth form is clearly a better adaptation to the more severe drought period which is experienced by shallow-flooded plants. Most *Brachiaria* species are also tufted, but the deeper-flooded *B. mutica* is, as would be expected, rhizomatous with nodal rooting. Yet again, *Echinochloa pyramidalis* is tufted, but *E. scabra* is rhizomatous with nodal rooting, like its common community associate *Vossia*.

3.6.1.2 *Leaves*

Leaf adaptations are more difficult to identify, but *Cyperus papyrus* is definitely one extreme case. Each umbel consists of several hundred small bracteoles arranged to form a perfect spherical radiation interceptor whilst retaining excellent provision for cooling. At the other end of the size scale, there are a few large-leaved swamp plants. *Mikania*, and some other climbers, have more than 50% of their leaves in shade and they need to be able to orientate them towards the light: only fairly large leaves would be suitable here. There are, however, a few large leaved monocotyledons, such as *Typha, Limnophyton, Crinum, Cyrtosperma, Suddia* and *Setaria*. Large leaves in full sun would normally raise a cooling problem; this does not arise with *Crinum*, which is found in swamp forest understorey. *Typha* and *Setaria* never occur in shade, but the strap-shaped leaves of *Typha* are held vertically to minimise heat load, and their large water content is guaranteed by the plant's habitat. *Setaria*, on the other hand, has a thin, highly lignified leaf which can withstand drying out without damage; this too is in keeping with the plant's habitat. Other grasses of the upper floodplain, particularly in nutrient-poor sites, such as *Brachiaria* and *Imperata*, also have hard, leathery leaves. Why *Typha* and some *Setaria* species have wide leaves at all is probably an irrelevant question, since both types can cope quite satisfactorily with the 'theoretical' problem of having large leaves exposed to a tropical sun.

Cyrtosperma, Limnophyton and *Suddia* all have much broader leaves than *Typha* and have all clearly evolved in shaded habitats, to which they are undoubtedly very well adapted. However, because all are permanent swamp species, they can also solve their heat-load problems in exposed situations by using the '*Typha*-solution': vertically-held leaves and a high transpiration rate.

3.6.1.3 *Roots*

Figure 3.16 and Table 3.5 may make the question of root adaptations appear complex, but the overall picture is quite simple: abundant adventitious rooting does confer adaptation to flooding and waterlogging. In fact, there is probably not a single significant example of a swamp forest tree in Africa which does not have stilts, adventitious roots or breathing roots (pneumorhizae).

Buttresses have been included in Fig. 3.16 and Table 3.5, because it has often been argued that they are more common in swamps than in dry forests. However, Richards (1939) observed that this is not, in fact, the case. Buttresses are particularly desirable for very large trees in high forest: all feeding roots in dense forest must be superficial, and this presents a stability problem which buttresses help to solve. Thus the 55-metre *Klainedoxa gabonensis*, a slow-growing, climax forest species, has some of the largest buttresses to be seen anywhere in Africa (Fig. 3.16, Types A(i) and A(ii)). The Fig. 3.16 Type (iii) buttress is developed by *Mitragyna* on mesic sites. *Pseudospondias*, which grows to 30 metres, probably has the largest buttresses of any African swamp forest tree. However, there appears to be a 'Catch 22' problem in swamp forests: Crawford (1983) has suggested that the key to waterlogging tolerance is very rapid root turnover. This requires a high growth rate and therefore a relatively low-density wood. But waterlogging also typically reduces plant stature and *limits* growth rates, because of oxygen shortage. Thus, within the genus *Syzygium*, the permanent swamp species *S. owariense* rarely exceeds 5 metres in height, whereas the seasonal swamp *S. cordatum* achieves 8 metres and *S. guineense* of gallery forest commonly grows to 12 metres.

Swamp substrates are also unstable and buttresses seem to be a good idea, although the enormous energy investment which must be built into them does not seem to be compatible with the other restrictions imposed by the swamp environment.

Most swamp trees usually do have relatively high growth rates and light wood (*Mitragyna, Maesopsis, Musanga, etc.*), so that 'skeleton buttresses' (= stilts) are quite adequate for support (Fig. 3.16D). They are also much more adaptable than buttresses to unstable substrates. By having a continuous production of new stilts, the adventitious root turnover is maintained at an adequately high level. Stilts are

Plate 3.10. Young 'aberrant' shoots of a 'stressed' papyrus rhizome. Normally, juvenile culms are devoid of leaves and do not open their umbels of photosynthetic bracteoles until they are almost two metres in height. However, they must be supplied by translocated carbohydrate from adjacent, photosynthesising mature culms. When these mature culms are damaged, as after fire, the young shoots must use stored carbohydrate from the rhizomes and develop their own photosynthetic capacity as early as possible. Under these stress conditions, *Cyperus papyrus* develops conventional sedge-like leaves. These aberrant shoots do not grow to more than a metre in height, but they do supply photosynthate for 'conventional' culms which develop later from the rhizome. (Photo: K. Thompson.)

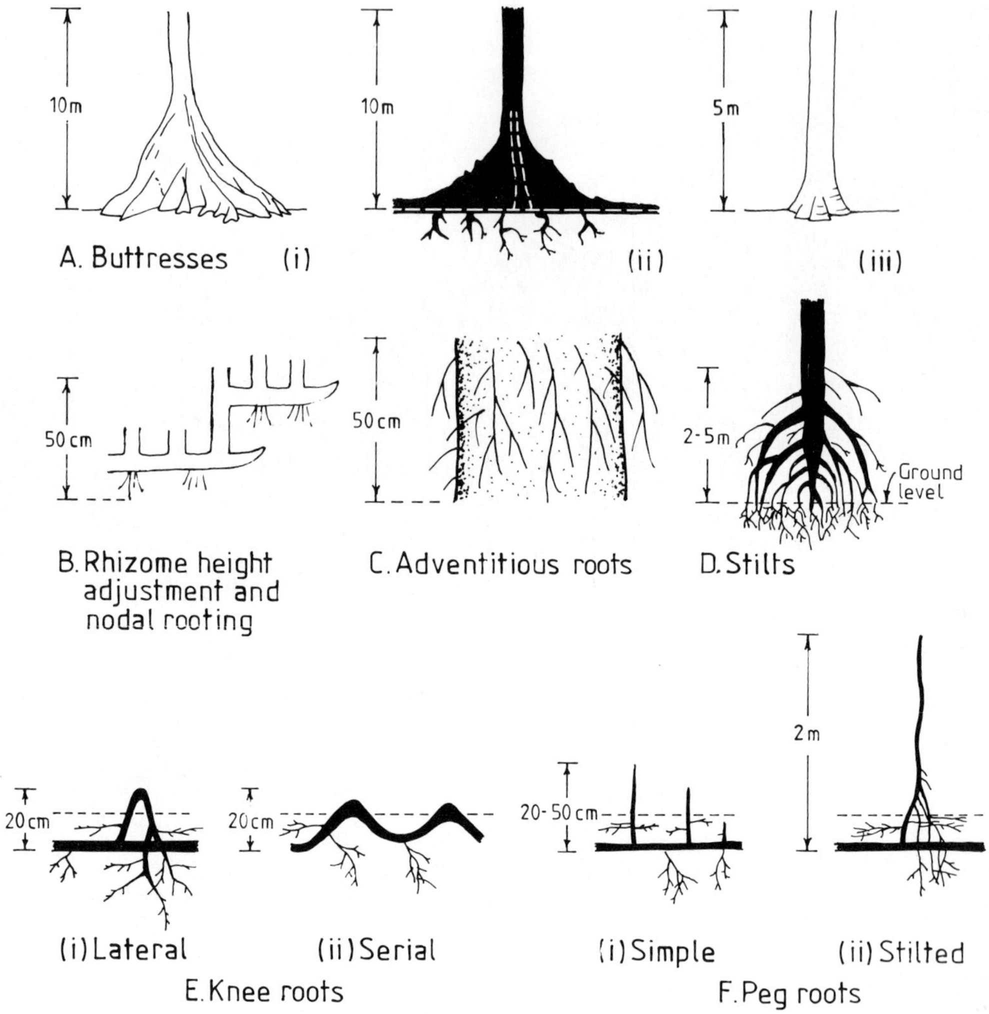

Fig. 3.16. The major types of morphological modifications to the stems and roots of (mainly woody) plants adapted to wetland habitats:

(A) (i) and (ii) represent heavily buttressed trees such as *Klainedoxa* or *Ficus*; (iii) *Mitragyna* buttress type on mesic sites.

(B) *Cyperus papyrus* growth form: the new rhizome develops as an outgrowth of the vertical culm in response to prolonged deep flooding.

(C) *Afroseralisia afzelii* trunk, showing small adventitious roots. These do not grow into stilts as do the stem outgrowths of D.

(D) *Uapaca* and *Xylopia staudtii* type stilting.

(E) Two types of knee roots, as found in species such as *Symphonia globulifera* or *Mitragyna stipulosa.*

(F) (i) *Anthocleista* and *Voacanga* type

(ii) *Xylopia staudtii* type.

(A. redrawn after Richards, 1957; C–F redrawn after Longman and Jenik, 1974).

often 'aeration organs' (aerial adventitious roots) in the early stages of their development (as in *Macaranga* and *Bridelia*), but maintain their growth. Several non-swamp forest trees also form aerial adventitious roots, such as *Afroseralisia*, in the Sapotaceae (Fig. 3.16C), and *Metrosideros* (Myrtaceae), but the sole purpose in these instances is likely to be absorption of water vapour or rainwater and they do not continue to grow until they form stilts.

The stilt habit is best developed in *Pandanus*, which is not surprising, since *all Pandanus* roots are adventitious. The success of the palms in wetland habitats probably also results from the ease with which they can develop adventitious roots. Among the dicotyledons, *Uapaca* and *Xylopia staudtii* are often almost entirely supported by stilts. *Culcasia striolata* has been included in Table 3.5 to show that herbs, too, may be stilted when the need arises. When the need (for support and aeration) does *not* arise, as on mesic sites, many normally stilted species (e.g. *Macaranga, Musanga*) will fail to develop these structures.

There is an alternative solution to the continual replacement of the rooting system from above: the true roots can grow *upwards* in order to reach aerated zones. Most people are familiar with the peg-roots of mangroves such as *Avicennia* and *Sonneratia*, but in fact many fresh-water swamp trees also develop these pneumorhizae (Fig. 3.16F). The pneumathodes of *Phoenix* are similar to pegs, except that only the white root tip protrudes above the soil. Both knee and peg roots are formed in response to a fluctuating water-table, but this is much more obvious with the serial knee-roots (Fig. 3.16E Type (ii)) which, as they grow, have endeavoured to keep within the aerated zone without at the same time drying out. Some species always have pneumorhizae; others, such as *Anthocleista, Symphonia* and *Xylopia*, do not develop them on mesic sites. *Ludwigia repens*, a fully aquatic and often free-floating emergent herbaceous dicot, develops erect spongy roots from the nodes which float to the surface as pneumorhizae; anchorage roots, where present, are quite different anatomically (*cf.* the mud roots of papyrus).

Table 3.5. Stem and root modifications of the more important wetland trees of Africa. (Data from a variety of sources, but particularly Longman and Jenik, 1974). Refer also Figure 3.16.

Species	Status			Annual or short-lived perennial	Buttresses	Stilts	Adventitious roots		Peg roots	Knee roots	Pneumathodes
	Herb	Tree/Shrub									
		Monocot	Dicot				All roots	Additional on aerial stem			
Afroseralisia afzelii			★					★			
Ancistrophyllum secundiflorum		★					★	★			
Anthocleista nobilis			★						★		
Bridelia micrantha			★			★			★		
Culcasia striolata	★					★	★				
Ficus congensis			★		★	★		★			
Grewia coriacea			★		★	★					
Macaranga barteri			★		★		★				
Mimosa pigra			★	★							
Mitragyna ciliata			★							★	
M. stipulosa			★		★					★	
Musanga cecropioides			★			★					
Pandanus candelabrum		★				★	★				
Phoenix reclinata		★					★				★
Pseudospondias microcarpa			★		★						
Raphia farinifera		★					★		★		
Sesbania sesban			★	★							
Spondianthus preussi			★			★					
Symphonia gabonensis			★							★	
Uapaca guineense			★		★	★			★		
Voacanga thouarsii			★						★		
Xylopia staudtii			★		★	★			★		

A further type of root adaptation found among aquatic emergents is that of dimorphic root structure. Many swamp plants show this, but *Cyperus papyrus* is a particularly good example (Tadros, 1940): roots suspended in water are large, white and fleshy, whereas plants rooted in alluvium develop the small, dark, wiry 'mud roots' which are much more similar to the roots of non-aquatics growing in soil. Root dimorphism is obviously related to nutrient uptake, since non-aquatics cannot be persuaded to develop 'water roots' under any circumstances.

3.6.2 *Anatomical adaptations*

A further alternative to root modifications as a means of effecting aeration of tissues below the water table is the proliferation of air spaces and channels within the stem. Most true swamp trees have light (*Ceiba, Musanga*) or medium (*Mitragyna, Spondianthus, Pseudospondias*) weight wood. Very few indeed have wood as dense as *Xylopia* (0.7 kg m^{-3}: almost four times the weight of the light woods). The heavier woods all have stilts or pneumorhizae and slower growth rates, and *Xylopia* has both stilts and peg roots (Table 3.5). Trees with light woods often have stilts too, but they are very fast growing and often tall (*e.g. Musanga* may attain 30 metres). Almost all of the medium weight woods are also coarse grained, which implies large xylem vessels or abundant xylem parenchyma or both, and most of the woods recorded as being 'medium to heavy' overall have relatively light sapwoods.

What this analysis means is that all swamp trees have effective channels for internal aeration. The ultimate woody example is *Aeschynomene,* which has an enormous stem pith, thus explaining its capacity to endure one metre of permanent flooding. With greater flooding depths, hydrostatic pressure fills the air spaces in the pith. *A. aspera* was once used in the manufacture of pith helmets. *A. uniflora* is light enough to be a common sudd component in Zaire. *Sesbania* also has exceedingly well-developed aerenchyma, as also has the herbaceous dicot *Ludwigia*.

The very well-developed aerenchyma tissues of *Typha* leaves and *Scirpus* and *Cyperus papyrus* culms perform a similar aeration service for the submerged roots and rhizomes of these plants. However, it has been puzzling to contemplate how gaseous diffusion can occur across the restriction of the basal intercalary meristems (in the case of papyrus, located at the junction of vertical culm with horizontal rhizome in Fig. 3.6) of these species: normal meristems just do not have inter-cellular air spaces. Conway (1938) certainly found one solution by blowing down the *senescent* culms of *Cladium* to demonstrate an unrestricted gaseous access to the submerged rhizomes. The diffusion channels down papyrus culms are probably complete after the basal meristem ceases activity (approximately one month after culm extension commences). Until a juvenile culm so matures, oxygen diffuses down the adjacent mature culm, and acropetally along the submerged rhizome to supply the young growing tissues.

Cyperus papyrus too is unable to stand prolonged deep flooding (compare *Aeschynomene*). *Phragmites*, however, may tolerate up to 4 metres of flooding for limited periods (Denny, 1973), but the hollow stems of the grasses may offer an advantage over aerenchymas. Although nodal blockages appear to constitute diffusion restrictions ('bulkheads'), they actually divide the diffusion pathway into flooding-resistant sectors, so that grasses are likely to have much greater flooding tolerance than sedges such as papyrus with continuous diffusion pathways.

3.6.3 *Physiological and metabolic adaptations*

3.6.3.1 *Anaerobic respiration*

There is probably no such thing as an emergent aquatic plant without the capacity for anaerobic respiration. Crawford (1982, 1983) has reviewed the state of our knowledge on the physiological aspects of flooding tolerance, and he points out that, although structural modifications are of primary importance to all wetland plants, they must also have metabolic back-up. This involves the short-term accumulation of ethyl alcohol, or one of a range of organic acids, as the currency of an 'oxygen debt' which must be 'worked off' later.

As far as can be seen at present, the oxygen debts of higher plants can only be cancelled by supplying molecular oxygen. Various micro-organisms can substitute nitrate or ferrous iron, but there is, as yet, no evidence that any higher plants can do this. Anaerobic respiration simply does not provide enough ATPs to keep higher plants going for long. Many African wetland plants have exceedingly high growth rates, so they obviously do not have an oxygen debt problem; they clearly have access to molecular oxygen through morphological and anatomical modifications. However, a rhizome section of *Cyperus papyrus* completely submerged in water with a very low dissolved oxygen content will form new culms (Thompson, unpublished data), so that appropriate metabolic pathways are certainly present in this plant to enable some anaerobic respiration to take place. The 'igapo' forests of the Amazonian basin are flooded for up to 9 months of the year to a depth of up to 8 metres (Gessner, 1968). Some smaller trees are completely submerged and others only have their crowns protruding from the water. Oxygen diffusion pathways down the trunk would not be enough in this extreme case, and metabolic adaptations or, at least, large carbohydrate reserves are obviously needed by submerged tissues. Gessner noted very slow growth rates and also frequent dormancy; he did not contemplate anaerobic metabolism. Nor does Wolfgang Junk (personal communication), who is currently researching, in the Amazon, this interesting phenomenon of deep-flooded trees. Feely (1965) observed that African *Acacia albida* can withstand several months of deep inundation, provided that its foliage remains above water. He also noted that young trees grow at over one metre per year, so that they can achieve the height needed to survive the flooding regime.

3.6.3.2 *Germination*

Most aquatic plants have very specialised seed adaptations and germination requirements, but there is not time to review here the complete range. Instead, a few illustrative examples will be taken.

Many species have seed which is very long-lived (at least several decades under anaerobic conditions), since germination requirements cannot often be met by the normal habitat of the mature plants. Thus the seed of most reeds (such as *Cyperus papyrus, Typha domingensis*) and many floating plants (e.g. *Pistia, Eichhornia*) will only germinate on nutrient rich, wet mud substrates. This reflects the high nutrient requirements of seedlings, but more importantly it shows that plants have adapted well to an environment which, at least in the long term, is unstable. The fully aquatic habitat is, compared with most soils, nutrient poor, so that the mud substrate is necessary for seedling development. Adapting to this also means that if a water body dries up, or recedes, the aquatic species still has a phase of its life cycle which is well adapted to the environment which prevails. Nevertheless, the critical phase is always that of the juvenile stage, and mature plants have much greater stress tolerance than their seedlings. Thus *Typha* will grow on (is tolerant of) alkaline substrates, as too will *Aeschynomene, Paspalidium, Paspalum repens* and *Acacia xanthophloea,* but none of these species will establish itself from seed on saline soils.

One example will also be taken from seasonal wetlands. *Themeda triandra* is a fire-adapted species; Lock and Milburn (1971) showed that *Themeda* seed needs a red light (phytochrome) stimulus before it will germinate. It does not get this stimulus unless and until the closed grass canopy above it is removed by fire. A burning regime (a common management practice on dark clay soils and alluvial floodplains) therefore encourages *Themeda* establishment, but only when the nutrient status is high. Once again, a species has managed to evolve two life-cycle phases which, between them, can cope with a variable environment which can eliminate the mature phase. In fact, overgrazed *Themeda* may be replaced by less palatable species such as *Sporobolus*. A fire event then closes the cycle of succession. Of course, the burning regime itself also raises substrate nutrient status.

Many wetland plants are known to be dispersed by animals. Thus *Acacia albida* seeds are passed intact through the gut of hippo (Feely, 1965) and *Kigelia africana* is dispersed by baboon.

3.6.3.3 *Nutrient and water availability*

Communities of *Cyperus papyrus* and *Phragmites* species have above-ground biomasses of between 2 and 5 kg m^{-2} (dry weight), *Vossia* and *Typha* from 1.5–2.5 kg m^{-2}, and *Acroceras* and *Cynodon* 100–500 g m^{-2} (Carmouze, Durand and Lévêque, 1983; Denny, 1984; Howard-Williams, 1973; Thompson, 1976*b*; Thompson, Shewry & Woolhouse, 1979). These community biomasses are determined largely by factors such as plant growth form and ecosystem stability.

Productivity, on the other hand, is determined (within the physiological capacity of each individual species) by factors such as temperature and nutrient availability. Thompson, Shewry & Woolhouse (1979) show, for papyrus growing in high nutrient and low nutrient sites, that biomass is unaffected, although crop growth rate may vary from 7–35 g m^{-2} day^{-1} (these are not peak rates; they are year-round sustained growth rates). The lower value was estimated for communities on the floodplain behind the levée, where the water regime is highly seasonal; individual plants were stunted but culm density was high. The higher productivities are achieved in sites with stable water regimes; individual plants are robust, but culm density is relatively low.

The same is true for forest communities: stature is limited by ecosystem stability and not by nutrient availability. Thus we may assume that the exceedingly well-developed nature of Type E (Dawkins & Philip, 1962) swamp forests in Zaire is due to the stability of the water regime and the lack of disturbance. Hamilton and Thompson (unpublished data) have studied a catena in Mabira Forest, Uganda, which appeared, on cursory examination, to assign an important role to nutrient status as a determinant of forest type and of stand biomass. The upper section of the catena was characterised by a 40 m canopy of the regional climax forest type, *Celtis/Chrysophyllum/Entandrophragma (angolense)*, but the lower section was dominated by a low (20 m) canopy of *Croton/Baikiaea/Lasiodiscus*. Examination of the geological map showed that the boundary between the two communities coincided with a fault line, which resulted in a switch from nutrient-poor kaolin soils (derived from granites) above, to nutrient-rich montmorillonite/illite soils (derived from amphibolite) below. However, just because a site has relatively high nutrient levels does not mean that the species present are responding to base status, and readers will by now have recognised that the dominants on the black clay soils are good indicators of seasonal swamp communities. The ecological determinant is, indeed, seasonal waterlogging and has nothing at all to do with nutrients.

The above forest example does not mean, however, that nutrient status does not select for species. It simply means that, in this case, waterlogging is more important. There are certainly a number of species, like *Cladium* or *Themeda*, known to require higher nutrient levels, and others, like *Miscanthidium*, known to be tolerant of low nutrient levels (they are probably 'nutrient conservers'). But because nutrients are not 'available' in the absence of water, water availability is a factor influencing both biomass *and* productivity. In practice, it does not limit the permanent swamp species, but it can have a major influence upon the performance of communities found on periodically-flooded substrates.

Thus, 'dry' savanna grasslands typically produce 0.5–3 t ha^{-1} yr^{-1} (above-ground dry matter), whereas humid zone mesophytic grassland achieves up to 12 t ha^{-1} yr^{-1} and edaphic grasslands up to 20 t ha^{-1} yr^{-1} (Braun, 1973; Thompson, 1976*b*; Menaut & Cesar, 1983). Bell (1982), however, argues that the community structure and biomass of African savannas is determined by the balance between water availability and soil nutrient availability, and that nutrient availability must be responsible for any vegetation patterns within areas receiving more than 400 mm of rainfall per annum. Bell does exclude floodplains (and, presumably therefore, all edaphic grasslands) from his hypothesis, so that he is partly correct with respect to species selection. However, as the statement stands it must be an over-generalisation, since the duration, frequency and seasonality of rainfall can either make 400 mm go a long way or, alternatively, nowhere at all. It is unlikely that nutrient status affects biomass, since poor soils support miombo woodland over much of the central African savanna zone, whereas the richer volcanic soils of Serengeti only support grasslands. Nevertheless, Bell does recognise the importance of grazing pressure as an ecological factor, and there are many floodplain and dambo grasslands which are dominated by poor quality herbage (*Sporobolus, Imperata,* small *Cyperaceae, etc.*) simply because over-grazing has lowered the nutrient status (*i.e.* exerted nutrient-stress on the palatable grasses to the point at which they are unable to compete with certain less-palatable species tolerant of low nutrient status).

Tinley (1983), however, takes the more conventional (and more often correct) view that water availability is the major determinant of plant communities throughout Africa. Nevertheless, as several examples in this chapter have shown, the situation is rarely as simple as it may seem. Interactions between nutrient status and water availability are complex, and each example must be studied individually in order to establish the relative importance of those two critical factors and plant adaptations to them.

3.6.3.4 *Photosynthesis*

The 'C_4 syndrome' (Laetsch, 1974) offers (in theory) a plant higher potential productivity and more efficient water and nutrient use (more carbon fixed per unit of water or nitrogen used) than is the case for 'conventional' C_3 metabolism. Therefore, most of the upper floodplain grasses would be expected to have C_4 photosynthetic metabolism in order to withstand the harsh dry-season conditions; most permanent swamp species should not be disadvantaged with C_3 metabolism. In practice, the picture is not so clear-cut. Shewry, Thompson and Woolhouse (unpublished data) showed, for several floodplains in the savanna zone of Zaire, that whereas the total *number* of C_4 species does not differ significantly between shallow-flooded and deep-flooded parts, the *coverage* (*i.e.* relative dominance; proportion of total biomass) of C_4 species is much greater on the upper floodplain than it is on the lower. Nevertheless, care must be taken in generalising, because the 'most aquatic of all' floodplain and swamp species, *Cyperus papyrus,* does have C_4 metabolism (Jones and Milburn, 1978). Indeed, it is one of only two or three permanent aquatic species in the world known to possess the C_4 photosynthetic pathway. Its phenomenal growth rate must owe something to its unique (for a freshwater aquatic plant) metabolism, but it is difficult to see how the most important benefit could be water conservation. Nutrient conservation is a more likely benefit for an aquatic plant with such a high growth rate. *Eichhornia crassipes,* another full aquatic, is one of the few plants in the world which can match the growth rate of *Cyperus papyrus.* The water hyacinth has C_3 photosynthetic metabolism but an exceedingly high rate of water uptake and loss. This not only enables the plant to cool (by transpiration) its large leaves, but also ensures a huge water throughput for nutrient extraction. The transpiration rate of a papyrus swamp is less than 20% of that of an *Eichhornia* sward, but the nutrient status of the water which papyrus absorbs is higher (*see* Chapter 6) and its nutrient conservation capacity is also presumed to be greater than *Eichhornia.*

Laetch (1974) further argued that the 'C_4 syndrome' should be a valuable adaptation to saline habitats, not only because of the 'physiological (osmotic) drought' experienced by alkaline swamp plants, but also because C_4 plants appear to operate 'differential ionic pumps' which may help to dispose of excess sodium. The fact that the C_4 pathway has now been identified in the very aggressive temperate estuarine aquatic emergent *Spartina anglica* may well transform Laetsch's hypothesis into a prediction. Indeed, it is not unreasonable to see the mildly saline dambos and upper reaches of African floodplains as being the most likely sites for adaptations like C_4 metabolism to evolve.

3.7 Evolution of aquatic emergent plants in Africa

Walter (1962) draws attention to the remarkable dominance of the monocotyledons over the dicotyledons on waterlogged and poorly-aerated soils. He attributes this dominance to greater tolerance of wet/dry cycles in the case of woody species, and to the monocot growth form in the case of herbaceous plants. These are very generalised explanations, and the present chapter has taken the question of adaptation to flooding much further. Woody monocots have xeromorphic features and adventitious roots; herbaceous monocots have well developed morphological and anatomical adaptations to flooding – as well as adventitious rooting. However, woody dicots should not be under-estimated; they have much broader ecological tolerances than the monocots. Thus there are many exclusively aquatic genera of monocots (all of the reeds, for instance), but very few, if any, woody dicot genera are restricted to aquatic habitats; even *Aeschynomene* has mesophytic species. There are, however, a very few genera of exclusively wetland or aquatic herbaceous dicots (*e.g. Ludwigia, Hydrocotyle*).

Monocotyledons are clearly more specialised than dicots. All of the grasses, for instance, belong to a single family (Section 3.3.1), but woody dicots alone are distributed among several 'family groups' (Section 3.3.2.2). Of course, woody plants are long-lived and must endure not only normal seasonal flooding events but also the occasional '50-year' flood. Herbaceous monocot communities are disrupted by such exceptional events (*e.g.* the Chad drought: Section 3.5.2), but quickly re-establish or adjust to the new hydrological regime. The total number of herbaceous monocots is very large, and each is adapted to fairly specific conditions, whereas most wetland woody dicots can endure considerable hydrological changes. Palms such as *Borassus* and *Hyphaene* are highly tolerant of prolonged waterlogging in savanna, but they do not have the tolerance of deep flooding events that riverine forest species such as *Rhus* and *Syzygium guineense* demonstrate. It is a pity that the Amazonian inundation forests (Gessner, 1968) have not been studied more intensively, because they probably hold the key to flooding adaptation in woody species, including most of the trees (Apocynaceae, Bignonaceae, Bombacaceae, Euphorbiaceae, Leguminosae, Moraceae, Myrtaceae, Polygonaceae, Urticaceae) which provide the important edaphic forest trees of Africa. There are inundation forests in Zaire, but the hydrological regime is much less extreme (it is less than 2 metres). It is clear from this survey of African edaphic forests, and from comparisons with the Amazon, that woody plants adapt more easily to flooding regimes than they do to dry seasons. The Amazon example also indicates that most of the permanent swamp forest species in Africa, and many of the seasonal swamp forest species too, could endure much more extreme flooding regimes than they do.

Looking now at the biogeography of individual species, it is quite clear that the Guinea/Congo zone has far more endemics than any other region of the African continent. This is not surprising; savanna species must be more adaptable and they are therefore more widely distributed. However, there are some peculiar distributions among the swamp forest trees: a remarkable number of species show distinct disjunctions in their ecological ranges. *Carapa procera, Croton, Gilbertiodendron, Rauvolfia, Symphonia globulifera* and *Syzygium guineense* are regular components of lowland swamp forest in Zaire, although they are also all quite widely distributed on free-draining soils above 1000 metres. *Dracaena, Ilex, Podocarpus* and *Rapanea* occur in seasonal swamps at about 1000 m altitude, and yet they are all more common above 2000 m on free-draining soils. No-one has satisfactorily explained these ecological disjunctions. Maybe ecotypes are involved, or maybe there are site similarities which have not been recognised.

Van Steenis (1981), in his monograph of rheophytes (plants adapted to flowing water habitats), lists about 12 families which contain a large number of rheophyte species. Fifty percent of these families (Araceae, Euphorbiaceae, Melastomataceae, Moraceae, Myrtaceae and Rubiaceae) are important wetland families in Africa. In his listing of genera which contain rheophytes, Van Steenis records the following which have been referenced in the present chapter: *Crinum, Cyperus, Eragrostis, Ficus, Diospyros, Garcinia, Ilex, Memecylon, Myrica, Pandanus, Rauvolfia, Rhus* and *Salix*. With the exception of *Crinum, Cyperus, Eragrostis* and *Pandanus,* all these genera are characteristic of riverine or gallery forests. Both the families and the genera listed are also major suppliers of non-rheophyte wetland species in Africa.

However, Van Steenis' monumental work is, in the context of the present chapter, more memorable for its omissions. It is maintained that no rheophyte exists in the Malvaceae (note, however, *Hibiscus diversifolius*, which is very effective on river levées), or in the Marantaceae (*Thalia geniculata*), or in the Polygonaceae (*see P. pulchrum* and *P. senegalense*). Moreover, Van Steenis does not recognise *Vossia* or *Ceratopteris thalictroides* as rheophytes. He includes some riverine and gallery forest species (such as *Diospyros natalensis*) but not others (such as *Salix subserrata* or *Syzygium guineense*). Van Steenis has established very rigid, and possibly rather arbitrary, definitions for his rheophytes. He may, therefore, by eliminating most 'seasonal rheophytes' (whilst accepting others arbitrarily), have sacrificed some opportunities to gain insight into rheophyte adaptations. In the present chapter, efforts have been made to incorporate the full spectrum of wetland plants in order that no such opportunity would be lost.

Another interesting group of species is the 'colonisers', since they do not respond to the 'stress' of waterlogging in the conventional fashion (reduced stature and growth rate). Species such as *Macaranga, Ceiba* and *Musanga* all have exceedingly high growth rates. Numerous herbaceous species also have phenomenal growth rates under flooding regimes. Obviously, such plants do not actually experience flooding stress at all. Their various adaptations remove the stress element, so that the plants are able to grow under optimal conditions.

It is little wonder that wetlands and floodplains come under such heavy management pressure in Africa. Their plant communities are an extremely valuable resource, not only for native wild animals, but also for domestic livestock. The conflicts which arise as a result are discussed in Chapter 8, but Bell (1982) adds a further dimension; he suggests that, since savanna grasses and large mammals have almost certainly co-evolved, the plants will have developed 'chemical defenses' (allelochemicals) to limit grazing intensity. Thompson and Hamilton (1983), on the other hand, suggest that floodplain grasses, such as *Cynodon* and *Themeda*, do not really need to evolve unpalatable features, because (up to a point) they actually grow *better* when grazed. Permanent swamp emergents do not need allelochemical defenses either, since they are already largely inaccessible to large herbivores and, in any case, their food value is limited by their high proportion of 'structural' chemical components.

3.8 Conclusions and acknowledgements

This chapter has been extraordinarily difficult to compile, because so many species are implicated and the information needed is so widely dispersed in the literature. Literature 'quality' is also, in general, poor, and interpretation of it would have been impossible without extensive personal field experience. There will be many errors of interpretation, but the aim has been to present a review which will establish a sound basis upon which others can build a better understanding of the ecology of wetland and floodplain emergents of Africa, and of other tropical regions of the world.

As usual, all the *really* hard work has been done by my typist, and I wish to thank Melody Bryan for her patience during alterations to every second paragraph.

3.9 References

Adejuwon, J.O., 1971. The ecological status of fresh water swamp savannas in the forest-zone of Nigeria. Journal of the West African Science Association, 16(2):133–154.

Anonymous, 1956. Reunion de spécialistes du C.S.A. en matière de phytogeographie à Yangambi. C.S.A. Publication 22, C.C.T.A., Londres.

Anonymous 1971. Fire in Africa. Proceedings of the annual Tall Timbers Fire Ecology Conference, Tall Timbers Research Station, Tallahassee, Florida.

Astle, W.L., 1965. The edaphic grasslands of Chambeshi Flats, Northern Province, Zambia. Kirkia: 37–50.

Bacalbasa-Dobrovici, N., 1971. The economic importance of the burgu (*Echinochloa stagnina* P. Beauv.) on the middle Niger in the Niger Republic. Hidrobiologia, Bucuresti, 12:41–46.

Barradas, L.A., 1951. Machongos reclamation. In: Proceedings of the Conference on Hydrology and Water Resources. East African High Commission, Nairobi.

Bell, R.H.V., 1983. The effect of soil nutrient availability on community structure in African ecosystems. In: B.J. Huntley & B.H. Walker (eds.), Ecology of Tropical Savannas. Springer-Verlag: 193–216.

Bell, R.H.V. & Grimsdell, J.J.R., 1973. The persecuted black lechwe of Zambia. Oryx, 12(1):77–92.

Berry, L., 1976. The Nile in the Sudan: geomorphological history. In: J. Rzoska (ed.), The Nile: Biology of an Ancient River. W. Junk, The Hague, Monographiae Biologicae 29: 11–18.

Bishop, W.W., 1958. Raised swamps of Lake Victoria. Geological Survey of Uganda Records (1955–56), 1–10.

Boughey, A.S., 1957. The physiognomic delimitation of West African vegetation types. Journal of the West African Science Association, 3(2):148–165.

Bouxin, G., 1976. Ordination and classification in the upland Rugege Forest (Rwanda, Central Africa). Vegetatio, 32:97–115.

Braun, M.H., 1973. Primary production in the Serengeti: purpose, methods and some results of research. Annales de l'Université d'Abidjan, (E)6:171–188.

Brenan, J.P.M., 1978. Some aspects of the phytogeography of tropical Africa. Annals of the Missouri Botanical Garden, 65(2):437–477.

Buddenhagen, I.W. & Persley, G.J., 1978. Rice in Africa. Academic Press.

Bunting, A.M. & Lea, J.D., 1962. The soils and vegetation of the Fung, east central Sudan. Journal of Ecology, 50:529–558.

Buringh, P., 1970. Introduction to the Study of Soils in Tropical and Subtropical Regions. Centre for Agricultural Publishing and Documentation, Wageningen.

Campbell, E.O., 1983. Mires of Australasia. In: A.J.P. Gore (ed.), Mires: Swamp, Bog, Fen and Moor. Elsevier, Ecosystems of the World Series 4B:153–180.

Camus, C., 1957. Problèmes posés par les papyrus au Lualaba Supérieur. Academie Royale de Sciences Coloniales, Bruxelles, Bulletin des Séances, 3:1164–1185.

Carmouze, J-P., Durand, J.R. & Lévêque, C., 1983. The lacustrine ecosystem during the 'Normal Chad' period and the drying phase. In: J-P. Carmouze, J-R. Durand & C. Lévêque (eds.), Lake Chad. W. Junk, The Hague, Monographiae Biologicae 53:527–560.

Chapman, J.D. & White, F., 1970. The Evergreen Forests of Malawi. Commonwealth Forestry Institute, Oxford.

Clayton, W.D., 1957. The swamps and sand dunes of Hadejia. Nigeria Geographical Journal, 1:31–37.

Clayton, W.D., 1958. A tropical moor forest in Nigeria. Journal of the West African Science Association, 4:1–3.

Conway, V.M., 1938. Studies in the autecology of *Cladium mariscus*. V. The distribution of the species. New Phytologist, 37:312–328.

Corner, E.J.H., 1966. The Natural History of Palms. Weidenfeld and Nicholson, London.

Crawford, R.M.M., 1982. Physiological responses to flooding. In: O.L. Lange *et al.* (eds.), Physiological Plant Ecology II. Springer-Verlag Encyclopoedia of Plant Physiology (NS), 12B:453–477.

Crawford, R.M.M., 1983. Root survival in flooded soils. In: A.J.P. Gore (ed.), Mires: Swamp, Bog, Fen and Moor. Elsevier, Ecosystems of the World Series 4A:257–284.

Dawkins, H.C. & Philip, M.S., 1962. Working Plan for Mpanga Forest for the period 7/7/60–30/6/65. Uganda Forest Department, Entebbe.

Dean, 1967. Grasslands of the Rukwa Valley. Journal of Applied Ecology, 4:45–57.

Denny, P., 1971. Zonation of aquatic macrophytes around Habukara Island, Lake Bunyoni, south-west Uganda. Hidrobiologia, Bucuresti, 12:249–257.

Denny, P., 1973. Lakes of south-western Uganda. 2. Vegetation studies on Lake Bunyoni. Freshwater Biology, 3:123–135.

Denny, P., 1984. Permanent swamp vegetation of the Upper Nile. Hydrobiologia, 110:79–90.

Deuse, P., 1966. Contribution a l'étude des tourbières du Rwanda et du Burundi. Treizième Rapport d'I.R.S.A.C., 1960–1964:53–115.

Drar, M., 1951. The problem of the Sudd in relation to stabilizing and smothering plants. Botaniska Notiser, 1951(1):32–46.

Duvigneaud, P. & Homès, P., 1955. Les *Xyris* du Bas-Congo et du Kwango. Bull. Soc. Roy. Bot. Belgique, 87:81–113.

Duvigneaud, P., 1958. La végétation du Katanga et des sols métallifères. Bull. Soc. Roy. Bot. Belgique, 90:127–286.

Eggeling, W.J., 1935. The vegetation of Namanve Swamp, Uganda. Journal of Ecology, 23:422–435.

Eggeling, W.J., 1947. Observations on the ecology of the Budongo rain forest, Uganda. Journal of Ecology, 34:20–87.

Eggeling, W.J., 1951. The Indigenous Trees of the Uganda Protectorate. (Revised Edition). Government Printer, Entebbe, Uganda.

El Din El Zarka, S., 1973. Kainji Lake case history. In: W. Ackermann *et al.* (eds.), Man-made Lakes, their Problems and Environmental Effects. American Geophysical Union, Geophysical Monograph 17, Washington D.C.

Environmental Council, 1984. Wetlands: a Diminishing Resource. (A report for the Environmental Council.) Water and Soil Miscellaneous Publication 58, National Water and Soil Conservation Organisation, Wellington, New Zealand.

Feely, J.M., 1965. Observations on *Acacia albida* in the Luangwa valley. Puku, 3:67–70.

Furness, H.D., 1981. The plant ecology of seasonally flooded areas of the Pongolo river floodplain with particular reference to *Cynodon dactylon* (L.) Pers. Ph.D. thesis, University of Natal.

Gaudet, J.J., 1977. Natural drawdown on Lake Naivasha, Kenya, and the formation of papyrus swamps. Aquatic Botany, 3:1–47.

Germain, R., 1965. Les biotopes alluvionaires herbeux et les savannes intercalaires du Congo equatorial. A.R.S.O.M.(N.S.), 15(4):237–244.

Gessner, F., 1968. Zur ökologischen Problematik der Überschwemmungswalder des Amazonas. Int. Revue ges. Hydrobiol., 53(4):525–547.

Gillman, C., 1949. A vegetation-types map (1:2,000,000 in colour) of Tanganyika Territory. Geographical Review, New York, 39:7–37.

Greenway, P.J., 1973. A classification of the vegetation of East Africa. Kirkia, 9:1–68.

Greenway, P.J. & Vesey-Fitzgerald, 1969. The vegetation of

Lake Manyara National Park. Journal of Ecology, 57(1):127–149.

Grimsdell, J.J.R. & Bell, R.H.V., 1976. The Bangueulu Basin of Zambia. Proceedings of a Symposium on the Okavango Delta and its Future Utilisation. The Botswana Society, Gaborone, Botswana: 257–262.

Hamilton, A.C., 1976. The significance of patterns of distribution shown by forest plants and animals in tropical Africa for the reconstruction of upper Pleistocene palaeo-environments: a review. Palaeoecology of Africa, 9:63–97.

Hamilton, A.C., 1981. A Field Guide to the Uganda Forest Trees. Makerere University, Kampala, Uganda (copyright A.C. Hamilton).

Hamilton, A.C., 1982. Environmental History of East Africa: a Study of the Quaternary. Academic Press, London.

Harker, K.W. & Napper, D., 1961. An Illustrated Guide to the Grasses of Uganda. Government Printer, Uganda.

Hedberg, O., 1964. Features of Afro-alpine ecology. Acta Phytogeographica Suecica, 49:118–128.

Heeg, J. & Breen, C.M., 1982. Man and the Pongolo Floodplain: a report of the Committee for Inland Water Ecosystems, National Programme for Environmental Sciences. South African National Scientific Programmes Report no. 56, C.S.I.R., Pretoria.

Hope, C.W., 1902. The 'Sadd' of the Upper Nile: its botany compared with that of similar obstructions in Bengal and American waters. Annals of Botany, 16:495–516.

Howard-Williams, C., 1973. Vegetation and environment in the marginal areas of a tropical lake (L. Chilwa, Malawi). Ph.D. Thesis, University of London.

Howard-Williams, C., 1975. Vegetation changes in a shallow African lake: response of the vegetation to a recent dry period. Hydrobiologia, 47:381–398.

Howard-Williams, C., 1979. The distribution of aquatic macrophytes in Lake Chilwa: annual and long-term environmental fluctuations. In: M. Kalk, A.J. McLachlan & C. Howard-Williams (eds.), Lake Chilwa. W. Junk, The Hague: 105–122.

Huntley, B.J. & Walker, B.H., 1982. Ecology of Tropical Savannas. Springer-Verlag, Ecological Studies 42.

Iltis, A. & Lemoalle, J., 1983. The aquatic vegetation of Lake Chad. In. J-P. Carmouze, J-R. Durand & C. Lévêque (eds.), Lake Chad: Ecology and Productivity of a Shallow Tropical Ecosystem. W. Junk, Monographiae Biologicae 53:125–143.

I.U.C.N., 1979. The distribution of protected areas in relation to the needs of biotic community conservation in West and Central Africa. (Draft). International Union for the Conservation of Nature and Natural Resources, Gland, Switzerland.

Jeník, J., 1971. Root system of tropical trees. 7. The facultative peg-roots of *Anthocleista nobilis* G. Don. Preslia, 43:97–104.

Jones, M.B. & Milburn, T.R., 1978. Photosynthesis in papyrus (*Cyperus papyrus* L.) Photosynthetica, 12:197–199.

Jonglei Investigation Team, 1954. The equatorial Nile project and its effects on the Anglo-Egyptian Sudan. I–IV. Report of the Jonglei Investigation Team, Sudan Government, Khartoum.

Keay, R.W.J., 1959. Vegetation map of Africa south of the Tropic of Cancer: Explanatory notes and map. Oxford University Press.

Killick, D.J.B., 1979. African mountain heathlands. In: R.L. Specht (ed.), Heathlands and Related Shrublands. A. Descriptive Studies. Elsevier, Ecosystems of the World Series, 9A:97–116.

Koechlin, J., 1961. La végétation des savannes dans le Sud de la République du Congo. Institut de Récherches Scientifiques au Congo Brazzaville, O.R.S.T.O.M.

Knapp, R., 1961. Die Vegetation von Afrika. Gustav Fischer Verlag.

Laetsch, W.M., 1974. The C^4 Syndrome: a structural analysis. Annual Reviews of Plant Physiology, 25:27–52.

Lamprey, H.F., 1975. The distribution of protected areas in relation to the needs of biotic community conservation in eastern Africa. I.U.C.N. Occasional Paper No. 16:1–83.

Langdale-Brown, I., Osmaston, H.A. & Wilson, H.G., 1964. The Vegetation of Uganda and its Bearing upon Land-use. Government Printer, Entebbe.

Launert, E., 1965. A survey of the genus *Leersia* in Africa. Senck. Biol., 46(2):129–153.

Lebrun J., & Gilbert, G., 1954. Une classification ecologique des forêts du Congo. Institut National pour l'Étude Agronomique du Congo Belge (I.N.E.A.C.), Série Scientifique, 63.

Lejoly J. & Mandango, A., 1982. L'association arbustive ripicole à *Alchornea cordifolia* dans le haut-Zaire. In: J.J. Symoens, S.S. Hooper & P. Compère (eds.), Studies on Aquatic Vascular Plants, Royal Botanical Society of Belgium, Brussels: 257–265.

Lind, E.M. & Morrison, M.E.S., 1974. East African Vegetation. Longmans.

Lind, E.M. & Visser, S.A., 1962. A study of a swamp at the north end of Lake Victoria. Journal of Ecology, 50:599–613.

Lock, J.M., 1973. The aquatic vegetation of Lake George, Uganda. Phytocoenologia, 1:250–262.

Lock, J.M., 1977. The vegetation of Ruwenzori National Park, Uganda. Bot. Jahrb. Syst., 98(3):372–448.

Lock, J.M. & Milburn, T.R., 1971. The seed biology of *Themeda triandra* in relation to fire. In: E. Duffy & A.S. Watt (eds). The Scientific Management of Plant and Animal Communities for Conservation. Blackwell Scientific Publications: 337–349.

Longman, K.A. & Jeník, J., 1974. Tropical Forest and its Environment, Longman.

Lubke, R.A., Raynham, G.L. & Reavell, P.E., 1981. Reassessment of plant succession on spoil heaps along the Boro River, Okavango Delta, Botswana. South African Journal of Science, 77:21–23.

Martin, A.R.H., 1960. The ecology of Groenvlei, a South African fen. 1. The Primary Communities. Journal of Ecology, 48:55–71.

Mavuti, K.M., 1982. Wetlands and shallow water bodies in Kenya: characteristics and general status of knowledge. In: Ecosystem Dynamics in Freshwater Wetlands and Shallow

Water Bodies. Proceedings of a SCOPE/UNEP International Scientific Workshop, Minsk-Pinsk-Tskhaltoubo, 12–26 July, 1981. Centre for International Projects GKNT, Moscow: 188–204.

Ménaut, J.-C., 1983. The vegetation of African savannas. In: F. Boulière, (ed.), Tropical Savannas. Elsevier 'Ecosystems of the World' Series, 13:109–149.

Ménaut, J.-C., & Cesar, J., 1983. The structure and dynamics of a West African savanna. In: B.J. Huntley & B.H. Walker (eds.), Ecology of Tropical Savannas. Springer-Verlag: 80–100.

Meusel, H., Jäger, E. & Weinert, E., 1965. Vergleichende Chorologie d. Zentraleurop. Flora.. Jena.

Michelmore, A.P.G., 1939. Observations on tropical African grasslands. Journal of Ecology, 27:282–312.

Milne, G., 1947. 'A soil reconnaisance journey through parts of Tanganyika territory, December 1935 to February 1936'. C. Gillman (ed.), Journal of Ecology, 35:192–265.

Mitchell, D.S., 1978. Freshwater plants. In: M.J.A. Werger & A.C. Van Bruggen (eds.), Biogeography and Ecology of Southern Africa. W. Junk, The Hague: 1113–1138.

Mohr, E.C.J. & Van Baren, F.A., 1959. Tropical Soils. N.V. Uitgeverij W. Van Hoeve, The Hague.

Monod, T. Les grandes divisions chorologiques de l'Afrique. CCTA/CSA no. 24.

Moore, H.E., 1973. Palms in the tropical forest ecosystems of Africa and South America. In: B.J. Meggers, E.S. Ayensu & W.D. Duckworth (eds.), Tropical Forest Ecosystems in Africa and South America: a Comparative Review. Smithsonian Institution Washington D.C.: 63–88.

Morgan, N.C., & Boy, V., 1982. An ecological survey of standing waters in northwest Africa. 1. Rapid survey and classification. Biological Conservation, 24(1):5–44.

Nielsen, M.S., 1965. Introduction to the Flowering Plants of West Africa. University of London Press.

Palgrave, K.C. Trees of Southern Africa. C. Struik Publishers, Cape Town.

Reeve, C.V., 1972. Rifting in the Kalahari. Nature, 237:95–96.

Reizer, C., 1974. Définition d'une politique d'amenagement des resources halieutiques d'une écosystéme aquatique complèxe par l'étude de son environment abiotique, biotique et anthropique, Le Fleuve Senegal Moyen et Inferieur. D.Sc. Thesis, Fondation Universitaire Luxembourgeoise, Arlon

Renvoize, S.A., Lock, J.M. & Denny, P., 1984. A remarkable new grass genus from the Southern Sudan. Kew Bulletin 39:455–461.

Richards, P.W., 1939. Ecological studies on the rain forest of Southern Nigeria. 1. The structure and floristic composition of the primary forest. Journal of Ecology, 27:1–61.

Richards, P.W., 1957. The Tropical Rain Forest: an Ecological Study. Cambridge University Press.

Robyns, W., 1936. Contribution à l'étude des formations herbeuses du district forestier central du Congo Belge. Mémoires de l'Institut Royale Colonial Belge, Section des Sciences Naturelles et Médicales, 5(1).

Robyns, W., 1947–1955. Flore des spermatophytes du Parc National Albert, Vols I–III. Institut des Parcs Nationale du Congo Belge, Bruxelles.

Schimper, A.F.W., 1903. Plant Geography upon a Physiological Basis, P. Groom & I.B. Balfour (eds.), Clarendon Press, Oxford.

Sheppe, W. & Osborne, T., 1971. Patterns of use of a floodplain by Zambian mammals. Ecological Monographs, 41(3):179–205.

Sita, P., 1970. Etude de la végétation de l'Ile M'Bamou (Stanley-pool). O.R.S.T.O.M., Congo-Brazzaville.

Smith, P.A., 1976. An outline of the vegetation of the Okavango drainage system. Proceedings of the Symposium on the Okavango Delta and its Future Utilisation. The Botswana Society, Gaborone, Botswana: 93–112.

Streel, M. La végétation tropophylle des plaines alluviales de la Lufira moyenne (Katanga meridional). Fondation de l'Université de Liège pour les Récherches Scientifiques en Afrique Centrale.

Symoens, J.J., 1953. Note sur la végétation des salines de Mwashya (Katanga). Bull. Soc. Roy. Bot. Belg., 86:113–121.

Tadros, T.M., 1940. Structure and development of *Cyperus papyrus*. Bulletin of the Faculty of Science, Fouad I University, Cairo, 20.

Terry, P.J. & Minto, J.D., 1970. Floating islands in Lake Victoria. Report of the Tropical Pesticides Research Institute, Arusha, Tanzania.

Thompson, K., 1973. The ecology of peatlands in East and Central Africa, and their classification for agriculture. In: R.A. Robertson (ed.), Classification of Peat and Peatlands. Proceedings of a Symposium of the International Peat Society, Helsinki: 60–73.

Thompson, K., 1974. Ecology, management and utilisation of aquatic and semi-aquatic vegetation in the Okavango Delta, Botswana: a review, with recommendations for further studies. Technical Report, UNDP Project BOT/506, Land and Water Development Division, UN/FAO, Rome.

Thompson, K., 1976*a*. Swamp development in the headwaters of the White Nile. In: J. Rzoska (ed.), The Nile, Biology of an Ancient River. W. Junk, The Hague, Monographiae Biologicae 29:177–196.

Thompson, K., 1976*b*. Primary productivity of African wetlands, with particular reference to the Okavango Delta. Proceedings of the Symposium on the Okavango Delta and its Future Utilisation. The Botswana Society, Gaborone, Botswana: 67–79.

Thompson, K. & Hamilton, A.C., 1983. Peatlands and swamps of the African continent. In: A.J.P. Gore (ed.), Mires: Swamp, Bog, Fen and Moor. Elsevier, Ecosystems of the World Series, 4B:331–373.

Thompson, K., Shewry, P.R. & Woolhouse, H.W., 1979. Papyrus swamp development in the Upemba Basin, Zaire: studies of population structure in *Cyperus papyrus* stands. Botanical Journal of the Linnaan Society, 78:299–316.

Tinley, K.L., 1982. The influence of soil moisture balance on ecosystem patterns in southern Africa. In: B.J. Huntley & B.H. Walker (eds.), Ecology of Tropical Savannas. Springer-

Verlag, Ecological Studies 42:175–192.

Tinley, K.L., Rosinha, A.J., Lobao Tello, J.L.P., & Dutton, T.P., 1976. Wildlife and wild places in Mozambique. Oryx, 13(4):344–350.

Tothill, J.D., 1948. Agriculture in the Sudan. Oxford University Press, London.

USAID, 1980. Environmental assessment: Burundi peat project. United States of America Agency for International Development, Report Peat II (695–0103).

Van Der Ben, D., 1959. La végétation des rives des lacs Kivu, Édouard et Albert. Exploration Hydrobiologique des Lacs Kivu, Édouard et Albert (1952–1954), 4. Institut Royal des Sciences Naturelles de Belgique, Bruxelles.

Van Rensburg, H.J., 1968. Multipurpose survey of the Kafue River Basin, Zambia. Vol. 4. The ecology of the Kafue Flats. UN/FAO SF: 35/ZAM.

Van Rensberg, H.J., 1971. Fire: its effect on grasslands including swamps. Proceedings of the Tall Timbers Fire Ecology Conference. Tall Timbers Research Station, Florida, II: 175–199.

Van Steenis, C.G.G.J., 1981. Rheophytes of the World. Sijthoff and Noordhoff, The Netherlands.

Vesey-Fitzgerald, D.F., 1963a. Annotated list of grasses collected in the Congo drainage basin of Northern Rhodesia and Tanganyika. Kirkia, 3:78–111.

Vesey-Fitzgerald, D.F., 1963b. Central African grasslands. Journal of Ecology, 51:243–274.

Vesey-Fitzgerald, D.F., 1970. The origin and distribution of the valley grasslands in East Africa. Journal of Ecology, 58:51–75.

Walter, H., 1962. Die Vegetation der Erde in ökologischer Betrachtung. Fischer-Verlag, Jena.

Welsh, R.P.H. & Denny, P., 1978. The vegetation of Nyumba ya Mungu Reservoir, Tanzania. Biological Journal of the Linnean Society, 10:67–92.

West, O., 1965. Fire in vegetation and its use in pasture management, with special reference to tropical and subtropical Africa. Commonwealth Agricultural Bureaux Publication no. 1/1965.

White, F., 1983. The Vegetation of Africa: a descriptive memoir to accompany the UNESCO/AETFAT/UNSO Vegetation map of Africa. UNESCO, Paris.

Author's address:
Department of Biological Sciences
University of Waikato
Private Bag
Hamilton, New Zealand

CHAPTER 4

Surface-floating aquatic macrophytes

DAVID S. MITCHELL

Detailed contents

Aquatic plants that float on the water surface occupy a distinctive habitat that requires particular adaptation. Furthermore, since they are not affected by water depth, they are not limited to a particular zone in the hydrosere but may occur wherever there is sufficient free water. Classically, however, surface-floating plants tend to dominate the offshore margins of the littoral vegetation continuum in those situations where they are not removed by wind or current.

Floating aquatic plants exhibit the anatomical characteristics of depauperate vascular systems and aerenchymatous tissue typical of vascular hydrophytes but, in addition, demonstrate positive bouyancy and, frequently, a distinction between upper non-wettable and lower wettable surfaces. These plants are, therefore, obligate aquatic plants which have features in common with euhydrophytes (see Chapter 2) that have floating leaves, such as *Nymphaea*, and with unattached plants that float below the surface, such as *Ceratophyllum*. Nevertheless, they are distinct from both these growth forms and occupy a particular ecological niche and thus merit consideration as a separate form. Floating islands of vegetation formed when emergent rooted plants, such as *Vossia cuspidata*, break free after growing out over the surface are not included in this growth form as they are typical emergents in spite of their floating habit.

Surface-floating aquatic plants are well adapted to survive in a habitat that poses particular constraints. In certain circumstances these survival mechanisms provide the means whereby plants with this growth form can cause some of the most spectacular and intransigent water weed problems. In this chapter, an examination will be made of the particular features of the surface-floating plant regime, the adaptations of the plants to survive and proliferate in the habitat, and the nature and extent of problems caused by floating aquatic plants in Africa.

Patrick Denny (ed.), The ecology and management of African wetland vegetation.
 ISBN 90 6193 509 1

4.1 Taxonomy

The floating plant habit is not common in the plant kingdom. Two families of ferns, Azollaceae and Salviniaceae, and one of flowering plants, Lemnaceae, consist solely of floating plants, though some members of the latter may also float submerged. Outside of these families there are relatively few species of surface-floating plants, and all belong to families where the majority of species are not of this growth form. For example, the notorious floating weed, *Eichhornia crassipes*, is the only species to exhibit this growth form in the family, Pontederiaceae. Surface-floating aquatic plants that have been reported to occur in Africa are set out in Table 4.1.

4.2 The water surface regime

4.2.1 *Definitions and terms*

Free-floating macrophytes were termed pleuston by Schroter and Kirchner (1896) whether they floated on, or below, the water surface. Gams (1918) and Hutchinson (1967) confined application of the term to surface-floating plants, while Luther (1949) reverted to the original use of the word and

Table 4.1. Surface-floating aquatic plants present in Africa.

Plant	Distribution in Africa
BRYOPHYTA	
Ricciaceae	
Riccia fluitans L.	pan-African
Ricciocarpus natans (L.) Corda	tropical & sub-tropical
PTEROPHYTA	
Azollaceae	
Azolla pinnata R. Br. var. africana (Desv.) Bak.	pan-African
A. nilotica Decne ex Mett. (e)*	eastern tropical & sub-tropical
A. filiculoides Lam. (a)**	South Africa
Salviniaceae	
Salvinia nymphellula Desv. (e)	western tropical
S. hastata Desv. (e)	eastern tropical & sub-tropical
S. molesta D.S. Mitchell (a)	Kenya, Tanzania, Zaire, Zambia, Botswana, Namibia, Zimbabwe, South Africa, Moçambique
Parkeriaceae	
Ceratopteris cornuta (Beauv.) Lepr.	tropical & sub-tropical
C. thalictroides (L.) Brongn.	Mauritius
C. richardii Brongn.	tropical & sub-tropical
ANTHOPHYTA	
Araceae	
Pistia stratiotes L.	tropical & sub-tropical
Pontederiaceae	
Eichhornia crassipes (Mart.) Solms-Laub. (a)	Egypt, Sudan, Tanzania, Zaire, Zambia, Moçambique, Zimbabwe, South Africa
Lemnaceae***	
Lemna spp.	pan-African
Spirodela spp.	pan-African
Wolffia spp.	pan-African
Wolffiella spp.	pan-African

*e = endemic to Africa
**a = alien introduction to Africa: distribution given in terms of countries from which incidences have been reported
***following the taxonomic treatment of Landholt (1980)

distinguished between acropleustophytes on the surface, mesopleustophytes between the surface and the bottom and benthopleustophytes on the bottom but not attached to it. Various other authors – notably Hejný (1960), Den Hartog and Segal (1964), Hogeweg and Brenkert (1969) and Hutchinson (1975) – also developed terms to classify aquatic plants with respect to their life forms with the consequence that there are now nearly as many terms to describe the group of plant species that float on the water surface as there are species that occupy the habitat! Therefore, there would be little benefit in suggesting another term to distinguish those species that grow exclusively in this habitat in the vegetative form and are especially fitted for it, from those that may routinely occupy it successfully but are not obliged to do so. Nevertheless, there is a need to recognise the existence of a group of particularly adapted plants and an apt, unambiguous term for them would be obligate acropleustophytes. This term is more narrow than the hydatoaerophyta of Hejný (1960), the pleustohelophytes of Den Hartog and Segal (1964), the free-floating hydrophytes of Sculthorpe (1967) and the acropleustophyta of Luther (1949), Hogeweg and Brenkert (1969) and Hutchinson (1975) in that it excludes those plants that are capable of existing for long periods below the water surface.

Plants which occur in Africa that belong to the obligate acropleustophytes are *Eichhornia crassipes,* all African species of *Azolla* and *Salvinia, Ricciocarpus natans,* all African species of *Spirodela* and *Lemna* (except for *L. trisulca*), and *Pista stratiotes. Riccia* and other genera of the Lemnaceae may float with their photosynthetic organs mostly below the water surface. Examples of obligate acropleustophytes are illustrated in Plate 4.1. In all cases it can be seen that there is clear distinction into non-wettable upper surfaces and wettable lower surfaces in direct contact with the water. It is this feature as much as any other that distinguishes these floating plants from others.

Plate 4.1. Examples of obligate acropleustophytes in Africa. (a) *Azolla pinnata* × 10; (b) *Lemna* sp. × 10; (c) *Salvinia molesta,* showing transition from primary invading leaf-form to the mat leaf-form. (Photos (a) and (b) L. Gallagher; (c) D.S. Mitchell.)

4.2.2 *Rigours of the water surface habitat*

Plants floating on the surface of the water have to contend with three main problems: (a) maintenance of positive bouyancy; (b) absorption of nutrients from the most dilute part of the ecosystem, namely the surface waters; (c) lack of anchorage so that they are at the mercy of wind and current.

They are also liable to physical damage from excessive wave action in a lake or turbulent flow in a river. Furthermore, some species develop growth forms when plants are crowded together, and these are readily upended by even mild winds and current when the population separates into individual plants.

Against these difficulties, plants occupying this niche benefit from being able to survive both rapid and extensive fluctuations in water level that eliminate attached plants, as they are not attached to the bottom. As a consequence, certain localities subject to such changes in water level are available for colonisation by floating plants free of competition from other vascular aquatic plants.

If all these factors are taken into account, it is obvious that surface-floating aquatic plants grow most successfully in calm sheltered waters where they are rapidly able to establish a complete blanket of cover and thereby provide anchorage for one another. Clearly, rate of population expansion is an important factor in these situations, and it is essential that nutrients are not limiting, at least in the initial stages of colonisation. Other factors that are important are the size of the colonisable area in relation to the size of the invading inoculum. The smaller the former and the larger the latter, the faster will be the rate of colonisation and the sooner the establishment of a stable, self-supporting, floating plant population.

4.2.3 *African waters suitable for surface-floating plants*

The major water bodies in Africa are lakes, which are especially large in the Rift Valley, and rivers with markedly seasonal flooding regimes. There is also a number of important and extensive swamps, such as the Okavango Swamps in Botswana, Bangweulu in Zambia, the fringing swamps of many East African lakes, the sudd region of the Upper Nile, and similar vegetation fringing Lake Chad. More ephemeral waters develop seasonally in the pans and vleis of southern Africa and savanna regions of other areas of Africa. Persistent stands of emergent aquatic plants develop in these, tiding over the dry season when the extent of free water may diminish to nothing, and, if rains provide sufficient run-off, expanding to cover areas of several hectares which may persist for several years during cycles of good rainy seasons.

The seasonal nature of the rains over much of the continent divides the year into a rainy season and an extended dry season. This has necessitated the conservation of water in man-made lakes where it is then available for a number of purposes including hydro-electricity. Some of the largest man-made lakes in the world have been built in Africa, and well known examples include Lake Kariba and Cabora Bassa on the Zambezi for hydro-electricity, Lake Nasser on the Nile for hydro-electricity and irrigation, Lake Volta on the Volta for hydro-electricity and the Kainji Reservoir on the Niger for hydro-electricity and flood control. Many of these schemes have become increasingly multipurpose in their function as they have developed and are now important for recreation, tourism, transport and fishery production (Lowe-McConnell, 1966; Obeng, 1969; Ackermann *et al.*, 1973). The development of industry and expansion of agriculture as well as the increasing demands from an exponentially expanding population have meant that there has been an exponential increase in the number of reservoirs being constructed in many African countries. Figure 4.1 depicts this pattern of development for Zimbabwe (Rhodesian Ministry of Information, Immigration and Tourism, 1973), where in addition to the 80 or so man-made lakes figured in the diagram, there were approximately 7000 farm dams in excess of 2 ha in area (Mitchell, 1974).

All these types of water body provide suitable habitats for floating aquatic plants, at least for some period of the year, though the extent of colonisable habitat will vary from time to time. In general, rivers are the least suitable water bodies for floating plants, because of the flushing effect of river flow, while the most suitable are man-made lakes in the early years of existence when nutrient content tends to be high, because attached macro-

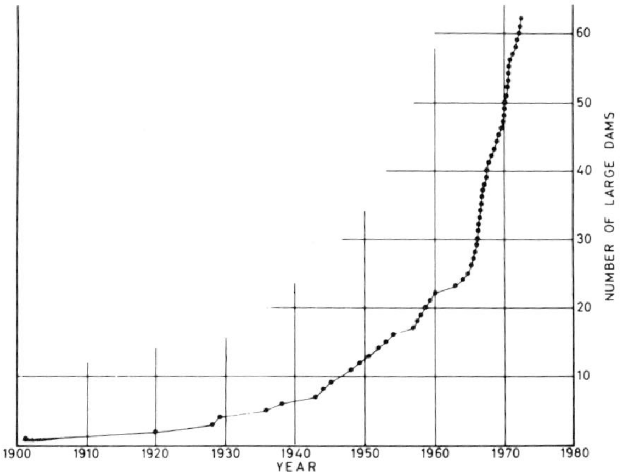

Fig. 4.1. The exponential increase in numbers of large dams in Zimbabwe from 1900 to 1973.

phytes are not able to contend with the fluxes in water level and offer no competition.

4.3 Adaptations and ecological strategies

4.3.1 *Non-wettable surfaces*

One of the most characteristic features of obligate acropleustophytes is their capacity to repel water from the upper surface of the leaf. This can be demonstrated readily by dropping water into the plants; it will run off immediately or bead up into spherical droplets because of surface tension effects. In addition, surfaces may be roughened by a sculptured epidermis, as in *Azolla*; by hairs, as in *Pistia*; or by multicellular papillae, as in *Salvinia molesta* where four uniseriate filaments on the apex of the papilla join distally to form a 'bird cage' – like arrangement (Plate 4.2).

That these surfaces are important in maintaining buoyancy is shown by the effect on the plants of adding an ionically positive detergent, such as calcium dodecylbenzene sulphonate. which causes *Azolla* and small *Salvinia* plants to sink within 2–3 minutes. Larger *Salvinia* plants and *Pistia* sink more slowly, while *Eichhornia crassipes* is relatively unaffected. This range in response is brought about by the differing extents to which foliage can be wetted by the reagent as surfaces become more vertically orientated. This effect of detergent chemicals has been employed in the development

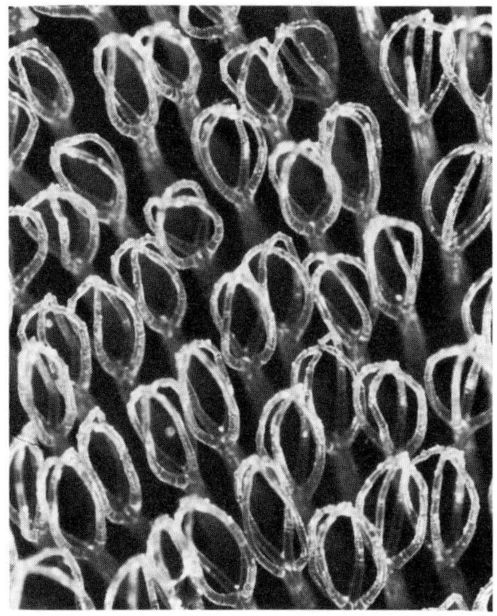

Plate 4.2. Rows of multicellular papillae surmounted by a 'bird-cage' of four uniseriate, distally united trichomes on the upper surface of the floating leaves of *Salvinia molesta* (× 10). The papillae repel water and trap air when pushed under water. (Photo L. Gallagher.)

of herbicides specific for these plants (Diatloff *et al.*, 1979).

It is noteworthy that, despite the importance of the water repellency of these surfaces to the ecology of these plants, very little has been done to study the nature of the phenomenon. The presence of a waxy chemical on the cuticle of these surfaces is assumed to be the cause, but nothing is known about the factors which promote its development on the upper surfaces of the plants and which prevent it on the lower surfaces of the same leaves. Furthermore, the extent to which the presence of projections from the leaf surface, such as has been described, enhance this effect has not been investigated. Even the most cursory examination of the plant *in situ* readily reveals that these hairs entrap air when plants are pushed below the water surface, and it has been assumed that this was due to their construction and arrangement (Sculthorpe, 1967). However, it is clear from the effect of the detergents mentioned earlier that the water repellent nature of the surfaces of these structures is more important, and, when this is destroyed, air is no longer trapped although the structure remains. Clearly more work is needed.

4.3.2. *Wettable surfaces and nutrient absorption*

The water-repellent nature of the upper surfaces of surface-floating plants is matched by the water-attractive nature of the lower surfaces of the leaves and any other submerged portions of the plants. Since the plants are not in direct contact with a nutrient rich substrate, nutrient absorption has to take place through these surfaces, and an intimate contact between water and plant is clearly required.

It is also obvious that, for effective absorption, the larger the absorptive area the better. Many plants have an increased surface area through the production of hairs and the development of much branched, or dissected, absorptive organs: *e.g.* the dense cover of lateral roots on the numerous adventitious roots of *Eichhornia crassipes* and the much dissected submerged axes (classically regarded as leaves) of *Salvinia molesta*. The lower surface of the latter plant is also covered with sharply pointed, uniseriate trichomes in which the cells appear to be dead. These trichomes are very dense on the root-like submerged axes of the plant. Whether they participate in nutrient uptake is not known, but it is clear that they assist in distributing water over the lower surface of the leaf when these are not in direct contact with the water. Some circumstantial evidence for their involvement in nutrient absorption is provided by the observation that they are more dense and longer in nitrogen deficient plants (Mitchell, 1970).

In some obligate acropleustophytes, however, roots are poorly developed or absent. Thus *Azolla* species generally have few roots, although those of *A. pinnata* are densely clothed with fine lateral rootlets. The reduced rooting surface in *Azolla* is supposedly compensated for by the presence of a submerged, hyaline leaf lobe. The other lobe of the leaf is green and floats on the water surface. The submerged lobe is presumed to absorb water and nutrients. The same trend for reduction in the extent of the rooting system is also seen in the Lemnaceae. *Spirodela* may have from 2 – 12 roots per frond, but *Lemna* has only one, while *Wolffia* and *Wolffiella* have none.

A number of authors (Steinberg, 1946; Landolt, 1957; White, 1936, 1937, 1940; Penfound and Earle, 1948; Musil and Breen, 1977; Mitchell, 1970; Cary and Weerts, 1983) have investigated the nutrient requirements of *Eichhornia crassipes, Salvinia molesta*, or members of the Lemnaceae. Work on the first two species has been stimulated by their weedy nature and on the latter largely by the ease and convenience with which plants can be cultivated in the laboratory, under axenic conditions if desired. These studies show that the growth rate of these plants is low in nutrient-poor conditions and greatly accelerated as nutrients become available in otherwise optimum conditions. Two features reflect the adaptation of these plants to the generally low nutrient status of their immediate environment and are noteworthy for this reason.

(a) The plants have a capacity to tolerate low nutrient status in the water for long periods even though growth has apparently ceased (*Salvinia molesta* has survived for over six months in distilled water).

(b) The plants have a capacity for the luxury uptake of phosphorus, *i.e.* the absorption and accumulation of phosphorus in excess of immediate needs.

Many of these plants also exhibit association with nitrogen-fixing micro-organisms. The symbiotic association between *Azolla* and the nitrogen-fixing, blue-green alga, *Anabaena azollae,* is well known (Eames, 1936; Kawamatu, 1965*a*, 1965*b*; Ashton and Walmsley, 1976). The alga is contained in cavities in the leaves of the plant, and the macrophyte is apparently dependant on its presence for normal growth. *Eichhornia crassipes* has been shown to be associated with the nitrogen-fixing bacterium, *Azotobacter chroococcum*, which becomes established in the mucilaginous coating of the base of the petiole (Iswaran *et al.*, 1973; Purchase, 1977). The association here is more casual, and it is not certain whether the host plant derives any direct benefit from the presence of the bacterium. A similar loose association between *Salvinia molesta* and the nitrogen-fixing, blue-green alga, *Nostoc,* has been reported by Mitchell (1970). In this case, the alga forms its colonies between the abaxial surfaces and along the margins of the more or less erect leaves of the mat form of the plant in stable populations. Whether these associations between aquatic plants and nitrogen-fixing organisms are casual or obligate, it is clear that nitrogen poor systems such as the surface waters inhabited by floating plants can be provided with nitrogen by this means.

4.3.3 *Adaptations to lack of anchorage*

The root systems of vascular plants serve two main functions: nutrient absorption and anchorage. Even when roots are well developed in these species, such as in *Eichhornia crassipes,* they do not penetrate a firm substratum at the bottom of a water body, although, exceptionally, they may become passively buried by sedimentation of organic and inorganic detritus. Thus these plants are capable of being moved by wind and current and, if the portion of the plant above the surface of the water exceeds that below it, of being turned upside down. Simple experiments, in which submerged portions of the plant are removed, reveal that these play an important role in stabilising the vertical orientation of those plants where the leaves project more or less vertically from the water surface, as in the case of *Eichhornia crassipes, Pistia stratiotes* and the mat form of *Salvinia molesta*. Plants such as *Azolla, Lemna* and the colonising forms of *S. molesta,* where the leaves float flat on the water surface, are relatively little destabilised by removal of 'root' systems.

Two factors are important in the formation of stable plant communities: (a) rapid colonisation of the suitable area available; (b) interlocking of individual plants in the community.

Studies of the growth rates of *Salvinia molesta* by Mitchell (1970), Gaudet (1973), Mitchell and Tur (1975), Cary and Weerts (1983) and Toerien *et al.* (1983) have shown that, under optimum temperature conditions (25–30°C), high light intensity, and non-limiting supplies of nutrients, increase in plant population is exponential, and the plant is capable of doubling in terms of leaf number in 2–3 days and occasionally in less than 2 days. Mitchell and Tur (1975) measured the growth rates in the field of different forms of the plant at different times of the year. They reported a mean value for doubling time of eleven days (n=73). The mean value for the mat form growing in crowded populations was 11.6 days and for the colonising form was 8.6 days. Mean doubling times (all forms) for winter/spring, spring/summer, and summer were 14.3, 13.2 and 8.1 days respectively.

Eichhornia crassipes, Azolla filiculoides and members of the Lemnaceae are similarly capable of remarkably rapid rates of growth for vascular plants. A doubling time, in terms of numbers of plants, of 6.2 days has been recorded for *E. crassipes* growing in sewage ponds (Bagnall *et al.,* 1974) and, in most tropical waters where nutrients are not limiting, doubling times of 10–12 days can be expected. *Azolla filiculoides* has been shown to have a doubling time of 3.4 days in good growing conditions and uncrowded populations, but this increased markedly to more than 200 days as the population aged and the plants became increasingly crowded together (Ashton, 1977).

Growth in the Lemnaceae is also exponential and is usually measured in terms of increase in dry matter or number of fronds in a population. Doubling times of about two days are regularly recorded in laboratory culture conditions, and similar rates of growth are likely to be possible in the field. Growth rate is affected by temperature, light intensity and the availability of nutrients and, as may be expected, both growth rates and optima differ between species. Thus Landolt (1957) recorded maximum doubling times at 24 °C and 100 foot candles continuous illumination with one per cent sucrose, and found that *Wolffiella gladiata (Wolffiella floridana)* doubled in 5 days, *Wolffia* spp. in about 2 days, *Lemna minor* strains in about 1.6 days, *Spirodela polyrhiza* strains in 1.5 – 1.3 days and *Lemna perpusilla* strains in 1.2 – 1.1 days.

That all these obligate acropleustophytes exhibit remarkably rapid rates of population growth is an obvious adaptation to the need for them to occupy the available colonisable area as quickly as possible and thereby stabilise the situation so that most other aquatic plants are unable to compete with

Plate 4.3. Pistia growing in the Kasinga Channel, Uganda. This plant is an excellent example of an obligate acropleustophyte. The upper leaf surface is covered in minute hains thus making it non-wettable, and the well-developed roots provide vertical stability. (Photo P. Denny.)

Plate 4.4. A floating mass of *Pistia* in the Kasinga Channel. The next day it was blown to the other end of the channel! (Photo P. Denny.)

them. In this way they can resist the effect of surface stress from wind or current.

As a further adaptation, *Eichhornia crassipes* (Center and Spencer, 1981) and *Salvinia molesta* (Loyal and Grewal, 1967; Mitchell, 1970) exhibit modifications in growth habit to suit the differing circumstances at different stages of the succession. Thus, in the initial stages, *Salvinia molesta* produces plants with small leaves about 1 cm across which float flat on the water surface, while *Eichhornia crassipes* grows as leaf rosettes, which may be less than 5 cm in diameter in disturbed conditions and which are close to the water surface with truncated swollen petioles. In these primary invading forms most of the plant matter produced is used to form photosynthetic tissue. As the area is colonised by a loose assemblage of plants or by plants encroaching on to it from the margin of stable fringing mats, another plant form is produced, which is larger and more robust. In *S. molesta* the leaves become keeled and reach a breadth of 5 cm but remain in contact with water. In *E. crassipes*, the leaves become larger and rise up to about 10 cm above the water surface, though still with bulbous petioles.

In all the above forms there is pronounced vegetative reproduction of new individuals: in *S. molesta* by profuse branching followed by breaking of the main stem at the node from which the branch originates and in *E. crassipes* by the production of stoloniferous offsets.

As the plants become crowded together, there are further changes in morphology and plants become even more robust. In *S. molesta,* internodes are shortened and the leaves overlap, folding along the midrib so that each half rises almost vertically above the water with the leaf tip becoming sharply incised, and the leaf developing an auriculate to oblong shape with a breadth of about 6 cm. In *E. crassipes* leaves of over 1 m in length are formed and the petioles lose their bulbous shape.

These sequences of morphological change form a continuum of modification to suit changing environmental conditions and are accompanied by an

Plate 4.5. The beautiful water hyacinth, *Eichhornia crassipes,* in flower in the Sudd of the Upper Nile. The plate-like leaves act as sails which aid the distribution of plants by wind. (Photo P. Denny.)

increase in leaf area index (Center and Spencer, 1981).

For both species, individual plants become interlocked with one another through the interweaving of their branches in *S. molesta* and through the entanglement of offsets in *E. crassipes*. As a consequence, the plants anchor one another and by expanding to fill the available surface become a fixed mat that is difficult to dislodge, thereby counteracting the disadvantage of not being fixed to the substratum.

4.3.4 *Reproduction and survival strategies*

It is apparent that obligate acropleustophytes depend primarily on rapid growth and vegetative multiplication for the establishment of stable communities. This has been seen as an adaptation to situations where there is a period or periods in the year when large quantities of plants are swept away by river floods and populations are virtually eliminated. This is particularly clear for *Salvinia molesta* and *Eichhornia crassipes,* both of which are native to South America where there is a number of large seasonally flooding rivers and few isolated standing waters (Mitchell and Thomas, 1972; Center and Spencer, 1981). Floating plant populations are loosened by rising water levels and swept away by the floods. However a few plants always remain, particularly in the upper reaches, and these provide the means of re-establishment.

Sexual reproduction in surface-floating plants is frequently linked with the extent to which the plants are crowded together. In *S. molesta* and *A. filiculoides* there is an increase in sporocarp production as the plants become crowded (Mitchell, 1970; Ashton, 1977), while, in contrast, *E. crassipes* flowers are often less frequent in tightly crowded populations, though flowering is also dependant on the time of the year.

Successful seed or spore set and germination is less common than in many other plants and, gener-

Plate 4.6. A close-up picture of the large, robust growth-form of *Salvinia molesta* from Lake Kariba. Life size. (Photo P. Denny.)

ally, it would be true to say that obligate acropleustophytes have a relatively low sexual reproductive capacity. *Salvinia, Azolla* and *Ceratopteris*, as ferns, exhibit alternation of independent gametophyte and sporophyte, the first two in addition being heterosporous. Furthermore, *S. molesta* is a sterile hybrid (Loyal and Grewal, 1966; Mitchell, 1970, 1972). Fertilisation in these heterosporous genera is also complex (Eames, 1936).

Eichhornia crassipes exhibits heterostyly, and there are many reports in the literature of poor seed production in the field in both native and alien habitats, which had been assumed to be due to the presence of only one style form. However Barrett (1980) has shown that poor seed set is not likely to be due to inherent sterility factors but to inefficient insect pollination and the requirements for seedling germination of high light, well oxygenated conditions and moisture (Hitchcock *et al.*, 1950) which are not often met in the field.

Sexual and asexual reproduction in the Lemnaceae have been the subject of considerable study, mostly under laboratory conditions, and relatively little is known about reproduction in the field. The flowers in the Lemnaceae have been markedly reduced to a single bilocular stamen in the male flower and a single carpel in the female. Inflorescences are also reduced: in *Spirodela* and *Lemna* to one female and two male flowers and in *Wolffia* and *Wolffiella* to one flower of each sex.

There is considerable information in the literature about the physiology of flowering and the various factors which can induce flowering to occur (see reviews in Hillman, 1961, and Sculthorpe, 1967). It is clear that daylength, temperature, the phytochrome system, the presence of kinetins and probably other factors, such as carbon dioxide concentrations, are important, but most of this work has not been applied to the ecology of the plant and the elucidation of its survival strategies. Flowering in the field appears to be comparatively rare and is only infrequently reported for Africa. However, the combination of the small size of the plants and the very reduced nature of the flowers probably means that the presence of flowers is often overlooked.

The Lemnaceae also produce specialised vegetative perennating structures called turions. In *Spirodella polyrrhiza* these are modified fronds which are devoid of air spaces, swollen, more darkly coloured (green or purple) and filled with starch. In species of *Lemna* and *Wolffia*, similar turions are produced, but these differ somewhat less from the normal fronds, while, in *L. trisulca*, hibernating fronds do not apparently differ morphologically from normal fronds. The turions are usually produced in response to the onset of adverse growth conditions and sink to the bottom until favourable conditions return. In central Europe, *Lemna* spp. perennate over winter as hibernating fronds, *Spirodella polyrrhiza* as turions and *Lemna gibba* as seeds (Rejmankova, 1982). Indeed Rejmankova (1976) has shown that in the spring about 80 per cent of the new fronds of the latter species have germinated from seeds.

4.4 Problems caused by surface-floating plants in Africa

Weed problems caused by floating aquatic plants have been reported from a number of localities in Africa (Table 4.2). The main problems have been caused by the plants which are alien to Africa and of the four species listed in the table only *Pistia stratiotes* is native to Africa.

Azolla pinnata var. *africana* and members of the Lemnaceae occasionally cover water troughs, ponds and small lakes and can cause a nuisance by inhibiting drinking by domestic animals, blocking pump intakes and tainting water (Wild, 1961; Jacot-Guillarmod, 1979).

It is clear from the table below that the water bodies most at risk are those where there is a high degree of human interference (man-made lakes, canals, *etc.*) and where the presence of the plant interferes with human use of the water, thereby constituting it to be a weed by definition. In most cases the plants act as primary colonisers of waters

Table 4.2. Problems caused by surface-floating plants in Africa.

Location	Type of waterbody	Nature of problem or interference	Source
Salvinia molesta			
Kenya			
Lake Naivasha	Rift Valley lake	Recreation	Gaudet 1976
Zaire			
Barrage d'Inga	Man-made lake	Water quality	van Himme 1973
Zaire River	River	Navigation	Little 1965*a*
Zambia			
Ndola Dambo	Town water supply	Recreation	personal observation
Zambia, Zimbabwe			
Lake Kariba*	Man-made lake	Fishing, navigation, recreation, spreads schistosomiasis	Boughey 1963 Mitchell 1969 Mitchell & Rose 1979 Marshall & Junor 1981 Hira 1969
Moçambique			
Cabora Bassa	Man-made lake	As above for Lake Kariba	Bond & Roberts 1978
Botswana, Namibia			
Chobe/Linyanti system*	River floodplain	Flow	Edwards & Thomas 1977
South Africa			
Swartvlei	River estuary	Flow, boating	Jacot-Guillarmod 1979
farm dams		Stock drinking	

Table 4.2. Continued.

Location	Type of waterbody	Nature of problem or interference	Source
Eichhornia crassipes			
Sudan, Egypt			
Nile River*	River	Flow, navigation, spreads schistosomiasis	Gay 1958, 1960 Pettet 1964 Obeid 1975 Batanouny & El Fiky 1975
Tanzania			
Pangani River	River		Wild 1961
Zaire			
Zaire River*	River	Flow, navigation	Dubois 1955 Wild 1961 Evans 1963
Malagarasy	River	Flow	Ramarokoto 1965 Wild 1961
Moçambique			
Cabora Bassa	Man-made lake	Navigation, increases evapotranspiration, spreads schistosomiasis	Bond & Roberts 1978
Incomati River	River		Wild 1961
Zimbabwe			
Makabusi River	River	Flow	Wild 1956, 1961
Sabi River	River	Flow	Bates & Phipps 1958 Gibbs Russell & Mitchell 1976
Lake McIlwaine	Man-made lake	Recreation	Jarvis *et al.* 1982
Senegal			Little 1965*b*
Mauritius			Robson 1976
South Africa			
Hartebeespoort Dam*	Man-made lake	Recreation, spreads schistosomiasis, increases evapotranspiration, obstructs flow	Scott *et al.* 1979
farm dams canals rivers			Jacot-Guillarmod 1979
Azolla filiculoides			
South Africa			
Verwoerd Dam	Man-made lake		Ashton 1977
farm dams		Stock drinking	Jacot-Guillarmod 1979
Pistia stratiotes			
Ghana			
Volta Dam	Man-made lake		Ewer 1965 Hall *et al.* 1969
Sudan			
Upper Nile	River	Flow	Wild 1961
Kenya			
Amboseli National Park	Pool		Wild 1961
Uganda			
Kasinga Channel	Natural channel	Boat movement (seasonal)	Denny, personal communication

*Serious problems are indicated with an asterisk

that are suitable for the establishment of aquatic vegetation and are, therefore, especially adapted to colonise open water as has been shown in previous sections. Their removal by some form of control merely restores the conditions they are most suited to exploit and in which they are readily dispersed. Mechanical and chemical control is, therefore, likely to be a continuing process and should not be undertaken without recognising this situation. Principles and techniques for the management of aquatic vegetation of all growth forms, including other water weeds, will be presented and discussed in Chapter 7.

4.5 Conclusions

Surface-floating aquatic plants occupy a particular niche for which they are especially adapted. The term, obligate acropleustophyte, is an appropriate one to distinguish the plants that cannot grow well in other situations. Because of their particular suitability for calm waters, their tolerance of water level fluctuations, and their capacity for rapid vegetative growth in conditions where nutrients are non-limiting, they are especially able to exploit the waters of new or enriched man-made lakes where they can be particularly troublesome as weeds.

4.6 References

Ackermann, W.C., White G.F. & Worthington, E.B. (eds), 1973. Man-made lakes: their problems and environmental effects. American Geophysical Union, Washington, D.C.

Ashton, P.J., 1977. Factors affecting the growth and development of *Azolla filiculoides* Lam. Proceedings of the second National Weeds Conference, South Africa. A.A. Balkema, Cape Town: 249–268.

Ashton, P.J. & Walmsley, R.D., 1976. The aquatic fern *Azolla* and its *Anabaena* symbiont. Endeavour, 35:39–43.

Bagnall, L.O., Furman, T. de S., Hentges, J.F., Nolan, W.J. & Shirley, R.L., 1974. Feed and fiber from effluent-grown water hyacinth. In: Wastewater use in the production of food and fiber-proceedings. EPA–660/2–74–041. Environmental Protection Agency, USA: 116–141.

Barrett, S.C.H., 1980. Sexual reproduction in *Eichhornia crassipes* (water hyacinth). II: Seed production in natural populations. Journal of Applied Ecology, 17:113–124.

Batanouny, K.H. & El Fiky, A.M., 1975. The water hyacinth *(Eichhornia crassipes* (Mart. Solms) in the Nile system, Egypt. Aquatic Botany, 1:243–252.

Bates, G.R. & Phipps, J.B., 1958. Water hyacinth and its control in Rhodesia. Proceedings of the First African Weed Control Conference: 135–139.

Bond, W.J. & Roberts, M.G., 1978. The colonisation of Cabora Bassa, Mocambique, a new man-made lake, by floating aquatic macrophytes. Hydrobiologia, 60:243–259.

Boughey, A.S. 1963. The explosive development of a floating weed vegetation on Lake Kariba. Adansonia, 3:49–61.

Cary, P.R. & Weerts, P.G.J., 1983. Growth of *Salvinia molesta* as affected by water temperature and nutrition. I: Effects of nitrogen level and nitrogen compounds. Aquatic Botany, 16:163–172. II: Effects of phosphorus level. Aquatic Botany, 17:61–70.

Center, T.D. & Spencer, N.R., 1981. The phenology and growth of water hyacinth *(Eichhornia crassipes* (Mart.) Solms) in a eutrophic north-central Florida lake. Aquatic Botany, 10:1–32.

Den Hartog, C., & Segal, S., 1964. A new classification of water-plant communities. Acta Botanica Neerlandica, 13:367–393.

Diatloff, G., Lee, A.N. & Anderson, T.N., 1979. A new approach for Salvinia control. Journal Aquatic Plant Management, 17:24–27.

Dubois, L., 1955. La Jacinthe d'eau au Congo Belge. Bulletin Agricole du Congo Belge, 46:893–900.

Eames, A.J., 1936. Morphology of Vascular Plants – Lower Groups. McGraw-Hill, New York.

Edwards, D. & Thomas, P.A., 1977. The *Salvinia molesta* problem in the northern Botswana and eastern Caprivi area. Proceedings Second National Weeds Conference, South Africa: 221–237.

Evans, A.C., 1963. The grip of the water hyacinth. New Scientist, 19:666–668.

Ewer, D.W., 1965. Biological investigations on the Volta Lake, May 1964 to May 1965. In: R.H. Lowe-McConnell (ed.), Man-made Lakes. Academic Press, London: 21–31.

Gams, H., 1918. Prinzipienfragen der Vegetationsforschung. Vierteljahresschrift der Naturforschenden Gesellschaft, Zurich, 63:293–493.

Gaudet, J.J., 1973. Standardised growth conditions for an aquatic weed, *Salvinia*. Hydrobiologia, 41:77–106.

Gaudet, J.J., 1976. Salvinia infestation on Lake Naivasha in East Africa (Kenya). In: C.K. Varshney & J. Rzoska (eds.), Aquatic weeds in South East Asia. Dr.W. Junk, The Hague: 193–209.

Gay, P.A., 1958. *Eichhornia crassipes* in the Nile of the Sudan. Nature, London, 182:538–539.

Gay, P.A., 1960. Ecological studies of *Eichhornia crassipes* (Mart.) Solms in the Sudan. I. Analysis of spread in the Nile. Journal of Ecology, 48:183–191.

Gibbs Russell, G.E. & Mitchell, D.S., 1976. Common aquatic plants on Rhodesian pans and lakes. Rhodesian Agricultural Journal, 73:13–17.

Hall, J.B., Laing, E., Hossain, M. & Lawson, G.W., 1969.

Observations on aquatic weeds in the Volta Basin. In: L.F. Obeng (ed.), Man-made lakes. The Accra symposium. Ghana Universities Press, Accra: 331–343.

Hejný, S., 1960. Oekologische Characteristik der Wasser- und Sumpfpflanzen in den Slowakische Tiefebenen (Donau und Theissgebiet). Verlag der Slowakischen Akademie der Wissenschafte, Bratislawa.

Hillman, W.S., 1961. The *Lemnaceae*, or duckweeds. A review of the descriptive and experimental literature. Botanical Review, 27:221–287.

Hitchcock, A.E., Zimmerman, P.W., Kirkpatrick, H. & Earle, T.T., 1950. Growth and reproduction of water hyacinth and alligator weed and their control by means of 2, 4-D. Contributions of the Boyce Thomson Institute. 16:91–130.

Hira, P.R., 1969. Transmission of schistosomiasis in Lake Kariba, Zambia. Nature, London, 224: 670–672.

Hogeweg, P. & Brenkert, A.L., 1969. Structure of vegetation: a comparison of aquatic vegetation in India, the Netherlands, and Czechoslovakia. Tropical Ecology, 10:139–162.

Hutchinson, G.E., 1967. A treatise on limnology. II: Introduction to lake biology and the limnoplankton. J. Wiley, New York.

Hutchinson, G.E. 1975. A treatise on limnology. III: Limnological botany. J. Wiley, New York,

Iswaran, V., Sen, A. & Sajne Apte, 1973. *Azotobacter chroococcum* in the phyllosphere of Water Hyacinth (*Eichhornia crassipes* (Mart. Solms). Plant and Soil, 39:461–463.

Jacot-Guillarmod, A., 1979. Water weeds in southern Africa. Aquatic Botany, 6:377–391.

Jarvis, M.J.F., Mitchell, D.S. & Thornton, J.A., 1982. Aquatic macrophytes and *Eichhornia crassipes*. In: J.A. Thornton (ed.), Lake McIlwaine: the eutrophication and recovery of a tropical African man-made lake. Dr.W. Junk, The Hague: 137–144.

Kawamatu, S., 1965*a*. Electron microscope observations on blue-green algae in the leaf of *Azolla imbricata* Nakai. Cytologia, 30:75–79.

Kawamatu, S., 1965*b* Electron microscope observations on the leaf of *Azolla imbricata* Nakai. Cytologia, 30:80–87.

Landolt, E., 1957. Physiologische und ökologische Untersuchungen an Lemnaceen. Berichte der Schweizerischen Botanischen Gesellschaft, 67:271–410.

Landolt, E., 1980. Key to the determination of taxa within the family Lemnaceae. In: E. Landolt (ed.), Biosystematische Untersuchungen in der Familie der Wasserlinsen (Lemnaceae). Vol. 1. Veröffentlichungen des Geobotanischen Institutes der Eidgenossichen Technische Hochschule, Stiftung Rübel, Zürich.

Little, E.C.S., 1965*a*. Occurrence of *Salvinia auriculata* Aublet on the Congo River. Nature, London, 208:1111–1112.

Little, E.C.S., 1965*b*. The world wide distribution of water hyacinth. Hyacinth Control Journal, 4:30–32.

Lowe-McConnell, R.H. (ed.), 1966. Man-made lakes. Academic Press, London.

Loyal, D.S. & Grewal, R.K., 1966. Cytological study on sterility in *Salvinia auriculata* Aublet with a bearing on its reproductive mechanism. Cytologia, 31:330–338.

Loyal, D.S. & Grewal, R.K., 1967. Some observations on the morphology and anatomy of *Salvinia* with particular reference to *S. auriculata* Aubl. and *S. natans* All. Research Bulletin of the Punjab University, Science, 18:13–28.

ence to *S. auriculata* Aubl. and *S. natans* All. Research Bulletin of the Punjab University, Science, 18:13–28.

Luther, H., 1949. Vorschlag zur einer okologischen Grundeinteilung der Hydrophyten. Acta Botanica Fennica, 44:1–15.

Marshall, B.E. & Junor, F.J.R., 1981. The decline of *Salvinia molesta* on Lake Kariba. Hydrobiologia, 83:477–484.

Mitchell, D.S., 1969. The ecology of vascular hydrophytes on Lake Kariba. Hydrobiologia, 34:448–464.

Mitchell, D.S., 1970. Autecological studies of *Salvinia auriculata* Aubl. Ph.D. Thesis, University of London.

Mitchell, D.S., 1972. The Kariba Weed: *Salvinia molesta*. British Fern Gazette, 10:251–252.

Mitchell, D.S., 1974. Water: the nations life-blood. Tobacco Forum, 1:19–23. Rhodesian Farmers Publications, Salisbury.

Mitchell, D.S. & Tur, N.M., 1975. The rate of growth of *Salvinia molesta (S. auriculata* Auct.) in laboratory and natural conditions. Journal of Applied Ecology, 12:213–225.

Mitchell, D.S. & Rose, D.J.W., 1979. Factors affecting fluctuations in extent of *Salvinia molesta* on Lake Kariba. PANS 25:171–177.

Musil, C.F. & Breen, C.M., 1977. The influence of site and position in the plant community on the nutrient distribution in, and content of *Eichhornia crassipes* (Mart.) Solms. Hydrobiologia, 53:67–72.

Obeid, M., (ed.), 1975. Aquatic weeds in the Sudan with special reference to water hyacinth. National Council for Research, Khartoum.

Obeng, L.E. (ed.), 1969. Man-made lakes. The Accra symposium. Ghana Universities Press, Accra.

Penfound, W.T. & Earle, T.T., 1948. The biology of the Water Hyacinth. Ecological Monographs, 18:447–472.

Pettet, A., 1964. Seedlings of *Eichhornia crassipes*: a possible complication to control measures in the Sudan. Nature, London, 210:516–517.

Purchase, B.S., 1977. Nitrogen fixation associated with *Eichhornia crassipes*. Plant and Soil, 46:283–286.

Ramarokoto, M.R., 1965. Water hyacinth in Madagascar. Working paper, FAO conference Quelea and water hyacinth control, Douala.

Rejmankova, E., 1976. Germination of seeds of *Lemna gibba*. Folia Geobotanica et Phytotaxonomica, Praha, 11:261–267.

Rejmankova, E., 1982. The role of duckweeds (Lemnaceae) in small wetland water bodies of Czechoslovakia. In: B. Gopal, R.E. Turner, R.G. Wetzel & D.F. Whigham (eds.), Wetlands: Ecology and Management. National Institute of Ecology, Jaipur: 279–403.

Rhodesian Ministry of Information, Immigration and Tourism, 1973. Rhodesia's Dams. Government Printer, Salisbury, Rhodesia.

Robson, T.O., 1976. A review of the distribution of aquatic weeds in the tropics and sub-tropics. In: C.K. Varshney & J.

Rzoska (eds.), Aquatic Weeds in southeast Asia. Dr W. Junk, The Hague: 25–30.

Schroter, C. & Kirchner, O., 1896. Die Vegetation der Bodensees. Lindau Bodensee Forsch, 9, T1 I:1–222.

Scott, W.E., Ashton, P.J. & Steyn, D.J., 1979. Chemical control of the water hyacinth on Hartbeespoort Dam, Water Research Commission, Pretoria.

Sculthorpe, C.D., 1967. The biology of aquatic vascular plants. Edward Arnold, London.

Steinberg, R.A., 1946. Mineral requirements of *Lemna minor*. Plant Physiology, 21:42–48.

Toerien, D.F., Cary, P.R., Finlayson, C.M., Mitchell, D.S. & Weerts, P.G.J., 1983. Growth models for *Salvinia molesta*. Aquatic Botany, 16:173–179.

Van Himme, M., 1973. Le probleme des salvinia au Barrage d'Inga (Zaire). Bulletin des Sciences, Academie Royale des Sciences d'Outre-Mer, Brussels, 1973: 534–551.

White, H.L., 1936. The interaction of factors in the growth of *Lemna*. VII. The effect of potassium on growth and multiplication. Annals of Botany, 50:175–196.

White, H.L., 1937. The interaction of factors in the growth of *Lemna*. XI. Nitrogen and light intensity in relation to growth and assimilation. XII. Nitrogen and light intensity in relation to root length. Annals of Botany (n.s.), 1:623–654.

White, H.L., 1940. The interaction of factors in the growth of *Lemna*. XV. On a rhythmic growth cycle of *Lemna* colonies associated with transference to a potassium free nutrient solution. Annals of Botany, 4:495–504.

Wild, H., 1956. Water hyacinth *(Eichhornia crassipes* (Mart.) Solms.) in Southern Rhodesia. Proceedings of second symposium on African hydrobiology and inland fisheries: 127–129.

Wild, H., 1961. Harmful aquatic plants in Africa and Madagascar. Kirkia, 2:1–66.

Author's address:
CSIRO Centre for Irrigation Research
Griffith NSW 2680
Australia

CHAPTER 5

The structure and functioning of African euhydrophyte communities

The floating-leaved and submerged vegetation

PATRICK DENNY

Detailed contents

The littoral vegetation consists of that part of the hydrosere which is permanently wet. It extends from the shore into the open water down to the limit of the euphotic zone (Chapter 1). The hydrosere is normally represented in time and space by an observed zonation of vegetation from emergent plants in the shallow water, to floating-leaved rooted plants and then totally submerged plants in the deeper water (Chapter 2). In Africa, and particularly in the tropics, the emergent macrophytes often dominate the vegetation, sometimes to the exclusion of other communities, giving rise to extensive swamps such as the Sudd in the Sudan and the Okavango Swamps in Botswana. Because of the importance of swamps in African freshwater ecosystems they will be considered separately (Chapter 6). This chapter considers the dynamics of the *euhydrophyte* communities, *i.e.* the floating-leaved and the totally submerged vegetation. The surface floating communities are discussed in Chapters 4 and 7.

5.1 Factors affecting distribution and zonation within a waterbody

Climate and environmental factors such as salinity, which may determine the geographical distribution of euhydrophytes, have been discussed in Chapter 2 so that the distribution within particular waterbodies can now be considered. There is a variety of factors which govern the pattern of vegetation, not the least of which is the availability of propagules (viable seeds and vegetative parts), their chance of becoming established and their ability to survive competition (Walker, 1970; Van der Valk, 1981). Water depth *per se* limits the zone of colonisation of floating-leaved plants as they are rooted in the substratum, and their floating-leaves must reach the water surface. Some plants, such as the floating-leaved potamogetons, extend their depth toler-

Patrick Denny (ed.), The ecology and management of African wetland vegetation.
 ISBN 90 6193 509 1

Plate 5.1. The penetration of light through an open canopy of papyrus. The climbers are *Mikania* and *Ipomoea*. Ground-cover plants must be shade-tolerant. (Photo K. Thompson.)

ance by shoot growth and elongation, whereas other plants elongate their leaf petioles. In Lake Bunyonyi, for example, petioles of *Nymphaea caerulea* exceed four metres in length (Denny, 1973). Other physical factors (wind exposure, light intensity, type of substrate, *etc.*) will determine the conditions for growth. The littoral region contains a variety of physical and physico-chemical habitats depending upon water depth and wind exposure. These are outlined below.

5.1.1 *Light*

The light climate can be divided into two parts: (a) the light reaching the water surface, and (b) the underwater light. Whilst floating-leaved plants will only be affected by the former, submerged plants will be affected by both. The general properties of light as an ecological factor are covered in Evans, Bainbridge and Rackham (1976) and need not be repeated here. For the particular characteristics of underwater light good accounts are provided by Kirk (1976; 1977; 1979) and Spence (1976; 1981).

Lake water is predominantly a shade habitat (Spence, 1981). Light attenuation in a well mixed water column follows Beer's Law and is a logarithmic function of depth. Rates of attenuation depend upon the solutes and the inorganic (silt) and organic (including phytoplankton) suspensions in the water, and absorption and scatter by these determines the underwater spectrum. In very clear (oligotrophic and calcareous) waters, red light has a higher attenuation than blue light, but in most lakes, where the Gelbstoff (dissolved yellow substances of organic origin) absorb significant amounts of the blue and ultra-violet wavelengths, the opposite is true. Kirk (1976) believes that the photosynthesis of macrophytes is markedly affected by the availability of blue light, which in turn, is dependent upon the concentration of Gelbstoff.

The depth to which the light intensity is still sufficient for the photosynthesis of phytoplankton is known as the euphotic zone, Z_{eu}. At its lower limit, light intensity is approximately equal to 1% of the subsurface photosynthetically active radiation (PAR). Talling (1965*a*; 1971) showed that in a wide variety of African lakes the depth of the euphotic zone may be determined most easily from the extinction coefficient (= attenuation coefficient) of the least attenuated wavelenghts of light, E_{min} (usually green light), according to the relationship: $Z_{eu} = 3.7/E_{min}$. In the open water of clear lakes such as Lake Tanganyika Z_{eu} may be as much as 20–30 m, but normally in African lakes it is from 5–10 m (*eg*. Lake Bunyonyi, $Z_{eu} = 7–8$ m, Denny, 1972*a*), and in algal-rich lakes, less than one metre. In Lake George, for example, $Z_{eu} = 0.7$ m (Ganf and Viner, 1973) and in the Ethiopian soda lakes, Kilotes and Aranguadi, $Z_{eu} = 0.15–0.49$ m (Talling *et al.*, 1973).

The depth of colonisation of submerged macrophytes, Z_c, is always less than the euphotic zone, and in certain circumstances, the empirical formula: $Z_c = 1.4/E_{min}$, can be applied (Spence, 1976). The precise value of Z_c will depend upon the spectral distribution of the light at Z_c and upon the plants under consideration.

The underwater light may determine the optimal depth of a particular species in a body of water, influence the position of plants in a zonation and affect the germination of seeds (Spence *et al.*, 1971). Sun – tolerant species occur near or at the surface of the water; shade species inhabit deeper water or exist in the shade of other species, and others adapt to local conditions (Spence and Chrystal, 1970*a*, 1970*b*). *Ceratophyllum demersum*, for example, is tolerant of low light and can be found at the limits of the euphotic zone (Denny, 1973), and *Lemna trisulca* grows below other Lemnaceae. *Potamogeton pectinatus*, on the other hand, has a high light requirement and is generally found in shallow water.

5.1.2 *Wave action and sediments*

The littoral can be divided conveniently into four zones: the splash zone, the wave-washed zone, the erosion and deposition terrace, and the profundal zone (Fig. 5.1). The splash zone is self-explanatory. The wave-washed zone is where surface travelling waves break onto the shore and is the region of greatest agitation. Coarse and fine particles are washed from between the stones and rocks, and a stony shoreline is often seen. A terrace is formed by the wave action to lakeward of the wave-washed zone. The sediments here are directly affected by surface waves to a water depth of half the wavelength (the wave-mixed depth), (Smith, 1979). Three processes take place along the terrace: (a) erosion, especially in the shallow water where strong wave action brings all but the coarsest particles into suspension, (b) sorting and elutriation, where turbulent action sorts the particles according to their physical characteristics, especially size and density, and (c) sedimentation of the particles in the deeper, less turbulent water. Particle deposition is graded according to the degree of turbulence, the larger, more dense particles being deposited before the finer ones. Below the wave-mixed

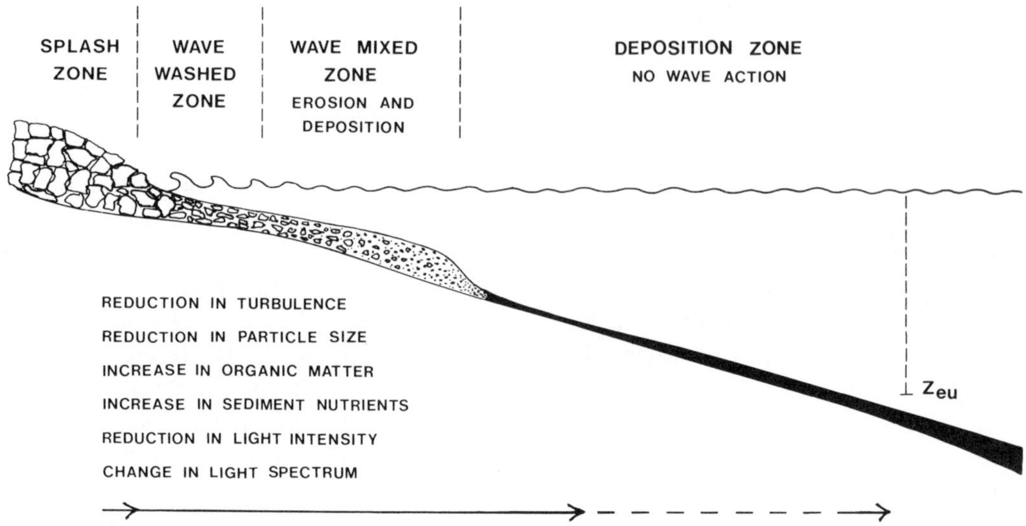

Fig. 5.1. A schematic diagram showing the influence of waves on the sediments of wind-exposed lake shores. Z_{eu} is the depth of the euphotic zone.

zone the coarser particles will have been deposited and the finest particles may settle out.

The net result of particle sorting by wave action and turbulence is a grading of the littoral sediments with depth. Rocks and stones occur in the wave-washed zone; gravel and coarse and fine sands grading towards silts are found along the terrace towards deeper water; and the fine silts and clays are deposited in the profundal zone. Organic matter does not have such a high density as the average inorganic material and a stable bed of organic particles always occurs in deeper water than a bed of similarly sized inorganic particles (Spence, 1982). Thus, the deposition of organic matter is associated with calm waters, and the levels of organic matter in the sediment tend to rise with increased water depth (Pearsall, 1917; 1920).

5.1.3 *Nutrient supply*

There are two potential sources of nutrient supply to euhydrophytes: the water and the sediment.

In a well mixed body of water solutes are uniformly distributed, but in stratified lakes large differences may occur between the epilimnetic and hypolimnetic water (Talling, 1969). The zone of colonisation in the littoral region is normally within the epilimnion, and nutrient deficiencies in the water are common. This may be aggravated by the vegetation itself, which reduces turbulence and diminishes the rate of exchange of water between the bed and the open lake. Where there is active photosynthesis of submerged vegetation and removal of free dissolved carbon dioxide, the pH value of the water rises, the carbon dioxide equilibrium shifts towards bicarbonate and carbonate forms, and carbon dioxide for photosynthesis may become chronically short, both on a diel and seasonal basis. This is enhanced by algal photosynthesis (Talling, 1976). Micro-stratification of water within the bed can lead to deoxygenation of the sediment surface and release of nutrients into the water above. Conditions near inflows are different. Water flowing over and through the vegetation will provide a continuously replaceable potential source of nutrients, but if the inflow has passed through a floating swamp, some ions may have been removed (Thompson, 1976; also *see* Fig. 6.5, p. 171).

The nutrients in the sediments are generally correlated with particle size and organic matter content. Sediments which are predominantly composed of coarse particles are nutrient-poor. The finer silts and clays, and organic particles, have a high cation exchange capacity. Also, the hydrous oxides of iron bind nutrient ions, especially anions such as phosphate. Thus, in the littoral region, the

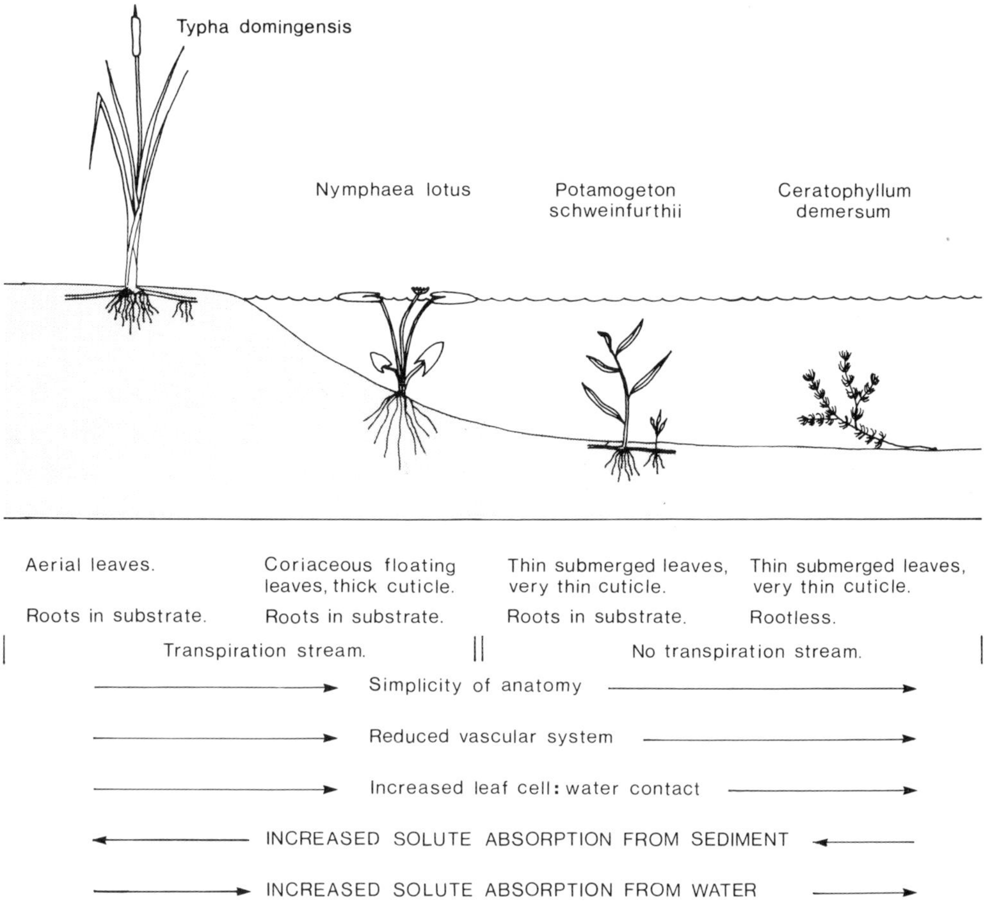

Fig. 5.2. A diagramatic representation of the main aquatic plant growth-forms showing trends in anatomy and morphology which affect the mode of solute absorption by the plants.

progression from coarse to fine particles, associated with reduction in turbulence (increasing depth), is accompanied by an enrichment of the sediment.

Experiments in Uganda (Denny, 1972*b*) have indicated that different euhydrophyte species have different abilities to absorb solutes from the substrate *via* the roots and from the water *via* the shoots, and observation which has been noted also in laboratory work (Welsh and Denny, 1979*a*; 1979*b*). Vigorously rooting, floating-leaved plants with thick cuticular membranes, well developed vascular tissues, and a transpiration stream are mainly supported by absorption of nutrients through the roots. Delicate submerged plants with a very reduced cuticle and vascular system, and frequently with less well-developed roots, can rapidly absorb solutes from the water through their shoots as well as utilise their roots; rootless plants, of course, absorb through their shoots alone (Fig. 5.2; Plate 5.2). Accordingly, the relative amount of solute absorption from sediment and from water depends upon the physiological, morphological and anatomical characteristics of the plants involved (Denny, 1980).

5.1.4 *Exposure*

Shores exposed to the wind are vulnerable to the full force of waves which may have developed over the entire fetch of the lake. Apart from their force in the splash and wave-washed zones (Smith, 1975), the magnitude of their wavelengths extends the depth of the wave-mixed zone so the erosion

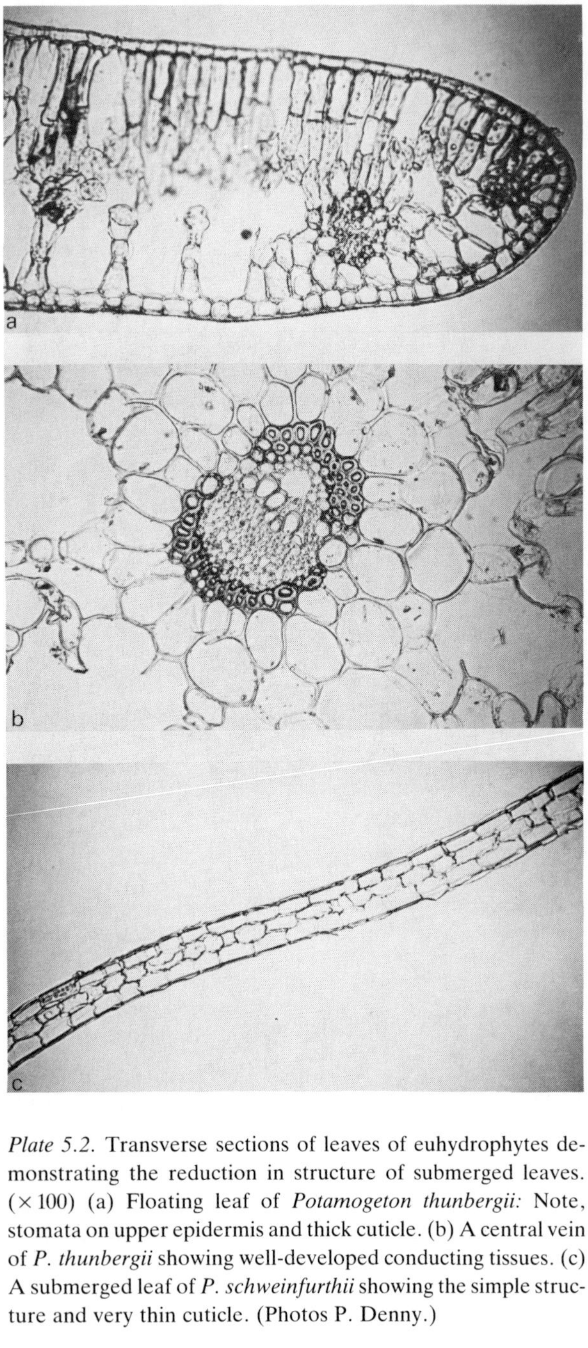

Plate 5.2. Transverse sections of leaves of euhydrophytes demonstrating the reduction in structure of submerged leaves. (×100) (a) Floating leaf of *Potamogeton thunbergii:* Note, stomata on upper epidermis and thick cuticle. (b) A central vein of *P. thunbergii* showing well-developed conducting tissues. (c) A submerged leaf of *P. schweinfurthii* showing the simple structure and very thin cuticle. (Photos P. Denny.)

terrace occupies a considerable part, if not all, of the euphotic zone. Lakes Tanganyika and Malawi, for example, are characterised by long stretches of rocky shore and broad stony and sandy beaches (Beadle, 1981). The wave-washed and terraced zones may be increased further by the change in water level due to wind-driven seiches and seasonal fluctuations. The effects of draw-down in man-made lakes are similar.

In sheltered bays and backwaters where turbulence is minimal the opposite trend is apparent, and the shores are largely regions of deposition rather than erosion. Soft, organic muds predominate and the accumulation of sediments, encouraged by the establishment of vegetation, further reduces turbulence. These are prime conditions for the development of swamp vegetation.

Water turbulence resulting from wind-induced currents in lakes, and flow in rivers and streams, has the same erosion, elutriation and deposition effects on sediments as the turbulence from surface travelling waves. Shores of erosion and deposition are found in lakes according to current velocity, and, at the bends of rivers, the inside bend usually provides a deposition zone for the establishment of vegetation whilst the opposite bank is being eroded.

5.1.5 *Discussion*

Which factor is the most important? Pearsall (1917; 1920) thought that the type of substrate determined the zonation of euhydrophyte vegetation, but a review by Spence (1982) shows it is not that simple. Although Spence's examples are taken from the temperate environment, his conclusions are pertinent to tropical systems. He argues that, of the vertical environmental gradients which change with depth of water – *i.e.* light, temperature, density and pressure – light is the single most important factor influencing the zonation of submerged plants. Turbulent motion is partially depth-dependent, but it is also dependent upon wind-exposure and water currents. The resolution of these forces determines the horizontal variations in sediment type and the extent of direct wave action. Turbulence and wave action account for the characteristic vegetation of sheltered and exposed shores. Spence concludes that, when the vegetation is mainly within the wave-mixed zone, the wave action/sediment-type interaction exerts the prime control on pattern and zonation. When the vegetation is mainly below the wave-mixed zone, the

Plate 5.3. Dense stands of papyrus on shallow levées in the permanent swamp of the Okavango Delta. This channel is deep, fast-flowing and heavily shaded: quite unsuitable for colonisation by euhydrophytes. (Photo K. Thompson.)

quality and quantity of light have precedence. Spence does not consider fully the implications of biological pressures such as grazing and interspecific competition. In the following pages it will be shown that interspecific competition, especially, can be a major regulator.

In African wetlands, three types of euhydrophyte communities may be recognized, *viz.* swamp communities, wind-exposed communities and open water communities. These divisions are purely artificial as each may grade into the others, but it is convenient to consider them separately.

5.1.5.1. *The swamp community*

In the next chapter Howard-Williams and Gaudet will describe different types of swamps drawing particular attention to the differences in the dynamics of floating swamps and rooted swamps. In both, the vegetation may be dense but there are often open pools and channels which provide a refuge for euhydrophytes. Propagules have to be distributed in the pools and perhaps, in the more isolated ones, birds are unwitting agents of dispersal. (The African wetlands attract many migratory species from Europe (Moreau, 1972)).

Pools and channels are likely to be protected from the rigours of wind exposure and surface wave action, but the plants may be under considerable competition for light and nutrients from emergent and surface-floating species. In the Sudd, for example, the dominant emergents *Typha domingensis, Phragmites karka* and *Cyperus papyrus*, form stands of vegetation up to 6 m tall, and the heads of papyrus culms hang over the channels

Plate 5.4. The native African species *Salvinia nymphellula* in its typical habitat within the *Echinochloa scabra* community fringing sheltered shores of an island on the Zaire River near Kisangani. There is little opportunity for submerged plant development in this community. (Photo K. Thompson.)

(Denny, 1984). Patches of 'open water' are largely covered by surface-floating plants, especially *Eichhornia crassipes*, so the light reaching the water surface may be only a fraction of the daylight intensity. A floating, rhizomatous mat of papyrus can spread from the rooted swamp over the water surface and suppress the growth of submerged plants below. The interface between the rhizomatous mat and the open water may appear suitable for euhydrophyte development, but the niche is often occupied by aquatic grasses such as *Vossia cuspidata*. The stagnant water within a floating

Plate 5.5. Lake Kisale, Kamulondo Depression, Zaire, is very shallow and still filling with sediment. This *Trapa Nymphaea* community is ideally suited to these conditions and excludes most submerged plants by shading. It represents a stage in the plant succession to papyrus swamp. (Photo K. Thompson.)

swamp will be substantially depleted of nutrients by the emergents (Chapter 6).

The competition is formidable; but, in shallow pools within the swamp, *Nymphaea lotus* and *Trapa natans*, together with submerged *Najas pectinata* and *Utricularia* spp. can be found. *Nymphaea* and *Trapa* occur in both mesotrophic and oligotrophic waters, their nutrient requirements being obtained from the organic detritus into which they are rooted. *Utricularia* can supplement its nutrients by the trapping and digestion of animals. The channels of the main rivers through the swamp provide a different habitat. The water may be quite fast flowing with high silt burdens resulting in an absence of beds of submerged plants, but at the edges of the swamp fringing the channels *Ceratophyllum demersum* and *Utricularia* spp. occur. These plants are typical of such an environment (see Chapter 2), can tolerate low light intensities, and thrive in slow flowing water.

At the outer edges of swamps towards the open lakewater, the euhydrophyte vegetation grades into an open water community. This can be recognised in Lake No (a lake within the Sudd region) where a zonation is discernible. *Nymphaea lotus* and *Trapa natans*, with an understorey of *Ceratophyllum demersum* and *Utricularia* spp., occur at the swamp edges, and beds of *Potamogeton* × *bunyonyiensis, P. schweinfurthii, P. pectinatus* and *Najas pectinata* stretch out into deeper water (Denny, 1984). However, within the swamp the pattern of euhydrophyte vegetation is determined mainly by the stands of emergents.

5.1.5.2 *The wind-exposed community*

Wind-exposed communities include littoral areas within the euphotic zone where the vegetation lies almost entirely within the wave-mixed zone. Typically, they occur on the exposed shores of lakes where a long fetch precipitates strong wave action, *eg*. the Rift Valley Great Lakes. The emergent and floating-leaved plant zones are absent, and, in the

Plate 5.6. The inflow region of Lake Bunyonyi showing the surrounding hills and the swamp vegetation. Fish traps are set around the swamp fringe and provide a thriving local fisheries. (Photo P. Denny.)

deeper water, sparsely distributed open communities of submerged plants predominate. Biomass and species diversity are poor, the most common plants being species of *Najas* and *Vallisneria* (Chapter 2; section 2.4).

This habitat has two obvious factors limiting the vegetation: direct wave action and wave-mixed interactions. The magnitude and force of the waves inhibits the colonisation of plants in the shallow water. (In temperate climates the zone may be occupied by small, deep-rooting, rosette plants such as *Littorella uniflora*, but an equivalent growth form seems to be absent from tropical Africa). The sediments of the deeper water are still susceptible to the erosion and particle sorting processes described in section 5.1.2, and nutrient-poor sandy silts predominate. However, the turbulent water above the sediment is well aerated and has a good carbon dioxide supply; in addition, turbulence over the submerged leaves ensures a continuously replaceable nutrient supply to the plants. Both *Vallisneria* spp. and *Najas* spp. have growth-forms suited to turbulent conditions (narrow or linear leaves and well-developed rooting systems for anchorage) and are well adapted for nutrient absorption through their leaves (Denny, 1980). Therefore, they need not be limited by the nutrient-poor substrate. Interspecific competition is minimal, the plants' main problems coming from the indirect effects of wind-exposure.

5.1.5.3 *The open water community*

The open water community occupies the littoral of lakes and rivers where swamps have not encroached and where some shelter is afforded. Contrasting with the wind-exposed community, the euhydrophyte vegetation is likely to be dense with a high biomass, and often has a rich flora. The

Plate 5.7. Cladium and *Commelina* at the water's edge of Lake Bunyonyi. Note how the flowering spikes of *Cladium* lean over the water and produce vegetative propagules in the axes of their nodes. (Photo. P. Denny.)

zonation of vegetation in the high altitude lake in Uganda, Lake Bunyonyi, is an example (Denny, 1971; 1972*a*; 1973). Lake Bunyonyi is a relatively small, steep-sided oligomictic lake with a nutrient-poor epilimnion and moderate flora. The shores are fringed with *Phragmites australis, Cladium jamaicense* or *Typha latifolia,* whilst *Nymphaea caerulea, N. lotus* and *Potamogeton* × *bunyonyiensis* provide a floating-leaved belt towards the open water. In the shallow water (0 to 2.5 m) the submerged vegetation is dominated by stands of *Potamogeton schweinfurthii, P. pectinatus* and *Hydrilla verticillata,* but in deeper water (2.5 to 5.5 m) their dominance is replaced by *Chara* sp. and *Ceratophyllum demersum*. Below this level, *Ceratophyllum* with some *Hydrilla* is found down to 7 or 8 m. *Utricularia vulgaris* (?) occurs amongst the beds of water-lilies, and clumps of *Lagarosiphon* sp. grow with the *Hydrilla*. The net result is a dense bed of vegetation extending from the reed swamp to the limit of colonisation (Fig. 2.4, Plate 5.6).

What determines the zonation of euhydrophytes? The wave-mixed zone is insignificant, and colonisation occurs predominantly in the deposition zone. According to Spence's hypothesis (Spence, 1982), light is the prime factor. Measurements on excised leaves of *Potamogeton* have shown that whilst rates of net photosynthesis of the floating leaves of *Potamogeton thunbergii* and semi-floating leaves of *Potamogeton* × *bunyonyiensis* rise with increasing light intensities (and higher temperatures), the leaves of *P. schweinfurthii* become light-saturated and damaged in bright light (Fig. 5.3) (Denny, unpublished). The light responses of these potamogetons correspond with their position in the zonation; *P. schweinfurthii* is more efficient at lower light intensities, and occurs in deeper water (Fig. 5.4). Spence and Chrystal (1970*b*) have made similar observations on shallow and deep water potamogetons from temperate climates. However, the depth ranges of all the submerged plants in Lake Bunyonyi sub-

ter; submerged macrophytes ~400–700 g m^{-2} dry organic matter and emergents are said to be from ~3000–10000 g m^{-2} dry organic matter (Westlake, 1975, Table 8.1) although maximal values of 5000 g m^{-2} are more likely. Differences are largely associated with an increase in the ratio of the support tissues compared with photosynthetic tissues.

5.2.1 *Biomass*

Estimates of biomass depend upon the methods used for obtaining data and the plants concerned. Most data exclude the root and rhizome portions and refer to shoot biomass. Measurements vary according to attributes such as ash content, and plants from calcareous regions (especially *Chara* species) require correction factors for calcium carbonate deposits. Biomass studies on submerged plants (normally, shoot biomass) suggest that a maximal dry weight value of ~300–500 g m^{-2} is typical for the temperate regions (Westlake, 1975). Specific examples can be obtained by reference to Westlake (1980, Table 5.2) but no data are included from Africa.

In sub-tropical South Africa, maximum values for *Potamogeton crispus* in the Pongolo River floodplain were estimated to be ~40–70 g m^{-2} dry weight (Rogers and Breen, 1980; 1982). Further to the south, in Lake Swartvlei, Howard-Williams (1978) estimated the maximum biomass of *Chara* to be ~140g m^{-2} (ash-free dry wt) and the value for *P.pectinatus* to be exceptionally high – in the order of ~2000g m^{-2} dry weight. In the nearby Wilderness Lakes system, values for *Chara* were some of the highest recorded for the Characeae – 850 g m^{-2} (dry weight) (~700 g ash-free dry weight) (Howard-Williams, 1980). In the same system, separate estimates for the shoots and roots + rhizomes of *Potamogeton pectinatus*, which gave values of ~900 and ~300 g m^{-2} respectively, provide some of the few data for below-ground biomasses. Dense beds of *Ceratophyllum* in Lake Sibaya (another southern African coastal lake) had maximum values exceeding 800 g m^{-2}, but values for *Myriophyllum spicatum* were around ~40–50 g m^{-2}, and for *P. schweinfurthii* ~60 g m^{-2} (Howard-Williams, 1979). The floating-leaved water-lily,

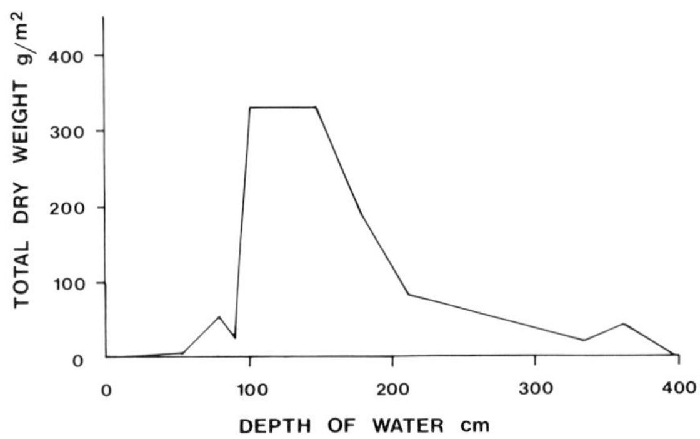

Fig. 5.6. Biomass/depth profile of submerged vegetation within the zone of colonisation of Loch Uanagan, Scotland. (Redrawn from Spence, 1982; Fig. 21.)

Nymphaea capensis was generally around 50 g m^{-2} although 128 g m^{-2} was recorded.

Correlations between biomass and environmental variables of Lake Sibaya showed that water depth and biomass were highly correlated; and plants such as *Nymphaea* spp. and *Myriophyllum* were not tolerant of exposure. The conflicting forces of wind-exposure (turbulence) and depth (light intensity), therefore, seem to govern the distribution of species and biomass values. Howard-Williams and Liptrot (1980) came to a similar conclusion for the *Potamogeton pectinatus*/*Chara* vegetation of Lake Swartvlei. In shallow water the chara only grew in more exposed sites behind the protection of *P.pectinatus* which reduced the effects of wave-action. Harbott (1982) noted that the extensive beds of *Potamogeton pectinatus* in the littoral regions of Lake Turkana were able to persist even though substantial fluctuations in water level occurred. This was achieved by rapid regeneration and growth of the plants at their optimal water depth.

The effect of depth on biomass is clearly demonstrated by Spence (1982) for Loch Uanagan in Scotland. The biomass/depth profile (Fig. 5.6) has a shape reminiscent of the depth profile for gross photosynthesis of a vertical column of phytoplankton in a lake where, near the surface, the high light intensities suppress photosynthesis. This may also occur with macrophytes, but Spence suggests that their decline in biomass in shallow waters is not normally due to light; it is caused by increased turbulence and wind-induced direct wave action (exposure). The biomass increases with depth to a maximum value (~400g m^{-2} dry weight, in this particular example) and then declines exponentially to Z_c, the decline being consistent with the reduction in irradiation (Spence, 1982).

5.2.2 *Photosynthetic rates*

Rates of net photosynthesis in a large variety of submerged plants (mainly from the temperate regions) usually lie between 2 and 10 mg C g^{-1} h^{-1} (dry weight) and a value of 4 is considered typical (Westlake, 1975). Specific examples can be obtained from Westlake (1980, Table 5.6), but there are no data from Africa. Rogers and Breen (1980) enclosed shoots of *Potamogeton crispus* from the Pongolo floodplain in 'light' and 'dark' bottles, and re-submerged them to their natural depth in the pans. Maximum photosynthesis rates were found to be ~7 mg C g^{-1} h^{-1} (~18 mg O_2). Similar experiments to estimate gross photosynthetic rates of *Chara globularis* in Lake Swartvlei gave values of ~3 mg C g^{-1} h^{-1} (~8 mg O_2) (Howard-Williams, 1978). 'Light' and 'dark' perspex tubes placed over *Potamogeton pectinatus* plants (plants + *Aufwuchs* + plankton) in Lake Turkana provided an estimate of gross photosynthesis of 2.6 mg C g^{-1} (~7 mg O_2) (Harbott, 1982).

The photosynthetic rates of excised, submerged leaves of *Potamogeton schweinfurthii*, semi-floating leaves of *P.* × *bunyonyiensis* (both from Lake Bunyonyi material), and floating leaves of *P.thunbergii*, were estimated in a Warburg apparatus and values of ~9–10 mg C g^{-1} h^{-1} (~24–28 mg O_2 for *P.schweinfurthii*, ~11–13 mg C g^{-1} h^{-1} (~30–36 mg O_2 for *P.* × *bunyonyiensis* and ~7–11 mg C g^{-1} h^{-1} (20–30 mg O_2) for *P.thunbergii* were obtained (Denny, unpublished). Rates were temperature sensitive (Fig. 5.3) and the high readings were associated with increased temperatures. This suggests that the plants do not have a high level of photorespiration. Evidence is accumulating that submerged plants can have unusual metabolic pathways which bear similarities to the C_4 pathway. Photorespiration seems to occur, but at levels very much lower than in terrestrial C_3 plants, and lower than rates of dark respiration (LLoyd, Canvin and Bristow, 1977). *P.schweinfurthii* has a photoactive uptake mechanism for bicarbonate ions through the plasma membrane, and a substantial bicarbonate pool exists within the vacuole (Denny and Weeks, 1970). This, and some preliminary metabolic studies, suggests it is not a normal C_3 plant. The data from the Ugandan potamogeton experiments are not directly comparable with the data using light and dark bottles as the techniques used were very different. However, all the results indicate that tropical species can have high rates of net photosynthesis.

There are few data expressed in terms of chlorophyll *a* content, but values quoted by Westlake (1975) are normally between 0.2 and 0.9 mg C mg^{-1} Chl *a* h^{-1}. These are in approximate agreement with the unit dry weight values assuming a mean chlorophyll content of leaves around 8 mg Chl *a* g^{-1} dry weight.

5.2.3 *Growth rates and annual production*

Photosynthetic rates may be high, but the photosynthetic tissues represent only a portion of the whole plant although all the plant will be respiring. Thus, photosynthetic capacity will not necessarily reflect the plant growth or crop growth rates.

Estimates of annual production can be made by recording growth at regular intervals throughout the year, or, more normally, by measuring crop growth rates (as changes in biomass during the growing season) and converting the data to an annual figure. It is not always possible to sample the biomass regularly through the growing season, but estimates of annual production can be made from a maximal seasonal biomass sample if the Production/Biomass (P/B) ratio of the species is known for the area.

Estimates of crop growth rates are sparse. Dry weight values for temperate submerged plants, which have usually been obtained by measured changes in biomass during spring and summer, range from ~2–10 g m^{-2} d^{-1} (Westlake, 1975). Data from the Pongolo River floodplains suggest that, in the (short) growing season, *Potamogeton crispus* has a maximum crop growth rate of <1 g

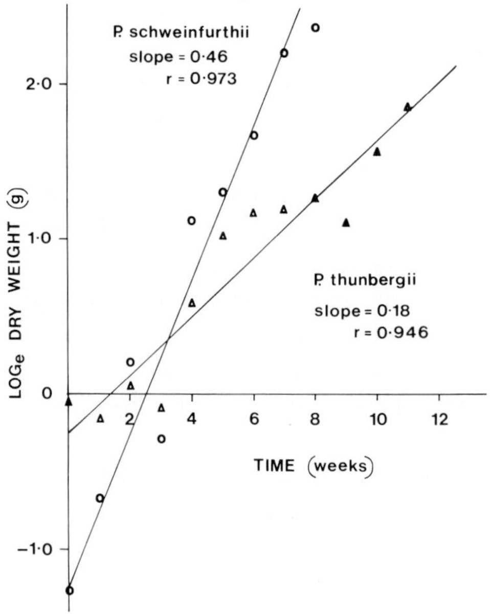

Fig. 5.7. Log plot of growth (measured as changes in [dry] weight with time) of two potamogetons in Uganda during their exponential growth phase. Uniform, young plants were planted in an experimental pond and were cropped at weekly intervals over 15 weeks. See text for details. (Denny, unpublished.)

m^{-2} d^{-1} (Rogers and Breen, 1980). This provides an estimated annual production of 40–50 g m^{-2} y^{-1}. The growth of *Potamogeton pectinatus* in Lake Swartvlei is also seasonal, having rates of ~16 g m^{-2} d^{-1} dry weight (Howard-Williams, 1978) during the period of maximum growth. This provides an annual production estimated of some 2500 g m^{-2} y^{-1} even though the winter production is negligible. By comparison, the annual production of *Chara globularis* is one tenth of this; around 260 g m^{-2} y^{-1}. In the Wilderness Lakes, mean production estimates for *P.pectinatus* and *Chara* (obtained from biomass values using the Lake Swartvlei P/B ratios of 1.22/1 and 2.0/1 respectively) are 415 and 231 g m^{-2} y^{-1} (Howard-Williams, 1980).

Experimental data for *Potamogeton schweinfurthii* and *P.thunbergii* from Uganda provide some data from tropical environments (Denny, unpublished). Individually weighed plants were grown in outdoor ponds at Makerere University, Kampala (see Denny, 1972*b* for details of local conditions). Samples were cropped at weekly intervals for 15 weeks, and the total biomass (shoots + rhizomes + roots) of each plant was measured. The increase in weight (final weight – original weight), expressed as a dry weight value, for a particular time period provides a measure of growth. After the first three weeks, which were taken as the period of adjustment, there was an approximately exponential increase in growth (Fig. 5.7). For *P.thunbergii* this continued for the full experimental period, but for *P.schweinfurthii* the stationary phase was reached after about 10 weeks. The Relative Growth Rate during the exponential phase of growth was ~0.18 g g^{-1} week $^{-1}$ for *P.thunbergii* and ~0.46 g g^{-1} week $^{-1}$ for *P.schweinfurthii*. Mean doubling times were ~4.0 and 1.5 weeks, respectively. The conditions for growth were good, and if it is assumed that the climate is relatively uniform over the year, then maximal annual growth rates (in the absence of plant competition) would be in the order of ~10 g g^{-1} y^{-1} for *P.thunbergii* and ~20 g g^{-1} y^{-1} for *P.schweinfurthii*.

If a biomass of 60 g m^{-2} dry weight for *P. schweinfurthii* in Lake Sibaya (Howard-Williams, 1979) is a normal density for this species, then a projected estimate of annual production (using the annual growth rate estimate) for Ugandan beds would be ~1200 g m^{-2} y^{-1}. This estimate is probably low as *P.schweinfurthii* is a tropical and equatorial species (Chapter 2) and in artificial ponds in Uganda, biomass values of 200 g m^{-2}, or more, were normal. This would give a projected estimate of annual production of 4000 g m^{-2} y^{-1} dry weight and is consistent with a doubling time of ~2 weeks, and a 12-month growing season. The carbon content of leaf tissues is 42–49% (Denny, unpublished) and thus, in terms of carbon fixed, the production would be ~1800 g C m^{-2} y^{-1}. The high annual production of *P.pectinatus* in Lake Swartvlei was achieved in only a seven month growing season; if it had a twelve month season, the rate could have more than equalled that of *P.schweinfurthii*. Evidence from Lake Turkana also suggests that *P.pectinatus* may have a high annual production (Harbott, 1982). Photosynthesis experiments on whole plants in the lake indicated a gross production value of up to 4.6 g C m^{-2} d^{-1}, and with a twelve month growing season this would

Table 5.1. Annual primary production estimates for a variety of aquatic plants. Unless otherwise stated, estimates are for net production. The variation in estimates is considerable and the data should be used only as an indication of production rates.

Plants	Annual production estimates		Climate/location	References
	$g\ m^{-2}\ y^{-1}$ C	$kg\ m^{-2}\ y^{-1}$ dry weight		
Algae				
Phytoplankton (general) (oligotrophic lakes)	<25	–	–	Westlake (1980, p. 246)
Phytoplankton (general) (meso/eutrophic lakes)	>100	–	–	Westlake (1980, p. 246)
Phytoplankton	730–1300	–	Sub-tropical Lake Chilwa	Westlake (1980, Table 5.19)
Phytoplankton	950 (gross)	–	Tropical Lake Victoria	Talling (1965b)
Marine grasses				
Thalassia	500–1500	–	Indian Ocean	Mann & Chapman (1975, Fig. 9.1)
Seaweeds				
Laminaria and Macrocystis	1000–2000	–	Temperate	Mann & Chapman (1975, Fig. 9.1)
Free-floating freshwater plants				
Eichhornia crassipes		1.5–4.4	Subtropical	Mitchell (1974, p. 45)
Emergent swamp plants				
Typha and Phragmites	~2000–2500[a]	4.5–7[b]	Temperate	Westlake (1975, Table 8.1)
Cyperus papyrus	~5000[c]	~12.5	Tropical	Thompson *et al.* (1979)
Submerged plants				
Submerged plants (general)	250–400	0.6–0.9[b]	Temperate	Westlake (1975, Table 8.1)
Potamogeton pectinatus	~1100[d]	2.5	South temperate Lake Swartvlei	Howard-Williams (1978)
Potamogeton crispus	~30[d]	0.05	Subtropical Pongolo floodplains	Rogers & Breen (1980)
Potamogeton schweinfurthii	~500–1800	1.2–4	Tropical Uganda	Denny (unpublished)

[a] Assuming 47 mg C (g organic weight)$^{-1}$
[b] Assuming 15% Ash
[c] Assuming 40 mg C (g dry weight)$^{-1}$
[d] Assuming 45 mg C (g dry weight)$^{-1}$

provide ~1700 g C $m^{-2}\ y^{-1}$ (≡ 3700 g dry weight assuming a carbon content of 45%). Net values would be somewhat less but of the same order of magnitude.

There may not be large differences between mean crop growth rates of tropical and temperate waterplants during their active growing seasons but, in tropical and sub-tropical regions, where rapid growth can proceed for a large part of – if not all – the year, the annual production values are significantly higher. This seasonal effect is also shown by Westlake (1975, Table 8.1) who suggested an annual production for submerged plants of ~2300 g in the tropics and ~600–900 g $m^{-2}\ y^{-1}$ dry weight (assuming ~ 15% ash content) in temperate regions.

Freshwater submerged macrophytes tend to have the reputation of being fairly low on the primary production scale (Talling, 1975; Westlake, 1975). Table 5.1 shows that this is not always the

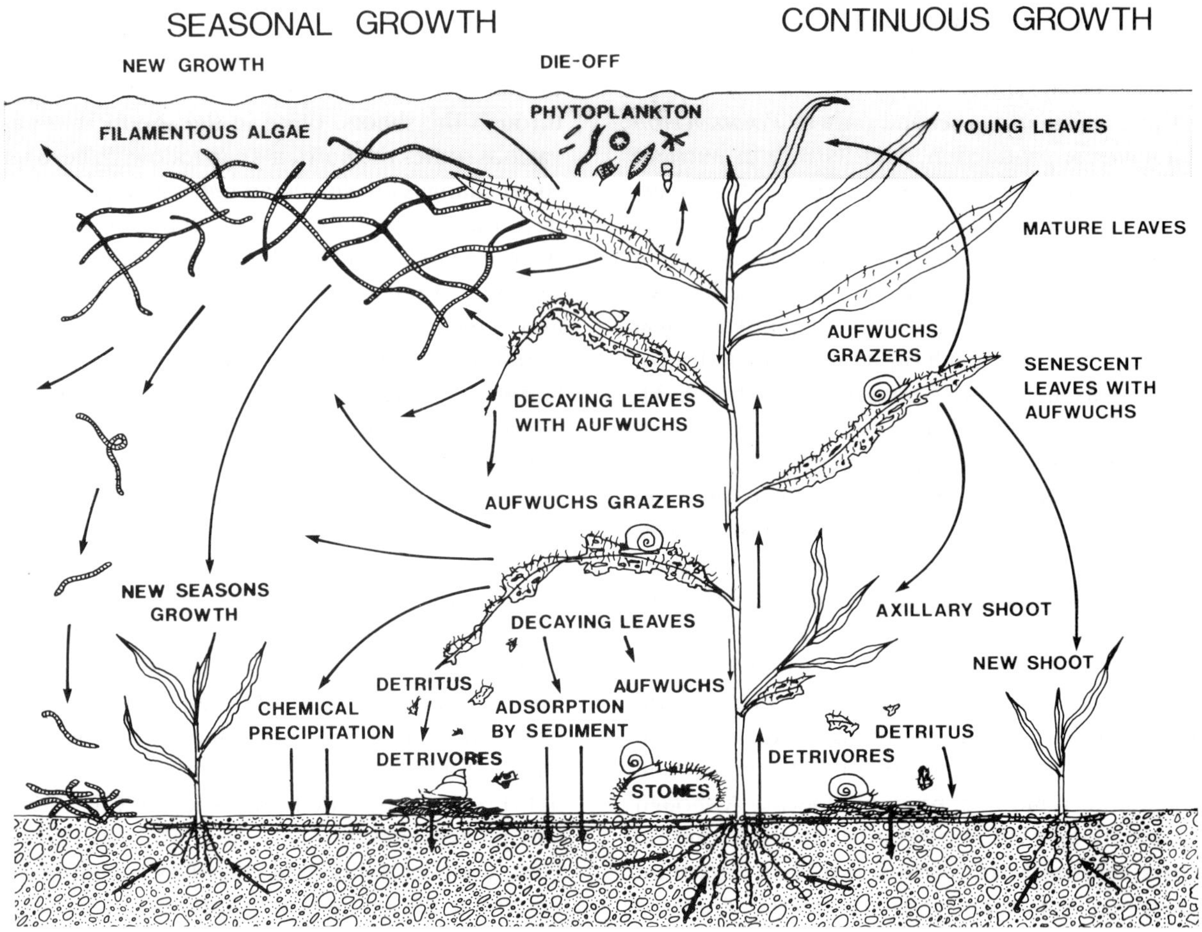

Fig. 5.9. A flow diagram showing the possible routes of nutrient cycling through euhydrophytes in warm, oligo/mesotrophic waters where there is either seasonal or continuous growth. The actual path will depend upon a variety of features including: – the ion in question; the type of plant; the proportions of components in the ecosystem and the trophic status of the water.

distribution. The plants are attached to the sediment and can only grow in relatively shallow water where the leaves of floating-leaved species can reach the surface, and where light penetration is sufficient for photosynthesis of submerged species. They normally colonise the littoral regions of lakes, but, in the tropics especially, they can be in intense competition with emergent plants from the swamps. Wind exposure, and the consequent turbulence and wave action, may further restrict them, and factors which affect the depth of the euphotic zone (such as inorganic suspensions in the water, and phytoplankton densities) determine their depth of colonisation. The nutrient status of the water influences species composition, but, as most of the plants can obtain nutrients from the sediment, it is unlikely to suppress colonisation directly. However, eutrophic waters can support dense growths of planktonic and epiphytic algae and these may shade out the submerged plants. The texture of the sediment must be suitable for anchorage and/or root growth, and its nutrient status affects plant growth.

Compared with the large expanses of African swamps and the broad areas of uncolonised lake bottoms (in which the littoral is often only a small fraction of the total surface area), the euhydrophyte zone must be considered small. Even in those shallow pools and running waters where the depth of water is suitable for colonisation, surface-floating plants may dominate. Therefore, what is the importance of floating-leaved and submerged plants to the aquatic ecosystem?

They are often labelled as troublesome water-weeds and purveyors of disease, but in the natural environment rarely are they a serious pest (certainly, not to the extent of surface-floating and emergent plants). They may invade waterholes and irrigation channels and, locally, require management. The extensive irrigation systems in Egypt and the Sudan are evidence of this and expensive controls are required to keep the channels clear (see Chapter 7). Of more social and economic importance is their association with water-borne diseases. In fast flowing waters their leaves and stems provide substrata for the larvae of *Simulium*, the adult flies of which transmit onchocerciasis (river blindness), and in quieter waters, they provide the favourite niche for the vector snails of schistosomiasis. The seriousness of these diseases and their effects cannot be overestimated, and management of waterplants around populated areas is one possible means of control. But unscrupulous clearance, particularly by chemical methods, may cause deoxygenation (with accompanying fish mortalities *etc.*), and release large quantities of nutrients into the water. If the main nutrient pool had been in the plants, the ecosystem may suffer permanent damage – like the indiscriminate lumbering of trees in some tropical rain forests.

An understanding of the functioning of the euhydrophyte community may ameliorate such potential hazards. Their occurrence around the perimeters of lakes and swamps, and in shallow channels, provides a physical stability so that the forces of wave action, turbulence and flow, are buffered. Particles in suspension settle out and build up a fine, nutrient-enriched sediment. Emergent swamp vegetation performs a similar function further up the waterplant zonation (see Chapter 6). In the absence of plants (or by their deliberate removal) the particles may get carried to the profundal, and in deep lakes, where circulating water rarely rises to the surface, become lost permanently. A parallel may be drawn with the 'slash and burn' land clearance practices which result in irreparable erosion of topsoil.

In floating swamps, if the biomass is large, and the water flow through the swamp small, the emergent plants act as an efficient filter, and water passing through becomes stripped of nutrients. However, sheet flow below the vegetation, particularly during river spate, can wash particulate detritus and solutes out into the open water beyond the swamp (Chapter 6). Euhydrophytes and their associated communities around the perimeter benefit from the nutrient enrichment (assuming the water is not too depleted of oxygen for survival), and retain some of these nutrients. Rooted swamps tend to cycle nutrients from the sediment into the water and the euhydrophyte communities in the channels through the swamp, and around the edge, once again, provide a nutrient buffer between swamp and open water. In other words, they provide a valuable *interface zone*.

The algal communities are an integral part of littoral. Consideration of phytoplankton is largely beyond the scope of this book, but its interactions in the littoral should not be overlooked. In oligotrophic waters where the density of algae is low, light to the submerged plants will not be intercepted significantly. There will be competition for soluble nutrients but as euhydrophytes can obtain nutrients from the sediment, interaction between the two will be minimal. In eutrophic waters, competition between the phytoplankton and euhydrophytes is intense and a high density of phytoplankton will dramatically reduce the euphotic zone. Consequently, the species diversity of euhydrophytes will be restricted and the spatial development of a zonation will be confined to a narrow band in shallow water. Floating-leaved plants, and strong-growing submerged plants with a thick leaf canopy just below the surface of the water, are unaffected (as long as their shoots can reach the surface), but deeper plants that are bottom-loving, will be eliminated. Alternatively, a canopy such as the *Potamogeton pectinatus* canopy of Lake Swartvlei, will shade the phytoplankton and reduce their growth.

The inter-relationship of the *Aufwuchs* in the littoral is intricate. If one includes all the surfaces available for colonisation by attached algae (the epiphyton, epilithon, epipelon, epipsammon and epixylon), the community must be considered a major primary producer. Productively values of ~800 mg m^{-2} d^{-1} organic dry weight (~385 mg C m^{-2} d^{-1}) have been measured on artificial substrata in Nyumba ya Mungu (a man-made lake in Tanzania) (Bowker and Denny, 1978). As the *Aufwuchs* covers all surfaces which receive light, and as its vertical distribution is on substrata throughout the euphotic zone, the integrated production down a water column may be high and, in some cases, equal or exceed euhydrophyte production. However, in dense stands where the annual production of euhydrophytes is high, the *Aufwuchs* must be considered the poor relative.

The relationship between the epiphytic algae and euhydrophytes is very complex. The growth of the epiphyton on shoots lags behind shoot growth, and, whilst the euhydrophytes produce new growth, the older and senescent tissues become colonised by algae. When the vegetation has reached the stationary phase of growth, the epiphytic algae may cover leaves of submerged plants, reduce the light available for photosynthesis, and enhance senescence. The relationship between 'host' and epiphyte need not concern us beyond that the euhydrophyte provides an extensive surface area for colonisation. In terms of nutrient dynamics, epiphytes present a large surface area for adsorption and absorption, moderate the phyllosphere around plant shoots and effectively act as a trap and buffer for nutrients to, and from, the shoots. The *Aufwuchs* community as a whole increases the effectiveness of nutrient retention and cycling within the littoral. Filamentous algae which are found amongst the vegetation increase this further.

Perhaps the most conspicuous benefit of euhydrophytes in the littoral (including interface) areas is in relation to secondary production, the plants offering refuge and food. A bed of vegetation provides protection for a host of animals. Zooplankton, for example, may take refuge amongst the plants during the daytime and emerge to graze the phytoplankton at night. Moss (*pers. comm.*) has suggested that the control of phytoplankton in small bodies of water can depend upon these zooplankters. Disappearance of the plants expose the zooplankters to greater predation, and grazing pressures on phytoplankton are reduced. An increase in phytoplankton increases light attenuation which in turn suppresses re-colonisation by euhydrophytes. Many fish hide and breed amongst the plants, and their fry are offered refuge – and a plentiful food supply. The abundance of fish-feeding birds around the shores of lakes, and along shallow waterways is ample evidence of a rich harvest (Plate 8.1, p. 214).

For an insight into the value of the euhydrophyte community in the food web consider some stages in the chain. The high primary production provides the energy base for decomposers, detritivores and grazers. The carbon loading may be increased further by trapping the allochthonous organic matter washed from the swamps *etc.* In the warm waters of Africa, heterotrophic microbiological activity is high and the breakdown of detritus into small, nutrient-rich, readily digestible particles, will be rapid (Howard-Williams and Davies, 1979). The bacteria and the comminuted detritus provide food for a large variety of invertebrates and the soft sediment is colonised by a range of mud-dwelling organisms. The importance of this vegetation (and the organic detritus from it) to the invertebrates is amply demonstrated by McLachlan (1974) in his review of some lake ecosystems in tropical Africa. Amongst the vertebrates, the food web constructed by Trewavas, Green and Corbet (1972) for fishes in the West Cameroon crater lake, Barombi Mbo, clearly shows the importance of organic detritus and associated bacteria as a fish food.

The *Aufwuchs* is heavily grazed by both invertebrates, such as snails, and vertebrates, especially fish fry. Indeed, in Barombi Mbo some fish appear to have become particularly adapted for specialised *Aufwuchs*-feeding (Trewavas, Green and Corbet, 1972). Analyses of gut contents of commercial fish in Nyumba ya Mungu Reservoir show that two species of *Sarotherodon*, *S. jipe* and *S. pangani*, are predominantly *Aufwuchs* grazers and when fish exclosures were placed around vegetation, the *Aufwuchs* community rapidly increased in biomass (Denny, Bowker and Bailey, 1978).

The euhydrophytes themselves house many invertebrates, especially insect larvae, which may graze the *Aufwuchs*, but few animals consume the healthy vegetation directly. Some fish, such as *Tilapia rendalli*, include euhydrophytes in their diet and Beadle (1981, p. 182) mentions that *Alestes dentex* consume quantities of *Potamogeton pectinatus* in Lake Turkana, but plant material generally seems to be relatively indigestable; at least, until a certain degree of decomposition has taken place.

Clearly, euhydrophytes, especially in the interface zone, can support a very rich ecosystem and the invertebrates particularly, occupy an important position in the food web. Predators, especially birds, congregate in the area and fish take full advantage of the habitat. In Nyumba ya Mungu, the dominant commercial fish were *Aufwuchs*-grazing riverine species, trapped in the man-made lake (Denny, Bowker and Bailey, 1978). The extensive interface zone between the open lake and the swamp provide the ideal conditions for these species. In Chapter 8, commercial development and exploitation of interface zones for fish will be considered further. The euhydrophytes and their associated *Aufwuchs* community with their high productivity, provide the energy base for this ecosystem, and the vegetation itself regulates the nutrient regime.

5.5 References

Barko, J.W. & Smart, R.M., 1980. Mobilization of sediment phosphorus by submersed freshwater macrophytes. Freshwater Biology, 10:229–238.

Beadle, L.C., 1981 (2nd. Edition). The Inland Waters of Tropical Africa. Longman, London.

Bowker, D.W. & Denny, P., 1978. The periphyton communities of Nyumba ya Mungu reservoir, Tanzania. Biological Journal of the Linnean Society, 10:49–65.

Denny, P., 1971. Zonation of aquatic macrophytes around Habukara Island, Lake Bunyonyi, S.W. Uganda. *Hidrobiologia* (Bucuresti) 12:249–257.

Denny, P., 1972*a*. Lakes of South-western Uganda, i. Physical and chemical studies on Lake Bunyonyi. Freshwater Biology, 2:143–158.

Denny, P., 1972*b*. Sites of nutrient absorption in aquatic macrophytes. Journal of Ecology, 60:819–829.

Denny, P., 1973. Lakes of south-western Uganda. ii: Vegetation studies on Lake Bunyonyi. Freshwater Biology, 3:123–135.

Denny, P., 1980. Solute movement in submerged angiosperms. Biological Review of the Cambridge Philosophical Society, 50:65–92.

Denny, P., 1984. Permanent swamp vegetation of the Upper Nile. Hydrobiologia, 110:79–90.

Denny, P., Bowker, D.W. & Bailey, R.G., 1978. The importance of the littoral epiphyton as food for commercial fish in the recent African man-made lake, Nyumba ya Mungu reservoir, Tanzania. Biological Journal of the Linnean Society, 10:139–150.

Denny, P. & Weeks, D.C., 1970. Effects of light and bicarbonate on membrane potentials in *Potamogeton schweinfurthii* (Benn.). Annals of Botany, 34:483–496.

Evans, G.C., Bainbridge, R. & Rackham, O., 1976. Light as an Ecological Factor: II. Blackwell Scientific Publications, Oxford.

Ganf, G.G. & Viner, A.B., 1973. Ecological stability in a shallow equatorial lake (Lake George, Uganda). Proceedings of the Royal Society of London B 184:321–346.

Harbott, B.J., 1982. Studies on algal dynamics and primary productivity in Lake Turkana. In: A.J. Hopson (ed.), Lake Turkana. Overseas Development Administration, London. Published by Institute of Aquaculture, University of Stirling, Scotland. 6 volumes. 1:109–161.

Howard-Williams, C., 1978. Growth and production of aquatic macrophytes in a south temperate saline lake. Internationale Vereinigung für Theoretische und Angewandte Limnologie, 20:1153–1158.

Howard-Williams, C. 1979. Distribution, biomass and role of aquatic macrophytes in Lake Sibaya. In: B.R. Allanson (ed.), Lake Sibaya, Dr. W Junk, The Hague: 88–107.

Howard-Williams, C., 1980. Aquatic macrophyte communities of the Wilderness Lakes: community structure and associated environmental conditions. Journal of the Limnological Society of southern Africa, 6:85–92.

Howard-Williams, C., 1981. Studies on the ability of a *Potamogeton pectinatus* community to remove dissolved nitrogen and phosphorus compounds from lake water. Journal of Applied Ecology, 18:619–637.

Howard-Williams, C. & Allanson, B.R., 1981. Phosphorus cyling in a dense *Potmogeton pectinatus* L. bed. *Oecologia* (Berlin), 49:56–66.

Howard-Williams, C. & Davies, B.R., 1979. The rates of dry matter and nutrient loss from decomposing *Potamogeton pectinatus* in a brackish south-temperate coastal lake. Freshwater Biology, 9:13–21.

Howard-Williams, C., Davies, B.R. & Cross, R.H.M. 1978. The influence of periphyton on the surface structure of a *Potamogeton pectinatus* L. leaf (an hypothesis). Aquatic Botany, 5:87–91.

Howard-Williams, C. & Liptrot, M.R.M., 1980. Submerged macrophyte communities in a brackish South African estuarine-lake system. Aquatic Botany, 9:101–116.

Kirk, J.T.O., 1976. Yellow substances (Gelbstoff) and its contribution to the attenuation of photosynthetically active radiation in some inland and coastal South-eastern Australian

CHAPTER 6

The structure and functioning of African swamps

CLIVE HOWARD–WILLIAMS and JOHN J. GAUDET

Detailed contents

In Africa the term 'swamp' generally applies to wet areas covered with tall herbaceous vegetation. Beadle (1974) calls such areas 'shaded swamps'. Swamps are basically aquatic systems, and the food web is largely aquatic, although limited interaction with surrounding terrestrial areas may occur. Therefore, the ecology of swamps is different from floodplains and other moist habitats which are only temporarily covered with water. Swamps are characterised by the presence of relatively tall vegetation. This tall vegetation has a significant effect on the ecosystem which is not found in other moist areas such as bogs, where the vegetation cover is short. In Africa herbaceous swamps can be defined as: flat areas which are flooded to a shallow depth either permanently, or for most of the year, and which are densely covered with herbaceous vegetation whose shoots rise out of the water to a height of more than one metre. Such swamps can be bottom-rooted or floating (Fig. 6.1).

In temperate regions swamp plants are normally bottom-rooted, good examples being the *Phragmites* grass marshes of Middle Atlantic USA or Eastern Europe, the *Typha* cattail or sedge marshes of Europe and Mid-western USA, the sedge marshes (*Cladium*) of Great Britain, the Everglades, *etc.* Tropical *Phragmites* and *Typha* marshes are structurally similar to their temperate counterparts. Floating swamps are more common in the tropics than elsewhere – for example, the floating grass meadows of South-east Asia and South America, or the *Vossia* swamps along African rivers (Sculthorpe, 1967; Junk, 1970). In Africa, floating *Vossia* or papyrus swamps (*Cyperus papyrus* L.) are structurally quite different from bottom-rooted swamps.

Patrick Denny (ed.), The ecology and management of African wetland vegetation.
 ISBN 90 6193 509 1

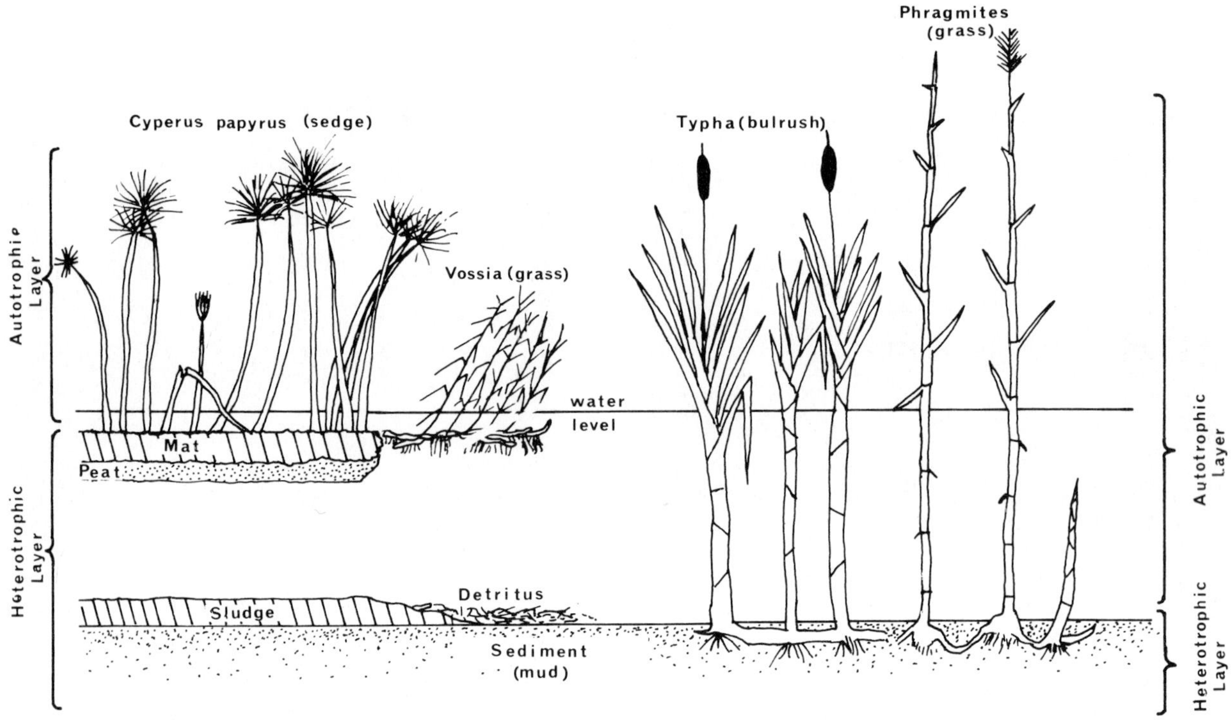

Fig. 6.1. Comparison of floating and rooted tropical swamps showing differences in general structure.

6.1 Basic structure of herbaceous swamps

Swamps occupy an ecological position between purely aquatic and purely terrestrial environments, and thus the structure of swamp ecosystems is rather different from both. Odum (1971) considers that ecosystems can be separated vertically into two layers, the autotrophic (producer) and the heterotrophic (consumer) layers. In swamps the bulk of the primary production occurs in air, and consumption and decomposition in water, thus producer and consumer organisms are physically separated (Fig. 6.1). This spatial separation between autotrophic and heterotrophic components leads in turn to a separation in time, because the foodweb has no direct access to the freshly produced material. There is, therefore, a timelag between production and the death and collapse of the produced material into the water where it can be attacked by aquatic organims. Such a space-time separation between ecosystem components is typical of systems in which the food web is based on detritus (Teal, 1962; Odum, 1971). Whilst this generalisation on trophic stratification is applicable to tropical swamps, this is not the case with modes of nutrient cycling, where we can recognise two very different systems reflected in the gross physical structure of two types of African swamp. The autotrophic component can either be rooted in the sediment (such as in swamps dominated by *Typha*, *Phragmites*, *Cladium* etc.) or rooted in a floating or almost floating mat (such as in swamps dominated by *Cyperus papyrus*, *Vossia* etc.). In the first type, the sediments can act as a source of nutrients for the overlying water column, whilst in the second type the sediments or bottom deposits act as a sink for nutrients from the water column. The two types of swamp (Fig. 6.1) are referred to here as *rooted swamps* and *floating swamps* respectively. Apart from showing different nutrient pathways, as will be discussed later, these two swamp types have a number of different environmental characteristics.

Plate 6.1. A rooted swamp. A stand of *Phragmites karka*, 6 m tall, rooted to the substratum in the Sudd region of the Upper Nile. (Photo P. Denny.)

Plate 6.2. A floating swamp. Workers from the Hydrobiological Research Unit of the University of Khartoum walking on a floating raft of papyrus in the Sudd. (Photo P. Denny.)

6.2 Swamp microclimate

6.2.1 *Light*

Light levels reaching the water surface have been shown to vary depending on water level and the state of growth of the swamp vegetation (Howard-Williams 1972), but seldom rise above 50%, and are often below 10% of the levels above the vegetation. Light penetration into swamp water is thus markedly reduced, and has a significant inhibiting effect on algal production in swamp water (Howard-Williams and Lenton, 1975). The humic substances present in many swamp waters give them a characteristic brown colour which further inhibits light penetration, especially at the blue end of the spectrum.

6.2.2 *Wind*

An analysis of the vertical changes in the airflow and air temperature in a *Typha* swamp in Lake Chilwa, showed that air temperature on average, dropped 3 °C from above the vegetation to the water surface (Howard-Williams, 1979*a*). Air flow, whilst often zero, was generally reduced to 18% of the value above the vegetation.

Floating swamps, typified by papyrus, show the same pattern, but to a greater degree because the canopy is generally more dense. Beadle (1974)

Plate 6.3. Floating swamp islands in the Kamulondo Depression, Zaire. The ones shown are almost entirely papyrus, but parts of the fringes are *Vossia* and *Phragmites* and *Typha* are minor components (see also Plate 3.5). Islands originate as fragments of lake-edge swamps (Plate 3.3) which detach during annual 4 metre floods and ground during dry season. Circular form developed by rolling and packing. (Photo P. Ichac.)

compares the interior of a papyrus swamp to that of a tropical forest with a closed canopy. The dense floating mat of rhizomes, roots and detritus (Fig. 6.1) on which the swamp develops inhibits most light penetration into the deeper swamp water. In such swamps also there is little or no photosynthetic production of oxygen by algae (Beadle, 1974).

6.2.3 *Evaporation*

The interior of tropical swamps is thus cool, dark and very still. These factors have a significant effect on evapotranspiration. Reduced solar radiation at the water surface, lower wind speeds and the ability of the plants to regulate transpiration by stomatal activity, result in reduced evapotranspiration in tropical swamp areas when compared with *equal areas* of open water. Thus in a rooted *Typha* swamp in Australia, Linacre *et al.* (1970) found that in dry weather evaporative losses were reduced to 80% of that of an adjacent lake, whilst Rijks (1969) found that papyrus reduced evaporation to 60% of that of open water in Uganda. Wilson and Dincer (1976) found that from water balance calculations over the Okavango Swamp area, the acceptable figure for swamp evaporation was 85% of that of open water.

6.3 Water movement

Water movements are obviously very different in rooted and floating swamps, and have a marked

Plate 6.4. Floating papyrus rhizomes extending beyond the swamp fringe in a sheltered location on Lake Naivasha, Kenya. Culms are always stunted where the root mat is poorly developed and the substrate nutrient status low. The culm 'stumps' locate the older parts of rhizomes. (Photo K. Thompson.)

effect on dissolved gases and nutrient levels. Most rooted swamps, no matter how dense, have some sort of system of channel(s) or network of small lagoons which play a critical role in the ecology of the swamp. Water flows are speeded up in these areas, and, due to the frictional resistance of the plants, flows are reduced markedly on either side of these more open sites. Horizontal water movements caused by wind drift and wind-induced mixing occur to a much greater extent in rooted than in floating swamps. Thus, Howard-Williams and Lenton (1975) found that lake water around the Lake Chilwa *Typha* swamp mixed with the swamp water to a distance of over 1 km from the lake edge. This increased zone of swamp-lake mixing had a marked influence on the oxygenation of the water

Plate 6.5. Part of a plot harvested for biomass estimation in a Lake George papyrus swamp, Uganda. As rhizomes continue to grow, culm aggregates break up and new groups form. The densest clumps are 'over-mature' and cast too much shade for continued growth. (Photo K. Thompson.)

and hence the movement of fish into and out of the swamp, and the transport of suspended solids into and out of the swamp. In a papyrus swamp the dense floating mat (Fig. 6.1) prevents such horizontal water movements, and causes vertical discontinuities in the water mass. In papyrus swamps in valleys the mat is so dense that inflowing river water may be canalised and the water directly under the mat will be quite stagnant. But in large floating lake side papyrus swamps, incoming river water spreads out under the swamp and a slow sheet flow results along the bottom. However, when the rains stop and drying commences, the water loss from large floating swamps may not be compensated by river input; as a result, lake water backs-up under the swamp. This infusion of lake water can be detected by monitoring conductivity profiles (Gaudet, 1979).

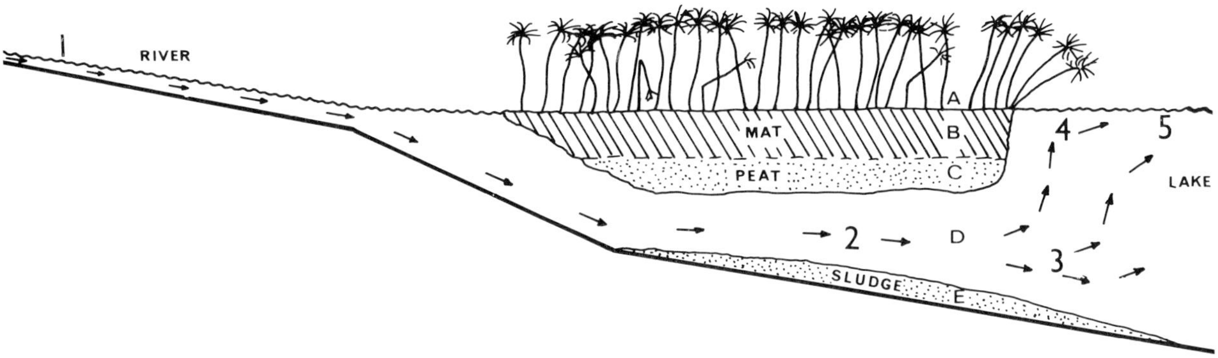

Fig. 6.2. Path of river water under a floating swamp on its way to a lake. A = surface water; B = mat water; C = peat; D = swamp water; E = sludge; 1, 2, 3, 4 and 5 are possible routes of water flow.

Following periods of heavy rain there is a gradual increase in throughflow and a seasonal flushing of swamps. This seasonal flushing is typical of African lake side swamps (*e.g.*, the *Typha* swamps on Lake Chilwa, Howard-Williams and Howard-Williams, 1978). In floating swamps following the rains, the river water becomes cooler, more dense, and comes in along the bottom of the swamp (Fig. 6.2) affecting conductivity profiles for some distance out into a lake. Gaudet (1976) showed that the same movement of swamp water resulted in the movement of large quantities of bottom sludge out of a papyrus swamp. In rooted swamps Howard-Williams and Lenton (1975) described how, following the onset of rains, water from the catchment piled up behind a *Typha* swamp, and as it flowed through it literally pushed in front of it the brown stagnant swamp water typical of lake side swamps in Lake Chilwa.

In floating swamps, although the major movement of water is along the bottom during throughflow, some dilution of the mat water occurs at the swamp surface because of cold rain water at the beginning of the rainy season. This input of colder water causes some mixing of surface water with the underlying swamp water. This sort of mixing was noted by Carter (1955) in Ugandan swamps and Gaudet (1979) in Kenyan swamps.

In rooted swamps vertical movement can occur because of temperature. This induces overturn because the water column is continuous from top to bottom. Such overturn would result in the redistribution of nutrients and oxygen in the water column each day in rooted swamps, whereas a pronounced stratification is typical of floating swamps (Gaudet, 1979).

6.4 Dissolved gases

6.4.1 *Oxygen*

The dissolved gases of most importance to the ecology of tropical swamps are oxygen, carbon dioxide, methane, nitrogen (Fig. 6.3) and hydrogen sulphide. Oxygen is introduced into the water by turbulence and by the activities of aquatic photosynthetic organisms. An appraisal of the physical structure of the swamps described in the previous sections shows clearly why in dense swamps it is unlikely that much oxygenation of swamp water will occur in either of these two ways. In tropical swamps another way oxygenation can occur is by water movements. Generally, oxygen in swamp water is rapidly reduced by consumer organisms which decompose the abundant organic matter produced. In closed canopy areas, the waters of tropical swamps are thus generally devoid of oxygen except for pulses of oxygenated water either from water driven into the swamp following onshore winds, or following rains when river water or rain enters the swamp. Thus, in floating papyrus swamps the oxygen concentrations vary from 0.9–4.6 mg dm^{-3} at the surface, to 0.0–2.4 mg dm^{-3} at the bottom (Carter 1955; Gaudet 1979). Oxygen is occasionally present but always at low concentra-

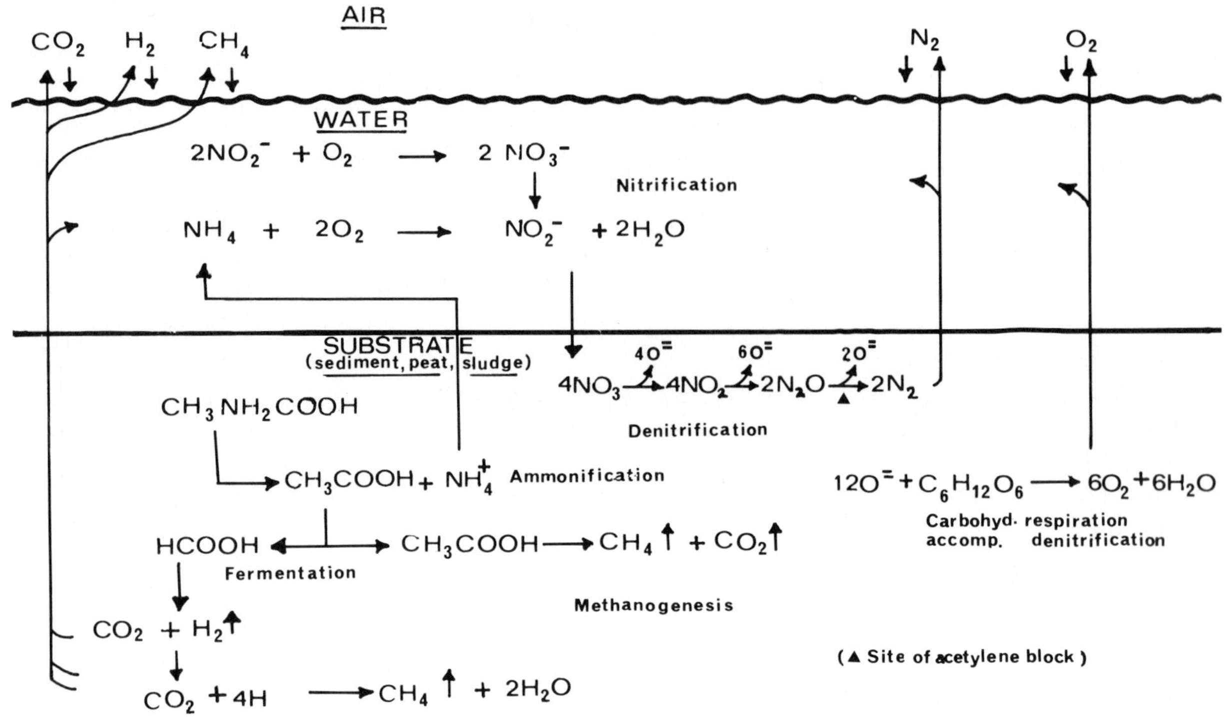

Fig. 6.3. General biochemical reactions common to swamps and the resultant degassing effect.

tions, with the percent saturation varying between 0 and 20% at prevailing temperatures. Similar conditions exist in the Nile swamps (18% saturation in swamp surface water, Rzoska, 1974). Some of this oxygen may have been physically transported into the swamp, but it could also be a product of papyrus photosynthesis passed by way of the continuous gas canal system to the roots. A similar oxygenation has been found in sediments in wetlands (Armstrong, 1975; Wium-Andersen and Andersen, 1972) and in sea grass beds (Oremland and Taylor, 1977).

The movement of oxygenated water and the periodicity of this movement plays an important role in swamp function as many organisms move in and out of swamps only when sufficient oxygen is present. Beadle (1974) points out that most reproduction of the inhabitants of floating swamps seems to occur during the rains when the concentration of oxygen increases in the swamp water. Fish spawning in the Lake Chilwa rooted swamps likewise occurs in the rains when open lake species are able to move through the swamp waters which, for the most part, are deoxygenated during the rest of the year. During the rains the swamp water is replaced by oxygenated water from the catchment (Howard-Williams, 1979*b*). A fairly comprehensive account of the adaptations by animals in order to live in deoxygenated swamp waters is given in Beadle (1974).

6.4.2 *Carbon dioxide*

The free-CO_2 content in tropical swamp water is greatly influenced by throughflow. Thus in a floating swamp on Lake Victoria the free-CO_2 varied from 27–58 mg dm^{-3}, while in a more stagnant valley swamp the free-CO_2 value rose as high as 75 mg dm^{-3} (Carter, 1955). These values dropped rapidly following rains, but such high CO_2 concentrations must render tropical swamps very difficult environments for aquatic faunal life, especially as high CO_2 conditions usually correspond with low or zero dissolved oxygen values.

Generally, the pH of swamp waters is lower than that of adjacent open waters in both tropical and temperate swamp communities (Carter, 1955; Beadle, 1974; Gorham, 1953; Howard-Williams, 1972; Gaudet, 1979; Dokulil, 1973). The drop in pH is most likely due to the fomation of dissolved carbon dioxide and humic acids formed from organic decomposition. It is not surprising that Milburn and Beadle (1960) found 148 mg dm^{-3} dissolved CO_2 (plus HCO_3^-) in a Ugandan papyrus swamp, after a dry period with little or no water exchange. In African papyrus swamps with dense peat deposits, the characteristic drop in pH has also been attributed to the release of H^+ ions (Visser, 1962), which have been exchanged from the decomposing peat for other ions in solution.

6.4.3 *Hydrogen sulphide*

Hydrogen sulphide is seldom encountered in tropical swamps. In a recent study of a papyrus swamp on Lake Naivasha in Kenya, Gaudet could not detect sulphide in the swamp water, but trace amounts of H_2S were found in a swamp gas sample (Gaudet, 1979).

During Talling's studies in the Sudd region on the Nile no H_2S was detected by smell (Rzoska, 1974). Hesse's work on the Lake Victoria papyrus swamps also failed to detect H_2S (Hesse, 1958). Beadle (1974) suggested that normally H_2S is not present in papyrus swamps in high enough concentrations to be detected except under special conditions, such as when large amounts of papyrus plants are submerged suddenly because of heavy rains. However, under normal conditions the absence of H_2S would indicate that microbial reduction would not be responsible for the decrease in SO_4^{2-} which occurs as water passes through a papyrus swamp. Hesse (1958), using inshore surface mud from Lake Victoria, where papyrus swamps are common, found that 53% of the added SO_4^{2-} was adsorbed. In addition to adsorption, S is also taken up and incorporated by papyrus plants, and is only slowly released during decomposition. As expected then, inshore muds near papyrus swamps have high inorganic S contents (Hesse, 1957). Gaudet (1976) also presented some evidence for both adsorption and incorporation in organic matter as mechanisms for $SO_4^{2}-$ depletion from papyrus swamp water. In order to determine if S was a limiting factor to H_2S production Gaudet (1979) added excess SO_4^{2-} to papyrus swamp sludge and tightly packed it into a glass cylinder. Within one week H_2S was produced in excess. Thus he assumed that a S-reducing microflora is present. It can also be assumed that the low redox potentials in the mat and bottom of papyrus swamps (Carter, 1955; Gaudet, 1979) would allow H_2S to be produced even though it was not detected in significant quantity. However, it may be that after formation this H_2S is quickly removed. Armstrong (1975) discussed several methods of sulphide removal in wetland soil by ferrous iron precipitation, sorption on clays, and conversion to elemental S by *Beggiatoa*. In the Lake Naivasha papyrus swamp, *Chromatium* sp. was seen in abudance (Widdowson, *pers. obs.*), and it contained well-developed S granules; possibly it fulfils a role similar to that of *Beggiatoa*.

6.4.4 *Methane*

In waterlogged conditions, swamp soils have been found to be a net source of methane to the atmosphere (Harris, Sebacher and Day, 1982). In papyrus swamps in Uganda occasional areas of intense decomposition were encountered by Carter (1955) where the redox potential was very low. It is likely that in such areas methane gas is produced, and occasionally enough is present to cause marsh gas explosions on seamers passing through such areas (on Lake Kyoga, Beadle, 1974; in the Nile Sudd, Rzoska, 1974). Visser (1964) also found methane during decomposition of papyrus under laboratory conditions. Swamp gas from a papyrus swamp on Lake Naivasha (Gaudet, 1979) contained N_2, CH_4, CO_2 and H_2S, in that order. The methane present may have been generated in local areas in sludge on the swamp bottom, or in peat and in underlying clay substrate. The underlying substrate had redox potentials of -71 to -150 mv, and is composed of a mixture of organic material and clay. There is little evidence that methane or any gas other than oxygen and carbon dioxide influences to any great extent, animal life in tropical swamps (Beadle, 1974).

6.4.5 *Nitrogen*

In gas samples from the papyrus swamp on Lake Naivasha, Gaudet (1979) found that nitrogen gas dominated the gas sample. Obviously it is readily available to swamp nitrogen-fixers. Papyrus roots, with adherent detritus, incubated with acetylene show strong ethylene production after 24 h (Gaudet, 1979; Widdowson, *pers. comm.*). Although no attempt was made to quantify these preliminary results, they are presumptive evidence for nitrogen-fixation. In a recent study, Viner (1982) found mean rates of N fixation for a papyrus swamp on Lake Naivasha to be only 0.0234 mg N m^{-2} day^{-1} (2 cm layer). This is several orders of magnitude less than calculated by Gaudet (1979) using a mass budget calculation. However, in spite of this discrepancy it is apparent that the amounts of nitrogen imported into a papyrus swamp in rivers, rain and runoff are insignificant in terms of the total annual uptake by papyrus and the microflora (Table 6.1).

Table 6.1. Annual import and export of nitrogen for the Kawaga Swamp (Gaudet, 1976).

Component	Tons	Nitrogen % of total
Plant and Detritus		
Papyrus stems	480.0	0.47
Mat	1221.8	1.21
Peat	15120.0	15.20
Sludge	81600.0	82.10
subtotal	*98422.7*	*98.98*
Water component		
Mat water	1.6	0.00
Swamp Water	12.0	0.01
subtotal	*13.6*	*0.01*
Annual imports		
Rivers	1.1	0.00
Rain	17.3	0.01
Runoff	8.3	0.00
import total	*26.7*	*0.01*
Annual exports		
Annual accumulation of sludge	990.0	0.98
Annual sludge loss	21.6	0.02
Through-flow	0.1	0.00
export total	*1011.7*	*1.00*
Total nutrient in swamp	98436 tonns	

Within a tropical swamp, competition must be quite keen between the emergent macrophytes and the autotrophic and heterotrophic microflora for fixed nitrogen. Most of the nitrogen uptake by papyrus probably goes on within the top 20–30 cm in the dense tangle of roots and detritus typical of the floating mat. It is possible that papyrus plants have adopted a system which localises the nitrogen fixation process, similar to the case in other macrophytes, where the rhizosphere is often a centre of intense nitrogen fixation (Patriquin and Knowles, 1972; Bristow, 1974). The cortex of a papyrus root undergoes a transition from a dense storage tissue to a loose fibrous tissue with age and, in addition, the cortex changes from a white to a dark red colour. As the roots mature, the breakdown of the storage tissue could allow a localized concentration of dissolved organic matter. This along with the release of O_2 by living papyrus roots (Tadros, 1940) would offer an ideal microhabitat, especially for micro-aerophilics like *Spirillum* spp. which are already implicated in nitrogen-fixation in macrophyte rhizospheres (Sylvester-Bradley, 1976). In a preliminary investigation similar spirilla have been isolated from papyrus roots (Widdowson and Mwaura, *pers. obs.*). The high Fe content of the root (Gaudet, 1975), the possible local sorbtion of SO_4^{2-} and a deposition of PO_4–P would all further enhance this microhabitat.

Deep in the mat at about 1.0 m there is less O_2 and no light. Here anaerobic nitrogen fixation may be possible. Nitrification would also occur in the mat water from 0–1.0 m depths and may be responsible for the increase in NH_3 seen at this level (Gaudet, 1979).

In African swamps the rate of nitrogen loss by denitrification has recently been determined by Viner (1982). A large amount of the fixed nitrogen must be lost via swamp gas bubbles by this process. Viner (1982) found maximum potential denitrification rates in papyrus mats on Lake Naivasha to be between 1.5 (lower surface) and 6.4 (upper surface) mg N m^{-2} day^{-1} (2 cm layer). Denitrification is an anaerobic process (Fig. 6.3), and although it obviously occurs in the upper floating mat, one would expect higher rates on the bottom in the sludge. Here on the swamp bottom a prevailing BOD, higher levels of NO_3-, and dissolved organic matter, would all favour denitrification (Anderson, 1977). Indeed, sediments from L. Naivasha showed potential rates of 12.9–17.4 mg N m^{-2} day^{-1} (2 cm layer), orders of magnitude higher than N fixation at the same sites (Viner, 1982).

Recent measurements of denitrification by a direct method in New Zealand wetland soils show rates from 1–4 mg m^{-2} N lost per hour in apparently aerobic soils (Howard-Williams and Downes, 1984). The process probably occurs on a matrix of anaerobic microsites in these soils.

Before accurate nitrogen balances can be constructed for tropical swamps, the losses due to denitrification must be determined. In Lake Okeechobee (Florida) 26–34% of the input is lost because of denitrification (Brezonik and Messer, 1978). Additional losses from solution could occur because of adsorption of NH_4 onto particles, and microbial immobilisation. These last processes, sorption and immobilisation, probably are more significant on swamp bottoms where particulate matter is continuously arriving. This detritus would serve as a substrate for both sorption and microbial growth.

6.5 Primary production

Wetlands are very productive ecosystems. For example, freshwater tidal marshes can sustain a dry weight production of 23 t ha^{-1} y^{-1} (Whigham *et al.*, 1978). Marsh production in temperate areas can vary considerably, from 15 to 84 t ha^{-1} y^{-1} in prairie marshes (Van der Valk and Davis, 1978) and 9 to 17 t ha^{-1} y^{-1} in sedge meadows (Bernard and Gorham, 1978). Bottom-rooted swamps in tropical areas show similar ranges in net production with 16–30 t ha^{-1} y^{-1} in an African *Typha* swamp (Howard-Williams and Lenton, 1975). Bottom-rooted tropical swamps then are no more productive than temperate swamps. There are two reasons for this. Firstly, the high tropical temperatures result in higher respiration rates. Gross production is thus higher in the tropics, but not net production. Secondly, much longer days during the growing season occur in temperate regions nullifying the apparent advantage of continuously high solar radiation in the tropics where little change in day length occurs through the year. Most of the emergent species in rooted swamps are C_3-plants, as is the case for the majority of aquatic macrophytes (Westlake, 1975). Tropical floating swamps are quite different, they are much more productive than rooted swamps. For example, net average production in papyrus is 48 to 143 t ha^{-1} y^{-1} (Thompson *et al.,* 1979). It is interesting that papyrus is a C_4-plant, the first freshwater macrophyte so classified (Jones and Milburn, 1978). Papyrus is thus an 'efficient' plant, with a low CO_2 compensation point, 'Kranz' anatomy, and high optimum temperature for photosynthesis. However, its photosynthesis is saturated at less than full sunlight (*i.e.*, it is shade tolerant) and has a low maximum rate of photosynthesis per unit leaf even at only 25 mg CO_2 dm^{-2} h^{-1} (compared to 44–69 mg CO_2 dm^{-2} h^{-1} for *Typha latifolia*, MaNaughton and Fullem, 1970). In comparison to *Typha*, however, the photosynthetic organ of papyrus (an umbel of finely dissected bracteoles) would have an estimated 10 × more photosynthetic area, and much of this would be arranged as a sphere at the apex of the culm. *Typha* leaves, on the other hand, are broad and vertical originating from the base of the stem (Fig. 6.1).

Both *Typha* and papyrus are characterised by a massive rhizome development. Howard-Williams (1973) found 52% of the maximum biomass was underground, while Thompson (1975) found 30–40% of the biomass of papyrus was root and rhizome material. Although this below-ground material is of considerable importance in nutrient cycling in swamp systems, as will become evident later, it is not generally available to consumer food webs in swamps. Thompson (1976) discussed the importance of the root and rhizome systems in maintaining the productivity of a tropical swamp plant such as papyrus. A considerable proportion of the carbon (Thompson suggests 50%) in the biomass of a mature stand of papyrus is 'recycled' within the plant by being translocated to the rhizome system as the old culms die off. This translo-

Plate 6.6. Papyrus valley swamp, 1850 m, Kigezi Region, Uganda. Papyrus is more productive than agricultural crops but doesn't yet have viable commercial use. Foreground is regenerating after harvesting for thatch. (Photo K. Thompson.)

cated material is then supplied to the young culms as they grow upwards. It must be remembered that not only is carbon recycled in this way within the plant, but mineral nutrients particularly phosphorus and nitrogen must also be recycled to an even greater degree (Bjork, 1967; Howard-Williams and Lenton, 1975; Bayly and O'Neill, 1972 and others).

Taking into account this internal carbon storage, Thompson (1979) suggests that the real net productivity of papyrus swamps is about 48–143 t ha^{-1} y^{-1}, whilst Howard Williams and Lenton (1975) give 16 t ha^{-1} y^{-1} for shoot production of *Typha domingensis* in the tropics. Underground production of *Typha* is generally between 30 and 90% of shoot production (Fiala 1973, Kaul and Vass, 1972). Total net productivity of *Typha* in the tropics is thus probably in the region of 25–30 t ha^{-1} y^{-1}.

It is interesting comparing these values with those of efficient crop plants (tropical swamps are generally monospecific stands, as are planted crops). Sugar cane in Swaziland can produce 63 t ha^{-1} y^{-1} and maize will produce 60 t ha^{-1} y^{-1} dry matter (Loomis and Gerakis, 1975). Thompson (1976) drew attention, however, to the important difference between crop plants and natural swamps. Papyrus can produce the same as crop plants, but *in low nutrient conditions*, whereas high yields in crop plants are only maintained by the addition of considerable fertiliser. In retrospect, this represents a considerable energy subsidy. If the amount of energy required to produce and distribute the fertiliser is subtracted from the produced potential energy of a crop, it is certain that tropical swamp plants under natural conditions will rate well above highly productive crop species in terms of production capacity.

In general in tropical swamps, producers other than the dominant swamp plant are of little significance in total production of dry matter. This is due largely to low light levels under the swamp canopy. A number of scramblers and vines such as *Ipomoea* spp. and *Vigna* spp. is commonly found at swamp edges as they are able to reach the canopy by twining around stems (Denny, 1984). *Utricularia* spp. are commonly found growing on the water surface in African swamps even in poor light conditions (where presumably its heterotrophic abilities confer an advantage, Howard-Williams and Walker, 1974). In the Lake Chilwa *Thypha* swamp Howard-Williams and Lenton (1975) found that chlorophyll *a* was almost undetectable in the water, as was photosynthetic oxygen production. Thus the contribution by various primary producers in tropical swamps must be similar to that of the extensive *Phragmites* reedswamp of the Neusiedlersee in Austria. Here Imhof (1973) reported that of the total net production of the swamp, 90% was contributed by the *Phragmites*, 6% by periphyton, 3% by planktonic algae and only 1% by *Utricularia*.

6.6 Herbivore activity in African swamps

Although the net production of African swamps is 8–10 × that of tropical grasslands (such as in Serengeti and Tsavo National Parks – Phillipson 1973, 1975) swamps do not support anything like the same herbivore populations. Among the few large herbivores adapted to swamp conditions are species of lechwe (*Kobus megaceros* and *K. leche*), and the situtunga (*Tragelaphus spekei*), all of which have greatly elongated hooves for walking across unstable substrates. The hippopotamus (*Hippopotamus amphibius*) is also found at and in swamp edges. However, none of these animals is widespread. Seasonally inundated landward margins of tropical swamps are, in contrast, very rich in wildlife, both in species and numbers of individuals. Henry Drummond, visiting the edges of Lake Chilwa in 1888, reported: 'nowhere else in Africa did I see such splendid herds of the larger animals as here' (Drummond, 1902). The Okavango Delta in Botswana represents a major African swamp area which is still relatively undisturbed by man. Patterson (1976) described the fauna of the swamp area, and he emphasised that, whilst only three species of large herbivore (hippopotamus, red lechwe and situtunga) inhabit the permanent swamps, the peripheral areas and channels and moist depressions support some 20 species of large herbivore ranging from the steenbok (*Raphicerus campestris*) to the elephant, with a total large animal biomass estimated at over 15 000 metric tons (Astle and Graham, 1976).

The same impression is obtained with respect to the avifauna of swamps, where swamp edges (both land and lake sides) are usually very rich in bird life. Few birds find the central areas of tropical swamps a favourable habitat other than for periodic feeding and nesting. Certainly, neither *Typha* nor papyrus is utilised to any extent by herbivores, except when the shoots are very young. Although Rzoska (1974) describes an amazing diversity of insects in papyrus swamps, the proportion of the vegetation actively grazed is very low. This raises two important questions:

(a) If, as we have already mentioned, swamps are so very productive at the autotrophic

Plate 6.7. Hippo are huge eaters, but only feed at night. During the day they wallow to keep cool. Above group, in Rwenzori Park, Uganda, will transfer one tonne of biomass to their pool each week as faeces. (Photo P. Denny.)

level, why don't they support a large consumer biomass?

(b) What happens to all the produced material? In answering the first question, the most important factors to be considered are physical environment, palatability and nutrient cycling.

(a) *Physical environment* – The unfavourable physical environment (dense vegetation, unstable substrates, permanent inundation) prohibits most large herbivores from entering swamps. Low oxygen levels, high CO_2 levels and low pH values, which are also prevalent, prohibit many fish and other aquatic organisms from utilising the produced material.

(b) *Palatability* – Swamp plants are generally very fibrous. For example, the cell wall fraction of *Typha domingensis* makes up over 70% of the dry weight of the plant (Polinisi and Boyd, 1972). For papyrus the values would most likely be higher, perhaps approaching that of another tropical sedge, *Scirpus cubensis*, which Howard-Williams and Junk (1977) found had a cell wall value of 83%. The protein content of emergent swamp plants is correspondingly low, that of papyrus culms being only about 5% (estimated from Gaudet, 1977) and *Typha* 4% (Polinisi and Boyd, 1972). Boyd and Goodyear (1971) specifically point out that *Typha latifolia* is seldom grazed because it is largely made up of indigestible material.

(c) *Nutrient cycling* – Thompson (1976) argues that the high productivity of African swamp plants is maintained in low nutrient conditions because of the provision of an efficient internal nutrient cycling mechanism. Thus, heavy grazing will remove nutrients and depress productivity. In addition, many African swamp areas are burnt each year (Lind, 1956; Lind and Visser, 1962; van Rensburg, 1971; Howard-Williams and Lenton, 1975; Patterson, 1976), often to remove dead vegetation and stimulate young growth which is more palatable to herbivores than the older material. Such burning, in nutrient limiting conditions, will also result in a long term depletion of many nutrients and hence decrease productivity. Although Thompson does not take the argument further, there may be a feedback mechanism operating here, where heavy graz-

Plate 6.8. The value of wetlands to wildlife: buffalo grazing floodplain grasses at the start of the wet season in Ngurdoto Crater, Arusha National Park, Tanzania. Although not a 'wetland ungulate' (like lechwe or reedbuck), buffalo make extensive use of floodplain grazing. (Photo K. Thompson.)

ing depresses productivity which, in turn, limits grazing. Thus in a system approaching a steady state, the amount of grazers will depend, not on the actual amount of plant biomass produced, but on the nutrient status of the swamp soil and water. This point is examined in more detail in Section 3.6.3.3.

This leads to the suggestion by Moss (1979) that evolution of the swamp ecosystem has moved in such a way as to eliminate any species which is able to change the structure of the swamp successfully.

6.7 African swamps as detritus-based ecosystems

Let us now consider the second question raised above, namely: What happens to the produced material? If it is not grazed, then there are at least four pathways open: it can be decomposed *in situ* by the activity of detritivores and microbial action; it can be burnt; it can accumulate as peat; or it can be taken out of the system to decompose elsewhere. In this respect, it is possible to compare rough estimates of major compartments and ob-

Plate 6.9. Warthog, zebra, oryx and wildebeest using artificial waterholes in Tsavo National Park, Kenya. The implications for both park and range management are clear: land itself is not in short supply, but water is needed for both grass and animals. (Photo K. Thompson.)

vious transfers in some recently studied wetlands (Fig. 6.4). When peatland, sedge marshes and papyrus swamps are compared in this manner, it is obvious that the floating papyrus swamps must produce large amounts of organic matter and, therefore, have a large export potential. The functional differences between rooted swamps and floating swamps are perhaps related to the differences in structure seen in Fig. 6.1, where bottom-rooted lakeside wetlands have small heterotrophic layers (10–50 cm) dominated by a large autotrophic layer (200–400 cm). Floating lakeside swamps combine both a very large heterotrophic layer (100–200 cm floating peat, 100–200 cm anaerobic swamp water, 0–1100 cm sludge) with a large overtopping autotrophic layer (300–500 cm). As a result, large amounts of detritus are available in floating swamps for use in detritus-based ecosystems. For

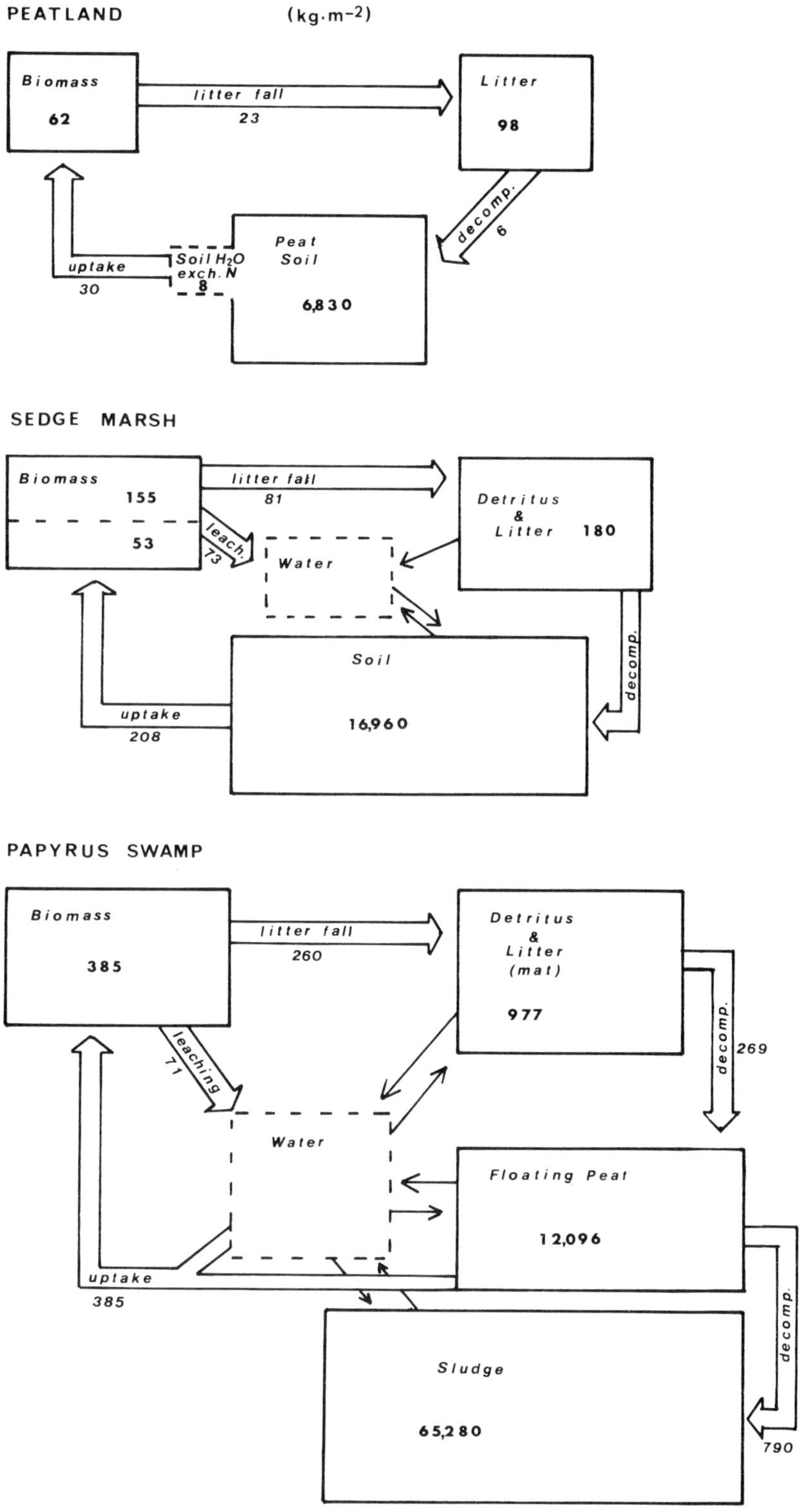

Fig. 6.4. Comparison of biomass stocks and transfers in different wetlands. Data from Gaudet 1976, 1977 and Gaudet (unpublished).

instance, there is increasing evidence that although a large consumer biomass may not be supported within a swamp, such a biomass can be supported once the swamp production has been transported to or beyond the swamp fringes. The Lake Chilwa swamps appear to be an ideal example of how organic material from a swamp provides both physical support and food for a major food web outside of the swamp (Howard-Williams and Lenton, 1975; Kalk and Schulten-Senden, 1977; Kalk *et al.*, 1979).

Why should swamp production support a food web once the plants have died when this production was not utilised when alive? There is much literature dealing with changes in composition of dead organic material (examples are given, for instance, in Dickinson and Pugh, 1974) showing that detritus often forms a better food source than the original living material. The plant fragments disintegrate into smaller and smaller pieces as a result of mechanical and microbial action, thus becoming increasingly available to small animals. Unpalatable polyphenols and tannins are rapidly leached from dead plant material in water (Howard-Williams and Junk, 1976) making much of the dead matter more acceptable. In addition the proportion of nitrogen (protein) increases with decreasing particle size (Odum and de la Cruz, 1967; Willoughby, 1974; Howard-Williams and Davies, 1979) thus rendering the small particles more nutritious. Gaudet (1976) shows clearly that the accumulated 'sludge' at the bottom of a papyrus swamp is enriched with nitrogen, and he suggests that this is most likely to occur because of anaerobic and aerobic N-fixation. In addition, and of some importance, is the fact that as decomposition progresses, the variability in the chemical composition of the original live plant material decreases significantly (Howard-Williams and Junk, 1976). Thus, irrespective of the original composition of the food source, it all ends up as a fairly homogeneous and often more palatable food base after decomposition. This would clearly give detritivores an advantage over grazers in a system where the original live plant matter is a highly variable and/or a basically poor quality food source.

Accumulated detritus, in the form of swamp sludge can be washed downstream out of a swamp (Gaudet, 1976). This serves to enrich the downstream ecosystems, not only with nutrients but with a valuable food source for fish, zooplankton, and benthos (Bourn, 1972; Howard-Williams and Lenton, 1975; Kalk and Schulten-Senden, 1977; McLachlan, 1974; 1975). Thus swamp-lake and swamp-river interfaces are areas of rich fishery potential in Africa (Ratcliffe, 1972; Bourn, 1972; Furse *et al.*, 1979; Welcomme, 1972; 1979; Beadle, 1974; Toews and Griffith, 1979).

6.8 Nutrient relationships in African swamps

Recent studies in African swamps by Gaudet in Kenya and Howard-Williams in Malawi have considered the nutrient dynamics of tropical swamps. One interesting facet of this work is the difference in nutrient pathways in floating and rooted swamps (Fig. 6.5).

The inflow of floating swamps is separated by the mat into two water types, mat water and swamp water. In turn, they both differ in general nutrient dynamics from the swamp water of rooted swamps. Mineral elements entering a floating swamp can:

(a) be immobilised by the swamp vegetation and incorporated into plant material;

(b) enter into trophic-dynamic metabolism;

(c) be lost by degassing (Fig. 6.3);

(d) be adsorbed on to the organic surfaces in the swamp; or

(e) flow directly through.

In rooted swamps, nutrients in the inflow are not incorporated directly into the plant biomass. The emergent aquatics of rooted swamps (*Typha*, *Phragmites*) have a dense root layer just beneath the sediment surface, where it is assumed that most nutrient uptake by the swamp biomass occurs. The sediment thus forms a direct nutrient source which is not the case in floating papyrus swamps.

In both swamps, once mineral elements are immobilised by the plant material, they can either be returned in solution to the mat water by rain leaching of the aerial shoots or elution and decomposition of dead shoots (Gaudet, 1977; Howard-Williams, 1979*b*), or they can also settle out in or on

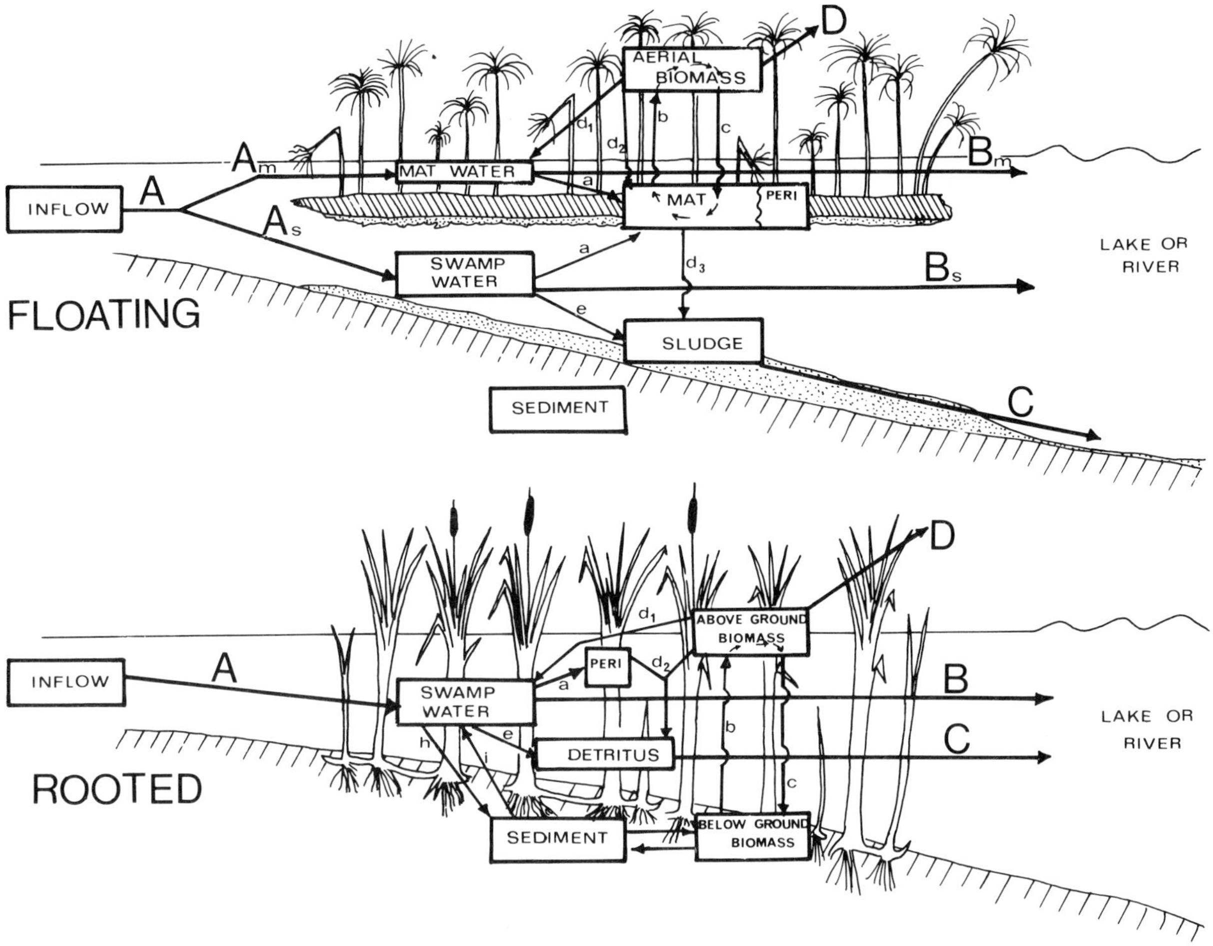

Fig. 6.5. General nutrient pathways in African swamp ecosystems (top: floating swamp; bottom: rooted swamp). Major compartments are shown as boxes. Two types of pathways are distinguished: those labelled with capital letters are throughput pathways (into and out of the swamp); those with small letters are internal swamp pathways. (Peri = periphyton).

Key: A = Nutrients in inflowing water: subscripts 'm' and 's' refer to mat and swamp water, respectively. B = Nutrients in outflowing water. C = Nutrients in the outflow of organic matter. D = Offtake: nutrients lost by grazing, burning or harvesting. a = uptake from the water into swamp plant or periphyton biomass. b = transfer from root/rhizome component to shoots. c = translocation from shoots back to root/rhizome compartment. d1 = transfer from shoot biomass to water (leaching, elution, decomposition). d2 = transfer from shoot biomass to detritus. d3 = transfer from floating mat to sludge. e = adsorption from swamp water onto sludge or detritus. n = loss to sediments from swamp water. i = enrichment of swamp water from sediments.

organic particulate material. This material accumulating on the bottom of the swamp is referred to as sludge (Fig. 6.5).

In both floating and rooted swamps, a further nutrient pathway occurs from the aerial shoots of the swamp plants, that is translocation back to the rhizome system – the internal recycling pathway (Bjork, 1957). When nutrient elements are in short supply in the system, this pathway may prevail as has been suggested for nitrogen in the Lake Chilwa swamps (Howard-Williams and Lenton, 1975). However, it is well known that emergent swamp plants can accumulate elements well in excess of the concentration in the environment (Bayly and O'Neill, 1972; Uhehlova *et al.*, 1973). Where elements are in excess of that required for growth they can accumulated in dead and dying tissues and be released back into the swamp water, such as is the case for sodium in Lake Chilwa (Howard-Williams and Lenton, 1975). In rooted swamps the vegetation can act as a 'nutrient pump' (Odum, 1971) by transferring mineral elements from the sediments to the swamp waters *via* the plant biomass. This does not happen in floating swamps.

Nutrients incorporated into the sludge of floating swamps can be slowly transported out of the swamp. The rate of sludge lost in this way from the Kawaga Swamp in Uganda was 2.4% of the total annual increment of sludge produced (Gaudet, 1976). Thus Gaudet suggests that from the point of view of nutrient relationships, papyrus swamps could be viewed as large holding tanks or septic tanks. A movement of detritus out of rooted swamps does occur also, but to date no quantitative figures are available.

Let us now consider the question of throughflow. In view of the general nutrient pathways outlined above, the waters flowing through a tropical swamp will obviously be altered in some way. The evidence available indicates that the major factor influencing the chemical composition of the throughflow is the rate of flow. Viner (1970) suggested that papyrus swamps around Lake George in Uganda had little effect on major ions entering the lake, because of high rates of throughflow. Gaudet (1978) found that in papyrus swamps where the throughflow is large, or where the swamp has been canalised, the swamp had little influence on salt (Na, K, Ca, Mg, Cl) content of the water. He suggested that the direct effect of swamps in trapping ions by organic sedimentation will only be felt where the water flow is much reduced, as in Kawaga Swamp in Uganda, or where water spreads out under a papyrus swamp, such as occurs in the Sudd.

Unlike the major mineral ions, papyrus swamps must have a large capacity to take up nutrient elements, such as nitrogen, phosphorus and sulphur in inflowing water.

Wastewater application to Michigan peatlands, Louisiana and New Jersey tidal marshes, Wisconsin *Typha* marshes, Florida cypress swamps and *Cladium* marshes in the Everglades, have all shown that wetlands have some ability to act as nutrient traps (Sloey, *et al.*, 1978). Also the recent work of D.F. Toerien (personal communication) showed that although nutrients from urban sewage effluent were removed in streams near Pretoria, the capacity for removal was greatly enhanced where reed beds (*Phragmites*) occurred in the stream. Decreases were particularly evident for NH_3-N, NO_2-N, NO_3-N and to a lesser extent PO_4-P. Water analyses show that the removal mechanism was not diminished, even during low water temperatures.

Much of this nitrogen removal is the result of denitrification (Fig. 6.3). In temperate wetlands denitrification may daily remove up to 350 g N. m^{-2}; and as much as 20 g P. m^{-2} may be immobilised during the growing seasons. However, seasonal washout may nullify the trapping effect of a wetland. For example, Harrison *et al.*, (1960) showed that large reedswamp areas subjected to urban runoff in South Africa, removed NO_3 during summer, but in winter, when the reeds (mostly *Phragmites australis*) died off, the capacity to remove nitrate deteriorated rapidly, although runoff was less.

The general idea of wetlands serving as traps has been criticised by Prentki, *et al.* (1978). They point out that the perennial nature of marshes composed of largely rhizomatous species may argue for strong nutrient conservation, but marsh plants also form very robust stands, suggesting that wetlands are very fertile habitats where little can be gained from saving nutrients. Their study of a *Typha* marsh near Lake Wingra showed that uptake from the soil and translocation could supply the annual P demand of this plant even during peak standing crop. Permanent burial (the ultimate sink for intercepted P) and denitrification (the ultimate sink for intercepted N) did not proceed at high enough rates to allow the Wingra marsh to act as a trap.

In rooted swamps the nutrient pump effect can in some cases enrich swamp water with nutrients while it is passing through the swamp. Howard-Williams and Lenton (1975) and Howard-Williams and Howard-Williams (1978) described how nutrients were lost from the swamp vegetation into swamp water at the edge of Lake Chilwa. This was particularly evident during high throughflow. Swamps therefore have the ability to deplete or enrich levels of nutrient elements in the water flowing through them depending on: (a) the type of swamp; (b) the rate of throughflow; and (c) the element concerned.

However, it is still not possible to cite a complete budget for a tropical swamp. The same is true for temperate areas. For example, Richardson, *et al.* (1978) point out that even the most complete nutrient budgets for bog wetlands are deficient. Most lack information on ground water input, while all lack information on N input by fixation or output by degassing (denitrification). Added to this is the problem that translocation is often ignored, yet this is a most important factor in internal nutrient cycles (Prentki, *et al.*, 1978).

6.9 References

Anderson, J.M., 1977. Importance of the denitrification process for the rate of degradation of organic matter in lake sediments. In: H.L. Golterman (ed.), Interactions between sediments and freshwater. W. Junk, The Hague: 357–362.

Armstrong, W., 1975. Waterlogged soils. In: J. Etherington (ed.), Environment and plant ecology. Wiley, London: 181–218.

Astle, W.L. & Graham, A., 1976. Ecological investigations of the UNDP in the Okavango Delta. Proceedings of the Symposium on the Okavango Delta and its future utilisation. Published by the Botswana Society, Gaborone: 81–91.

Bayly, I.L. & O'Neill, T., 1972. Seasonal ionic fluctuations in a *Typha glauca* community. Ecology 53:714–719.

Beadle, L.C., 1974. The inland waters of tropical Africa. Longmans, London.

Bernard, I.M. & Gorham, E., 1978. Life history aspects of primary production in sedge wetlands. In: R.E. Good, D.F. Whigham & R.L. Simpson (eds.), Freshwater Wetlands. Academic Press, New York: 39–51.

Bjork, S., 1967. Ecologic investigations of *Phragmites communis*. Studies in theoretic and applied limnology. Folia Limnologica Scandinavica 14.

Bourn, D.M., 1972. The feeding, diet and ecological relations of three species of fish of economic importance in southern Malawi. M.Sc. Thesis, University of Edinburgh.

Boyd, C.E. & Goodyear, C.P., 1971. Nutritive quality of food in ecological systems. Archiv für Hydrobiologie 69:256–270.

Bristow, J.H., 1974. Nitrogen fixation in the rhizosphere of freshwater angiosperms. Canadian Journal of Botany 52:217–221.

Carter, G.W., 1955. The papyrus swamps of Uganda. W. Heffer and Sons, Cambridge.

Denny, P., 1984. Permanent swamp vegetation of the Upper Nile. Hydrobiologia, 110:79–90.

Dickinson, C.H. & Pugh, G.J.F., 1974. Biology of plant litter decomposition. Academic Press, New York, Vols. 1 and 2.

Dokulil, M., 1973. Planktonic primary production within the *Phragmites* community of Lake Neusiedlersee (Austria). Polskie archiwum hydrobiologii 20:175–180.

Drummond, H., 1902. Tropical Africa. 11th Edition. Hodder and Stoughton, London.

Fiala, K., 1973. Growth and production of underground organs of *Typha latifolia L.* and *Phragmites communis Trin.* Polski archiwum hydrobiologii 20:59–66.

Furse, M.T., Kirk, R.C., Morgan, P.R. & Tweddle, D., 1978. Fishes: distribution and biology in relation to change. In: M. Kalk, A.J. McLachlan & C. Howard-Williams (eds.), Lake Chilwa. W. Junk, The Hague: 175–208.

Gaudet, J.J., 1975. Mineral concentrations in papyrus in various African swamps. Journal of Ecology 63:483–491.

Gaudet, J.J., 1976. Nutrient relationships in the detritus of a tropical swamp. Archiv für Hydrobiologie, 78:213–239.

Gaudet, J.J., 1977. Uptake and loss of mineral nutrients by papyrus. Ecology 58:415–422.

Gaudet, J.J., 1978. Effect of a tropical swamp on water quality. Internationale Vereinigung für theoretische und angewandte Limnologie 20:2202–2206.

Gaudet, J.J., 1979. Seasonal changes in nutrients in a tropical swamp: North swamp, Lake Naivasha, Kenya. Journal of Ecology 67:953–981.

Gorham, E., 1953. Chemical studies on the soils and vegetation of waterlogged habitats in the English lake district. Journal of Ecology 41:345–360.

Harrison, A.F., Keller P. & Dimovic, D., 1960. Ecological studies on Olifantsvlei near Johannesburg. Hydrobiologia 15:89–124.

Harris, R.C., Sebacher, D.I. & Day, F.P., 1982. Methane flux in the Great Dismal Swamp. Nature, London 297:673–674.

Hesse, P., 1957. The distribution of sulphur in the muds, water and vegetation of Lake Victoria. Hydrobiologia 10:29–39.

Hesse, P., 1958. Fixation of sulphur in the muds of Lake Victoria. Hydrobiologia 11:171–181.

Howard Williams, C., 1972. Limnological studies in an African swamp: seasonal and spatial changes in the swamps of Lake Chilwa, Malawi. Archiv für Hydrobiologie 70:379–391.

Howard Williams, C., 1973. Vegetation and environment in the marginal areas of a tropical African Lake (L. Chilwa, Malawi). Ph.D. Thesis, University of London.

Howard-Williams, C., 1979*a*. The aquatic environment. II: Chemical and physical characteristics of the Lake Chilwa swamps. In: M. Kalk, A.J. McLachlan & C. Howard-Williams (eds.), Lake Chilwa. W. Junk, The Hague: 81–90.

Howard-Williams, C., 1979*b*. Interactions between swamp and lake. In: M. Kalk, A.J. McLachlan & C. Howard-Williams (eds.), Lake Chilwa. W. Junk, The Hague: 231–245.

Howard-Williams, C. & Davies B.R., 1979. Rates of dry matter and nutrient loss from decomposing *Potamogeton pectinatus* in a south temperate coastal lake. Freshwater Biology 9:13–21.

Howard-Williams, C. & Downes, M.T. 1984. Nutrient removal

by stream bank vegetation. In: R.J. Wilcock (ed.) Land treatment of wastes. Water and Soil Miscellaneous Publication no. 70. Govt. Printer, Wellington, New Zealand, pp 409–422.

Howard-Williams, C. & Howard-Williams, W.A., 1978. Nutrient leaching from the swamp vegetation of Lake Chilwa, a shallow African lake. Aquatic Botany 4:257–267.

Howard-Williams, C. & Junk, W.J., 1977. The chemical composition of Central Amazonian aquatic macrophytes with special reference to their role in the ecosystem. Archiv für Hydrobiologie 79:446–464.

Howard-Williams, C. & Lenton, G.M., 1975. The role of the littoral zone in the functioning of a shallow tropical lake ecosystem. Freshwater Biology 5:445–459.

Howard-Williams, C. & Walker, B.H., 1974. The vegetation of a tropical African lake: classification and ordination of the vegetation of Lake Chilwa (Malawi). Journal of Ecology 62:831–854.

Imhof, G., 1973. Aspects of energy flow by different food chains in a reed bed. A review. Polskie archiwum hydrobiologii 20:165–168.

Jones, M.B. & Milburn, T.R., 1978. Photosynthesis in papyrus (*Cyperus papyrus* L.) 12:197–199.

Junk, W.J., 1970. Investigations on the ecology and production biology of the 'floating meadows' (Paspalo-Echinochloetum) on the Middle Amazon. I: The floating vegetation and its ecology. Amazoniana 2, 449–495.

Kaul, V. & Vass, K.K., 1972. Production studies of some macrophytes of Srinagar lakes. In: Z. Kajak & A. Hillbricht-Ilkowska (eds.), Productivity problems of freshwaters. Polish Scientific Publishers, Warsaw: 725–731.

Kalk, M. & Schulten-Senden, C.M., 1977. Zooplankton in a tropical endorheic lake (Lake Chilwa, Malawi) during drying and recovery phases. Journal of the Limnological Society of Southern Africa 3:1–7.

Kalk, M., McLachlan, A.J. & Howard-Williams, C. (eds.), 1979. Lake Chilwa – studies on change in a tropical ecosystem. W. Junk, The Hague.

Linacre, E.R., Hicks, B.B., Sainty, G.R. & Grauze, G., 1970. The evaporation from a swamp. Agricultural Meteorology 7:375–386.

Lind, E.M., 1956. Studies in Uganda swamps. Uganda Journal 20:166–176.

Lind, E.M. & Visser, S.A., 1962. A study of a swamp at the north end of Lake Victoria. Journal of Ecology 50:599–613.

Loomis, R.S. & Gerakis, P.A., 1975. Productivity of agricultural ecosystems. In: J.P. Cooper (ed.), Photosynthesis and productivity in different environments. Cambridge University Press, Cambridge: 145–172.

McLachlan, A.J., 1974. Recovery of the mud substrate and its associated fauna following a dry phase in a tropical lake. Limnology and Oceanography 19:74–83.

McLachlan, A.J., 1975. The role of aquatic macrophytes in the recovery of the benthic fauna of a tropical lake after a dry phase. Limnology and Oceanography 20:54–63.

McNaughton, S.J. & Fullem, L.W., 1969. Photosynthesis and photorespiration in *Typha latifolia*. Plant Physiology 45:703–707.

Messer, J.J. & Brezonik, P., 1978. Denitrification in the sediments of Lake Okeechobee. Internationale Vereinigung für theoretische und angewandte Limnologie 20:2207–2216.

Milburn, T.R. & Beadle, L.C., 1960. The determination of total carbon dioxide in water. Journal of Experimental Biology 37:444–460.

Moss, B., 1979. The Lake Chilwa ecosystem – a limnological overview. In: M. Kalk, A.J. McLachlan & C. Howard-Williams (eds.), Lake Chilwa. W. Junk, The Hague: 401–415.

Odum, E.P., 1971. Fundamentals of ecology. W.B. Saunders Company, Philadelphia.

Odum, E.P. & de la Cruz, A.A., 1967. Particulate organic detritus in a Georgia salt marsh. In: G.D. Lauff (ed.), Estuaries. American Association for the Advancement of Science, Washington, D.C.: 383–388.

Oremland, R.S. & Taylor, B.F., 1977. Diurnal fluctuations of O_2, N_2 and CH4 in the rhizosphere of *Thalassia testudinum*. Limnology and Oceanography 22:566–570.

Patriquin, D. & Knowles, R., 1972. Nitrogen fixation in the rhizosphere of marine angiosperms. Marine Biology 16:49–58.

Patterson, L., 1976. An introduction of the ecology and zoogeography of the Okavango Delta. Proceedings of the Symposium on the Okavango Delta and its Future Utilisation. Published by the Botswana Society, Gaborone: 55–60.

Phillipson, J., 1973. The biological efficiency of protein production by grazing and other land based systems. In: J.W.G. Jones (ed.), The biological efficiency of protein production. Cambridge University Press, Cambridge: 217–235.

Phillipson, J., 1975. Rainfall, primary production and 'carrying capacity' of Tsavo National Park (East), Kenya. East African Wildlife Journal 13:171–201.

Polinisi, J.M. & Boyd, C.E., 1972. Relationships between cell-wall fractions, nitrogen and standing crop in aquatic macrophytes. Ecology 53:484–488.

Prentki, R.T., Gustafson, T.D. & Adams, M.S., 1978. Nutrient movements in lakeshore marshes. In: R.E. Good, D.F. Whigham & R.L. Simpson (eds.), Freshwater Wetlands. Academic Press, New York: 169–194.

Ratcliffe, C., 1972. The fishery of the lower Shire River area. Malawi Fisheries Bulletin No. 3. Fisheries Dept. Lilongwe Malawi.

Richardson, C.J., Tilton, D.L., Kadlec, J.A., Chamie, J.P.M. & Wentz, W.A., 1978. Nutrient dynamics of northern wetland ecosystems. In: R.E. Good, D.F. Whigham & R.L. Simpson (eds.), Freshwater wetlands. Academic Press Inc., New York: 217–241.

Rijks, D.A., 1969. Evaporation from a papyrus swamp. Quarterly Journal of the Royal Meteorological Society 95:643–649.

Rzóska, J., 1974. The Upper Nile swamps, a tropical wetland study. Freshwater Biology 4:1–30.

Sculthorpe, C.D., 1967. The biology of aquatic vascular plants. Edward Arnold, London.

Sloey, W.E., Spangler, F.L. & Fetter, C.W., 1978. Management of freshwater wetlands for nutrient assimilation. In: R.E. Good, D.F. Whigham & R.L. Simpson (eds.), Freshwater Wetlands. Academic Press Inc., New York: 321–340.

Sylvester-Bradley, R., 1975. Isolation of acetylene reducing spirilla from the roots of *Potamogeton filiformis* from Loch Leven (Kinross) Journal of General Microbiology 97:109–132.

Tadros, R., 1940. The daily changes in the internal atmosphere of *Cyperus papyrus* and the ventilation of submerged organs. Bulletin Faculty of Science of Fouad I University, Cairo 20:31–66.

Teal, J.H., 1962. Energy flow in the saltmarsh ecosystem in Georgia. Ecology 43:614–624.

Thompson, K., 1976. Swamp development in the headwaters of the White Nile. In: J. Rzoska (ed.), The Nile. W. Junk, The Hague: 177–196.

Thompson, K., 1976. Primary productivity of African wetlands, with particular reference to the Okavango delta. Proceedings of the Symposium on the Okavango Delta and its Future Utilisation. Published by the Botswana Society, Gaborone: 67–79.

Thompson, K., Shewry, P.R. & Woolhouse, H.W., 1979. Papyrus swamp development in the Upemba Basin, Zaire: studies of population structure in *Cyperus papyrus* stands. Botanical Journal of the Linnean Society 78:299–316.

Toews, D. & Griffith, J.S., 1979. Empirical estimates of potential fish yield for the Lake Bangweulu System, Zambia, Central Africa. Transactions of the American Fisheries Society 108:241–252.

Ulehlova, B., Husak, S. & Dvorak, J., 1973. Mineral cycles of reed stands of Nesyt fishpond in southern Moravia. Polski archiwum hydrobiologii 20:121–129.

van der Valk, A.G. & Davis, C.B., 1978. Primary production of prairie glacial marshes. In: R.E. Good, D.F. Whigham and R.L. Simpson (eds.), Freshwater Wetlands. Academic Press Inc, New York: 21–37.

Van Rensburg, H.J., 1971. Fire: its effect on grasslands including swamps. Proceedings of Tall Timbers Fire Ecology Conference 11:175–199.

Viner, A.B., 1970. Ecological chemistry of a tropical African lake. Ph.D. Thesis, University of London.

Viner, A.B., 1982. Nitrogen fixation and denitrification in sediments of two Kenyan lakes. Biotropica 14:91–98.

Visser, S.A., 1962. Chemical investigations into a system of lakes, rivers and swamps in S.W. Kigesi, Uganda. East African Agriculture and Forestry Journal 28:81–86.

Visser, S.A., 1964. A study of the decomposition of *Cyperus papyrus* in the swamps of Uganda, in natural peat deposits as well as in the presence of various additives. East African Agriculture and Forestry Journal 29:268–287.

Welcomme, R.L., 1972. A brief review of the floodplain fisheries of Africa. FAO Committee for Inland Fisheries of Africa. Report CIFA/72/S. 17.

Welcomme, R.L., 1979. Fisheries ecology of floodplain rivers. Longmans, London.

Westlake, D.F., 1975. Primary production of freshwater macrophytes. In: J.D. Cooper (ed.), Photosynthesis and productivity in different environments. Cambridge University Press, Cambridge: 189–206.

Whigham, D.F., McCormick, J., Good, R.E. & Simpson, R.L., 1978. Biomass and primary production in freshwater tidal wetlands of the middle Atlantic coast. In: R.E. Good, D.F. Whigham & R.L. Simpson (eds.), Freshwater Wetlands. Academic Press Inc., New York: 3–20.

Willoughby, L.G., 1974. Decomposition of litter in freshwater. In: C.H. Dickinson & G.J.F. Pugh (eds.), Biology of plant litter decomposition. Academic Press Inc., New York: 659–681.

Wilson, B.H. & Dincer, T., 1976. An introduction to the hydrology and hydrography of the Okavango Delta. Proceedings of the Symposium on the Okavango Delta and its Future Utilisation. Published by the Botswana Society, Gaborone: 33–47.

Wium-Andersen, S. & Andersen, J.M., 1972. The influence of vegetation on the redox profile of the sediment of Grane Langso, a Danish *Lobelia* lake. Limnology and Oceanography 17:948–952.

Authors' addresses:
Clive Howard-Williams
Department of Scientific and Industrial Research
Taupo Research Laboratory
P.O. Box 415, Taupo
New Zealand

John J. Gaudet
USAID, East Africa Regional Economic Development Services Office
P.O. Box 30261, Nairobi
Kenya

CHAPTER 7

African aquatic weeds and their management

DAVID S. MITCHELL

Detailed contents

7.1 Definition and nature of aquatic weed problems

7.1.1 *Definition*

The simplest definition of a weed is a plant that is a nuisance. Thus, an aquatic weed is an aquatic plant which interferes with the use of water, or in some other way constitutes a nuisance to man or hazard to human welfare. However, in contrast with single purpose systems such as an agricultural crop in which weeds are readily identified as such, water bodies frequently have more than one use, and assessment of the weediness of a plant may be confused when it interferes with one use, such as navigation, while promoting another, such as fish production (Mitchell, 1974, 1978). The value judgments made in these situations are often one-sided and based on incomplete understanding of the nature of the system, the role of the plants in it and the long-term implications of the relationships involved in these situations. Sometimes, as in the case of *Eichhornia crassipes* in the Nile (see section 7.2.2) or *Salvinia molesta* on Lake Kariba (see section 7.2.5), there is general agreement that the plants in question are weeds, though it should be noted that even the most detrimental plant invasion can have some benefits and that adjustments to problems that are intransigent or too expensive to control are always possible.

In other situations opinions may differ markedly

Patrick Denny (ed.), The ecology and management of African wetland vegetation.
 ISBN 90 6193 509 1

and assessments are controversial. For example, proposals have been made for the use of the water in the Okavango Swamps for agriculture and mining in Botswana. Alternative sources of water are not available (most of the country is semi-desert), and the importance and value of the water is unquestionable. Water balance studies indicate that large volumes of water entering the swamp through the Okavango River are dissipated by seepage (perhaps promoted by seismic activity) or by evapotranspirative loss (Thompson, 1974). Both are notoriously difficult to measure, and in the absence of reliable data it is very tempting to advocate the removal of some of the vegetation in order to cut down on water loss. (That this may have the opposite effect, as explained in section 7.3, is seldom appreciated.) Consideration also has to be given to the value of the Okavango Swamps to the climate, tourist potential and wild life of Botswana. In such situations a balanced assessment of the weediness of plant populations is very difficult and decisions are consequently usually made on a socio-economic basis with little regard to ecological considerations. Thus the excavation of channels to drain water from the swamp for various purposes avoids assessment of the effect of the plants on water supply (and probably leads to their ultimate demise anyway).

In general, a useful guide to sound assessment may be obtained from consideration of the normal distribution of the plant in question. If it is native to the area where it is causing a problem, there have to be good reasons for considering it a weed and any proposal to control the plant will have to take account of the probability that its weediness is very localised and that re-invasion will occur from uncontrolled populations nearby. For example, emergent swamp plants which may colonise an irrigation channel and interfere with flow in the channel are clearly weeds in that situation. However, the same plants in a swamp being managed as a nature reserve in the vicinity cannot be considered weeds. In such a situation, control methods for the plant in the irrigation system will obviously need to be confined in their effect to the channel or drain being treated. By contrast, the presence of a plant that is alien to the area could be treated as a potential weed even before it causes problems and drastic measures aimed at its eradication would be justified, especially if it is one of the species, such as *Eichhornia crassipes*, which is known to be a widespread problem. In such a situation the assessor is in the position of having to justify why the plant is not considered a weed and handled accordingly.

Normally a proper and full assessment of the weed status of an aquatic plant in a particular situation requires that rigorously collected, quantitative data of an ecological as well as a socio-economic nature be obtained. This must be objectively examined and a decision reached that is justifiable and open to continuing re-examination as the situation changes, as set out in section 7.6 (Mitchell, 1979*a*, 1979*b*).

7.1.2 *The nature of weed problems*

Aquatic weeds cause a variety of problems in Africa that are broadly similar to those caused by aquatic weeds elsewhere in the world as described by Guscio *et al.* (1965), Holm *et al.* (1969), Little (1968, 1969) and Mitchell (1974, 1976, 1978). In Africa, aquatic weeds interfere with water flow in rivers, canals and drains, thereby imperilling irrigation schemes and slowing drainage of water from floodlands; impede the movement of boats for transport, fishing and recreation; interfere with various methods of catching fish; compete with rice in paddy systems; degrade water quality by adding taints and odours to the water and by decreasing dissolved oxygen content; alter the flora and fauna of aquatic ecosystems by providing new habitats, removing others and by affecting the light climate in the water; favour the spread of diseases such as malaria and schistosomiasis by providing habitats for the intermediate vectors of the parasites causing these diseases; threaten engineering structures such as bridges, weirs and devices to control and measure water flow, especially when large mobile mats of aquatic vegetation move around the water body; block pump intakes; impair the access of stock and wild life to drinking water; decrease the capacity in usefulness of a reservoir by occupying useful volume and by increasing water loss through evapotranspiration, although these effects are often less than first thought; and entrap sediment and thus cause a progressive decrease in the capacity of a reservoir, etc. However it should be stressed that aquatic weeds also have some benefits and uses (see section 7.5), and a balanced view should be taken.

It is very difficult to estimate the annual cost of aquatic weeds in Africa. The cost to human society of the problems described above from uncontrolled plant populations would be impossible to calculate. Some indication of the cost may be obtained by examining the amount of money that society has been prepared to spend in control measures. Five examples can be cited. In the Sudan approximately U.S. $ 1.5 million was being spent annually controlling *Eichhornia crassipes* in the Nile in the early 1960s (Holm *et al.*, 1969). Similarly, about U.S. $ one million (fifty million Belgian francs) were spent on attempts to control the same plant in Zaire (Congo) river in 1956/57 (Lebrun, 1959). In both these cases, herbicides were used together with limited manual and mechanical control. In neither of them was control successful, although considerable quantities of weed were destroyed. In 1977 *Eichhornia crassipes* on Hartebeespoort Dam, South Africa, was successfully controlled with herbicides by a firm specialising in weed control, under contract to the government at the cost of R220 000. Previously, manual and mechanical methods to control the weed had been used for a number of years at a total cost of R600 000–R1 000 000. The cost in decreased use of recreational facilities and reduced land values as a consequence of the weed infestation was estimated to be of the order of R2 000 000 over about 20 years (Bruwer, 1979). The chemical control of *Salvinia* on Lake Kariba, was investigated and a number of herbicides was screened. Field trials with paraquat and sodium arsenate, the two chemicals that were shown to be most effective against the weed, confirmed that they could be used at a cost of U.S. $ 4.5–9 per acre and U.S. $ 0.85–1.8 per acre, respectively (Hattingh, 1962, 1963). However if the whole infestation was to be controlled by even the least expensive method, the cost would have been $ 350 000 and this was not considered feasible, especially as the area of the plant then appeared to be declining. *Salvinia molesta* in the Chobe River system, Botswana, was sprayed with paraquat and glyphosate at a cost of R200 000 in a pilot control project in 1974 and 1975 (Edwards and Thomas, 1977). Again control was only partially successful.

These figures can only give an indication of the cost of water weeds in Africa, and in any case only represent a small part of the total adverse effect of water weeds to man in the continent. It should also be noted that as development proceeds, human populations increase, and the competition for water resources becomes more acute, so will the adverse effects of water weeds increase and their cost to human society rise. An indication of the extent to which costs could rise is given by the U.S. $ 43 million cited by Holm *et al.* (1969) as the estimated annual loss from *Eichhornia crassipes* in the states of Florida, Alabama, Mississippi and Louisiana in 1956.

Most of the research into aquatic weed problems has been directed towards means of controlling unwanted vegetation rather than towards understanding the ecological basis of the problem or to understanding the effect of the plants on the waterbody and its use. However, amongst the latter, studies have been carried out on the effect of aquatic plants on flow in channels and drains, on water loss through evapotranspiration, and on relationships with invertebrate vectors of disease. Each of these will be reviewed in turn.

The relationship between vegetation and resistance to flow is complex as it depends on the cross section of the channel, depth of water, density of plants, morphology of plants, stiffness of plant stems and volume of submerged plant material. Manning's formula is used to calculate a 'coefficient of roughness' or 'resistance coefficient', n, as follows:

$$\bar{v} = \frac{1}{n} A.R.^{2/3}\ S^{1/2} \quad \text{,where}$$

$\bar{v}$ = average flow ($m^3\ s^{-1}$)
n = coefficient of roughness
A = cross-sectional area (m^2)
R = hydraulic radius: A divided by the wetted perimeter of the cross-section in m
S = slope of the water level.

However, deviations from this formula occur in the presence of dense submerged vegetation. For example, Pitlo (1978) showed that in the presence of a dense growth of *Ranunculus circinatus*, $\bar{v}$ was proportional to S and not to $S^{1/2}$. Also, resistance to floods will change suddenly in communities of some submerged plants with slack stems, such as *Potamogeton natans*, which at a certain velocity is pulled under the water (Pitlo, 1982).

More empirical observations in the field have yielded estimates of a reduction in design flow rate of up to 97% by submerged plants, while *Eichhornia crassipes* populations were reported to reduce flow in large canals by 60% and in small canals by 80% (Bogart, 1949). Clearly these are close to maximum values, and a great deal will depend on the shape and size of the canals and the mass and morphology of the plants. Graham (1973), for example, suggested mean flow losses of about 50% in irrigation supply channels in Victoria, Australia. It appears that no estimates have been made of such losses in Africa, but there are no reasons to believe that they would be any different.

The effect of water plants on evapotranspirative loss is less clear. On the one hand there is evidence of considerable quantities of water lost from wetland systems. For example, of 29 000 million cubic metres of water entering the swamps of the Nile sudd just below Bor in Sudan, only 14 000 million cubic metres leave the swamp at Lake No (Moghraby, 1982). Obviously, not all of the difference can be accounted for by evapotranspirative loss, but a large proportion of it must be due to this process.

In Europe there is a number of measurements to indicate that aquatic plants lose considerable quantities of water by evapotranspiration. Burian (1971) measured an average consumption, during the 190 day period of main growth, of 1 metre of water by 1 square metre of the reed, *Phragmites australis*, in the reed swamps fringing the Neusiedler See, Austria. Over the same period the open water of the lake lost 70 mm of water and the free water in the swamp itself lost 400 mm.

In contrast with the above data, Rijks (1969) found that a papyrus swamp in Uganda lost 60% of the water being lost from an adjacent open water surface at the same time. Linacre *et al.* (1970) obtained similar results in Australia. Evapotranspiration from a swamp was compared with that from a lake nearby, and it was found that water lost from the swamp was half to a third of that from the lake during periods of low atmospheric humidity but about the same after rain. It was considered that the lesser loss of water from the swamps during the dry periods was due to physiological resistance to water loss by the plants, a higher reflectivity of the plants and the shelter from wind that the stands of vegetation afforded to the water surface.

For the most part observations referred to in the preceding paragraph have been derived from indirect measurements involving the use of complex formulae in which evaporation is derived from related climatic factors such as fluxes in temperature and humidity. Penman (1963) has reviewed the use of such formulae and should be consulted for further detail.

Other workers have made direct measurements of evaporation from communities of aquatic plants. Thus Brezny *et al.* (1973) obtained the following evaporation: evapotranspiration ratios in Rajasthan, India, – 1.00:1.26 for *Eichhornia crassipes*, 1.00:0.93 for *Pistia stratiotes*, 1.00:0.99 for *Trapa natans*, 1.00:1.13 for *Ipomea aquatica*, 1.00:1.52 for *Typha latifolia* and 1.00:2.41 for *Cyperus rotundus*. Other workers have shown that evapotranspiration losses from *Eichhornia crassipes* communities can be higher than evaporation losses by 3.7 times (Timmer and Weldon, 1967), 4–5 times (Little, 1967) and 6.6 times (Penfound and Earle, 1948), while similar comparisons with *Salvinia molesta* showed that evapotranspirative losses from a weed mat were about the same as evaporation losses from open water (Little, 1967; Mitchell, 1970).

Probably the most serious adverse effect of aquatic weeds on human welfare in the long term is the extent to which they harbour the invertebrate agents which are vectors for disease in man and domestic animals. The principal diseases are schistosomiasis, malaria and fascioliosis. Unfortunately, the connection between the presence of these diseases in the human community and the aquatic vegetation in nearby water bodies is not always clear, because in all cases the intermediate vector does not appear to be dependent on the presence of the aquatic vegetation. However, the vectors are clearly favoured by calm water conditions which also promote the growth of plant communities which then further stabilise water conditions.

Schistosomiasis infected at least 150 million people in 1965 (Jordan and Webbe, 1969), and the number must be considerably higher now. The disease is caused by a worm which infests the blood vessels of the bowel or bladder. It has a complex life cycle in which eggs excreted by the human host enter water, develop into mobile miracidia and infest aquatic snails belonging to the genera *Bulinus*, *Biomphalaria* and *Oncomelania* (the last is not present in Africa). After further development in the snail, large numbers of motile cercariae are produced by the snail hosts, and these infect people who come into direct contact with the water. A number of studies has shown that the snails have preferences for certain vegetation. Dawood *et al.* (1965) showed that in Egyptian irrigation systems the snails prefer *Potamogeton crispus* followed by *Eichhornia crassipes* and then *Panicum repens*. In Ghana, Paperna (1968) showed that the numbers of *Bulinus* snails was much higher when the submerged plant, *Ceratophyllum demersum* was present, while *Biomphalaria* snails were mostly associated with marginal and emergent plants. The snails appear to live in the microhabitats which are provided by the aquatic vegetation. They lay their eggs on the surface of the plants, and they obtain food both from the plants and from grazing epiphytic algae and detritus. However, the relationship between the snails and the vegetation is complex, and it is difficult to define clear causal links (Jordan and Webbe, 1969). It is obvious that more research is needed to clarify this situation. However, whatever that may show, it is likely that the incidence of the disease will increase whenever human populations, particularly with low standards of hygiene and sewage disposal, come into regular contact with water in which there are prolific stands of aquatic vegetation. These situations occur in irrigation systems and in a number of other engineering schemes concerned with the development of water resources in Africa. It should be noted that the floating aquatic plants, *Eichhornia crassipes* and *Salvinia molesta* exacerbate the problem by their mobility. Snails which are present on the plants are spread through a water body when the plants are moved by wind and current. Young *Biomphalaria* snails were observed by Hira (1969) to be clinging to the submerged portions of *Salvinia molesta* on Lake Kariba. However, the writer's observations indicate that *Eichhornia crassipes* is a much preferred host to *Salvinia molesta* for the snails concerned. Thus the former plant is especially likely to be a nuisance in this regard.

7.2 Types of water weed problems

7.2.1 *Classification*

There is a variety of ways of classifying water weed problems so as to make worthwhile comparisons and draw on existing experiences. One approach is to arrange problems in terms of the growth form (*i.e.* submerged, floating, emergent, *etc.*) of the principal plants involved. Many control measures are related to the plants' position in the water, and this approach has the benefit of facilitating comparisons between like problems. Table 4.2 displays weed problems caused in Africa by surface floating plants and provides an example of this approach. The similarity between the reported incidents is apparent, and ecological aspects of the problems have much in common.

Another approach is to recognise that water weeds are weeds primarily in relation to their interference with water use and to classify weed problems in terms of their interference with recreation, water flow, the use of water for human consumption, *etc.* or the spread of disease. This approach can take only little account of ecological factors and generally becomes a catalogue of events that are difficult to relate to one another. For this reason this approach is of limited value.

A third approach is to classify weed problems in terms of the nature of the water body (*e.g.* farm dam, man-made lake, river, channel, drain, *etc.*). As in the first approach there is a number of ecological factors that are common to these water bodies, and this raises the possibility of predicting

Plate 7.1. Introduced to Zaire River little more than 30 years ago, water hyacinth now infests over 1500 km of major waterways. It eliminates, riverine fringe plants like *Echinochloa*; results in increased bank erosion, navigation hazards (drifting mats) and depleted fish stocks. (Photo K. Thompson.)

the occurrence, course and outcome of weed infestations in one water body on the basis of what occurred in another that is similar, although, obviously, account has to be taken of local site conditions and of the availability of weed inocula. Both the first and third approaches facilitate the development of weed management programmes which are related to the ecology and biology of systems that are themselves the result of evolutionary adaptation of both plant and ecosystems (see Chapter 1, section 1.2). The following sections summarise the range of problems that have been reported to occur in Africa in terms of the types of water body that have been infested.

7.2.2 *Weeds in rivers*

Most of the rivers in Africa are liable to annual floods owing to the very seasonal nature of the rainfall over much of the continent. For most of the year, therefore, these systems provide a hostile environment for plant growth and aquatic vegetation is usually depauperate and reduced to a few emergent species such as *Phragmites mauritianus* which is able to colonise sandy substrates and resist mechanical damage during floods. Sand bank rivers typical of this type do not have weed problems, at least in those middle reaches which are characterised by extensive ranges in water level resulting from scouring floods being interspersed with isolated, progressively shrinking pools between rocks and sand banks. Some euhydrophytes may occur in the pools but seldom constitute a weed problem and indeed enhance a limited biological productivity.

The weed problems that do occur are, however, clearly generated in the non-flood periods, though they may only be manifested as problems during the flood season. Thus *Eichhornia crassipes* populations which build up rapidly in the quiet backwaters and swampy areas of the Zaire (Congo) River are loosened by rising water levels and flushed out onto the main river by increasing currents. Lebrun (1959) estimated that, despite an expenditure of 50 million Belgian francs in 1956/57 on attempts to control the weed in the nursery areas, approximately 150 tonnes of the plant were still passing through Kinshasha during the floods of the following year, threatening engineering structures and making navigation hazardous as it did so. The situation in the Nile River is similar, though here the population of the plant builds up in the immense swamp regions of the Sudd in the Upper Nile (Rzoska, 1974; Denny, 1984).

Eichhornia crassipes has been reported to infest a number of other rivers in Africa, such as the Pangani River, Tanzania, the Incomati River, Moçambique, the Sabi and Makabusi Rivers, Zimbabwe, the Kafue River, Zambia, and the Swartkops and Vaal Rivers in South Africa (Wild, 1961; Scott *et al.*, 1979; Jacot-Guillarmod, 1979). While occasionally a nuisance, it does not appear to be a major problem in such rivers unless these flow into a man-made lake (see 7.2.5).

The infestation of the Chobe (1) Linyanti system by *Salvinia molesta* is also promoted by flood action, but in a different way. The Chobe is a small river draining a large floodplain adjacent to the Zambezi above the Victoria Falls. When the Zambezi floods, the Chobe River reverses its flow and the weed is carried upstream. Between floods it completely covers the river, making navigation impossible.

By contrast, the infestation of the Bree River near Cape Town, South Africa, by *Myriophyllum aquaticum* (Jacot-Guillarmod, 1979) is exacerbated more by pollution of the river than by floods. For the most part this river is slow-flowing through clear sandy beds. *Myriophyllum aquaticum* has infested those reaches of the river where there is nutrient rich inflow. The plant is an attached emergent that has the capacity to grow out from the bank over the water. Rafts of floating material eventually break free and are carried down the river where they interfere with fishing and other activities as well as fouling otherwise clear reaches.

7.2.3 *Weeds in canals and drains (artificial rivers)*

The most intensive construction of canals and drains is usually related to the movement of water to and from irrigation areas; in Egypt, for example, there are approximately 44 000 kilometres of irrigation canals (Van der Bliek *et al.*, 1982). However, canals may also be constructed to provide a passage for boats and/or to redistribute water from one region to another, such as the Jonglei Canal to by-pass the Sudd in the Sudan for both purposes. Similarly, drains may also be constructed to drain water from agricultural areas liable to flooding (usually these are 'reclaimed' from wetlands by drainage to begin with) or to drain water from a swamp for use in industry as in Botswana.

Generally flows are too fast and the water too deep for the development of major weed problems in the larger canals, even if the substratum was suitable for the establishment of plants. The most serious problems occur in the smaller canals of irrigation systems, especially if the water is clear so that there is sufficient light penetration for the growth of submerged plants and there are periods during which flow is slow or absent so that the plant populations may establish themselves. Thus problems in Egyptian irrigation canals are reported to have increased since 1975 when much of the silt load previously carried in the water began to be deposited in Lake Nasser (El Gharably *et al.*, 1982), and the light climate for submerged plants was consequently improved. Irrigation supply canals are generally kept full of water and thus provide a stable environment for colonisation by submerged aquatic plants, the most troublesome in Egypt, for example, being *Potamogeton pectinatus, P. crispus, P. nodosus* and *Ceratophyllum demersum.* Filamentous algae, such as *Cladophora*, can become associated with these weed beds and increase the resistance to the flow of water. Emergent plants, such as *Typha domingensis*, can colonise the canal banks but will only invade the deeper water in neglected supply canals after some time. Free floating weeds, such as *Eichhornia crassipes*, are seldom a problem while flows are maintained, unless there is an invasion of a large number of plants into the system from another source.

In contrast with supply canals, flow in irrigation drains is generally slower, and the system is likely to be stagnant more frequently. Also water is more shallow and nutrient and salt levels are higher in drains than in supply canals. Such an environment is more favourable for emergent weeds such as the grasses, *Phragmites australis, Echinochloa* spp., *Paspalum* spp. and *Panicum* spp. as well as *Polygonum* spp., *Typha* spp. and *Cyperus* spp. Submerged weeds occur but may be shaded out, while free-floating plants are generally not swept away because of the absence of strong flows and so can accumulate in nuisance proportions.

7.2.4 *Weeds in natural ponds and lakes*

Standing (lentic) waters are more likely than flowing (lotic) waters to provide conditions for the development of nuisance quantities of aquatic plants. Higher temperatures; greater sedimentation contributing both to a finer, more nutrient-rich substratum and to clearer water and therefore to a better light climate for euhydrophytes; more sustained nutrient cycling systems; and a retention time both of water and free floating organisms that is several orders of magnitude longer than that of lotic systems; all contribute to the situation. The weediness of the problem plants obviously depends on the current uses being made of the water body, but the nature and extent of the problem depends on a number of factors including the morphology of the water body (*e.g.* presence and extent of shallow shelving shore lines, mean depth, presence and extent of areas sheltered from wave action, *etc.*); its limnological behaviour (*e.g.* frequency of mixing, level of biological productivity, *etc.*); and its hydrology (nature of water currents, water balance, and extent of water level fluctuation, *etc.*). The large Rift Valley lakes of Africa, Lakes Malawi, Tanganyika, Kivu, Turkana and Edward have only sparse development of aquatic macrophyte vegetation and no weed problems. However, a number of African lakes is associated with large swamps, and these interface with the lake in a number of ways. In most cases there is an interchange of water between the systems (Carter, 1955; Gaudet, 1975, 1976*a*, 1977, 1979*a*; Beadle, 1981) and, in areas sheltered from wave action where water is clear, there is the development of classical hydroseral zonation from the emergent plants of the swamp to the euhydrophytes in the lake. Large lakes associated with swamps in this way include Lakes Victoria, Chad, Chilwa, Bangweulu, Kyoga and Naivasha. The lake: swamp interface is important for the biological productivity of the lake (Beadle, 1981; and see Chapters 5, 6 and 8), for although vegetation may interfere with navigation, and areas of papyrus may break free and drift around as floating islands, the plants cannot be considered as weeds.

Under certain circumstances, however, there may be a marked change in the extent of the aquatic vegetation which interferes with well established patterns of human exploitation of the lake. For example, the excessively low levels of Lake Chad following the Sahelian drought of 1972/73 provided conditions for the establishment of extensive areas of aquatic vegetation and the extension of the fringing swamps. Okafor (1980) estimated

Plate 7.2. Development of 'Kariba Weed' (*Salvinia molesta*) on Lake Kariba. (Photo P. Denny.)

that 2000 square kilometres of the lake were then infested with aquatic vegetation in the period since the drought during which the lake varied from 6000 to 25 000 square kilometres in area. Okafor (1982) has reported on the problem in the Nigerian section of the lake where it drastically interferes with transport and fishing in one of the potentially most productive inland fisheries in Africa. Similar problems obviously occur throughout the lake (Beadle, 1981). Journeys which previously took one hour now take over four hours. Fishing is directly impeded and the delay in transporting fish from where they are caught to the points of export results in a high proportion of spoilage. The main plants involved are the emergent swamp species, *Cyperus papyrus, Echinochloa pyramidalis, Nymphaea guineansis, Phragmites australis, Typha domingensis* and *Vossia cuspidata*. The floating plant, *Pistia stratiotes*, has also increased in extent and is particularly troublesome on two intake channels for irrigation schemes: the fifteen kilometre channel at Baga feeding the 20 000 ha Baga Polder Project; and the 30 kilometre channel at Kirinowa feeding the 67 000 ha South Chad Irrigation Project. *Pistia* has completely covered the channels and is impeding flow (Okafor, 1982). Beadle (1981) has also noted that even at times of more 'normal' lake levels during the 1960s submerged vegetation, consisting mainly of *Potamogeton schweinfurthii, Vallisneria spiralis* and *Ceratophyllum demersum*, blocked the passage between many of the swamp islands that are a feature of the lake. There is also extensive development of euhydrophyte vegetation on a flat plain along the southern coast east of the Chari River, depending on the extent of inundation which in turn depends on the season. Some species together with *Nymphaea* spp. dominate the vegetation in this situation.

All these plants are native to Africa and in this case are exploiting opportunities presented by naturally occurring environmental changes. They can only be assessed as weeds because man's exploitation of the resource is less adaptable than the plants in that a decrease in utilisation or in fisheries production with the consequent decrease in income and lifestyle is unacceptable to a significant number of people (Maembe, 1981, reported that fish production in Lake Chad supported about 10 000 full time, part time and occasional fishermen).

Natural lakes can also be invaded by alien plants as exemplified by the infestation of *Salvinia molesta* on Lake Niavasha, Kenya (Gaudet, 1976*b*). The plant was first reported to be present in 1963 and initially spread in the vicinity of the yacht club where it interfered with recreational boating. The plant was vigorously controlled with herbicides and for a couple of years was not reported on the lake. However, in 1969 further control measures had to be undertaken in another part of the lake where the weed was found growing inshore and on the lakeward side of the papyrus stands. Further attempts to control the plant have not succeeded in eradicating it, and it can now be considered a permanent constituent of the lake's vegetation. The same situation has occurred with the same plant in the much smaller and more remote Lake Liambezi which is part of the Chobe-Linyanti system (Edwards and Thomas, 1977).

Among the characteristic standing water bodies of the savanna and grassland regions of Africa are the shallow depressions which are filled with water to a lesser or greater extent depending on the time of the year and the total annual rainfall for a particular year. These are variously called pans, or dambos, or some other term. They provide ideal opportunities for the growth of aquatic vegetation and for the most part this is not regarded as a problem. However, in some situations the vegetation can become so dense that it can apparently interfere with the access of stock and wildlife to the water.

7.2.5 *Weeds in farm dams and man-made lakes*

Much of the development of Africa's resources has involved the construction of dams across rivers, whether for the conservation of water to ensure supply for agricultural, industrial and domestic purposes or for hydro-electricity. Among the lakes created by these dams are some which are amongst the largest in the world, namely Kariba, Kainji, Nasser, Volta and Cabora Bassa, but there is also a considerable number which, while not as spectacularly large, is nevertheless big by normal standards – *e.g.* Verwoerd and Strijdom dams in South Africa, Kyle in Zimbabwe, Kafue in Zambia, Barrage d'Inga in Zaire and Nyumba ya Mungu in Tanzania, *etc.* Besides these there are countless others of lesser size. In addition many agricultural enterprises throughout Africa, whether communal or single owner, have required the construction of small catchment dams and other forms of small water reservoir (see section 4.2.3). All these waters can provide good conditions for plant growth, and, since the constructors or owners did not envisage aquatic plant communities in their original concepts, the plants are unwanted and regarded as weeds, even when a problem has not been identified.

Farm dams are frequently relatively shallow, and thus encourage plant growth. However, depending on the climate, many are also liable to excessive fluctuations in water level which has the opposite effect. Thus it is usually only in those situations where water levels are reasonably stable, at least for most of the growing season, that submerged and emergent plants establish in sufficient numbers to constitute a problem. However if floating plants are present, especially those that are alien such as *Eichhornia crassipes* and *Salvinia molesta*, considerable difficulties may occur. The main problems are: interference with stock access for drinking, tainting of water, occupation of useful volume, evapotranspirative loss and fouling of pump intakes (see section 7.1.2).

The growth of aquatic plants in man-made lakes has been described in general by several authors including Little (1969) and Mitchell (1973), while Gaudet (1979*b*) has reviewed aquatic weeds in African man-made lakes. Other authors have reported on the vegetation of specific lakes and their catchment areas, such as the Volta (Hall *et al.*, 1969; Hall and Okali, 1974), Kainji (Cook, 1968), Lake Nasser (El Hadidi, 1976; Entz, 1976, 1980)

Plate 7.3. A dense growth of *Pistia* on a windward shore of Nyumba-ya-Mungu, Tanzania. (Photo P. Denny.)

and Cabora Bassa (Bond and Roberts, 1978). In these lakes aquatic vegetation has generally established sparsely though a succession of species may occur as the lake matures (Gaudet, 1979*b*). However, the growth of *Salvinia molesta* on the first of the large lakes built in Africa, Lake Kariba, was a spectacular example of the invasion of a man-made lake by an aquatic weed (Schelpe, 1961; Boughey, 1963; Mitchell, 1969, 1970; Mitchell and Rose, 1979; Marshall and Junor, 1981). Increasing rapidly in the natal waters of the lake from the small areas first reported in 1959 to over 1000 square kilometres in 1962, the plant acquired a notoriety that it has since justified by increasing incidences of it invading suitable waters elsewhere in the tropical world. The plant principally interferes with navigation, recreation and fishing (though the fish fauna are favoured by its presence, Mitchell, 1976) and these effects are exacerbated when floating mats are colonised by emergent aquatic plants, such as *Ludwigia* spp. and *Scirpus cubensis*, forming a kind of sudd of floating islands (Boughey, 1963). It was also feared that the hydro-electric output of Lake Kariba would be adversely affected, and concern that similar explosive growth of aquatic weeds might occur in other man-made lakes then being planned for Africa was expressed in several quarters.

There was a number of factors that contributed to the early explosive growth of *Salvinia* on Lake Kariba, the most important being: the presence of a large proportion of calm water sheltered from strong winds, waves and currents; the plant's capacity to rise with the rising waters of the filling lake; the continuing supply of plant nutrients being made available from the decay of plants and animals progressively drowned by the rising waters of the lake; the absence of a complex aquatic flora and therefore of competition from established plants; and the absence of animals which grazed on the weeds themselves. However as the lake increased in size to reach full capacity in 1962, wave action increased in extent and in frequency, the proportion of sheltered areas suitable for the establishment of stable weed populations decreased, and the first marked lake drawdowns left large areas of *Salvinia* stranded. Over the next ten years there was a progressive decline in suitable habitat with the increasing break up of partially submerged trees which had previously anchored weed mats. Increasing competition from a complex of other aquatic plants which colonised the lake during this period, and a decrease in the availability of plant nutrients with a trend in the lake to increasing oligotrophy also had an adverse effect on the weed. During this period the areas of weed on the southern bank of the lake fluctuated at around 400 square kilometres (areas could not be measured on the northern bank but on the basis of previous measurements would have behaved in the same way and the total area of weed in the lake would have been about 750 square kilometres). In 1970 the biological control agent, *Paulinia acuminata*, an aquatic Acridid grasshopper was introduced to the lake from South America. Over the following two years populations of this insect increased, and in 1973 the area of weed on the southern bank had decreased to 77 square kilometres. It has remained at about this level or below ever since (Mitchell and Rose, 1979). It is considered that plants took longer to recover from the destructive action of seasonal floods in river inlets (a favoured habitat) and other agents such as wave action during a storm because of the decrease in the availability of plant nutrients and the presence of *Paulinia acuminata* which tends to feed particularly on young plant tissue (Mitchell and Rose, 1979, Marshall and Junor, 1981).

The invasion of Lake Kariba provides a spectacular and reasonably well documented example of an aquatic weed problem in an African man-made lake, even though the areas of plant eventually declined. However, it should be stressed that the rise and decline in the weed was due to particular factors present in Lake Kariba and to changes in these factors as the lake matured. The same events may not occur on other lakes and, particularly if nutrients are being supplied to a lake, populations of floating and submerged aquatic plants may continue to grow. For example, Bon Accord Dam near Pretoria, South Africa, which receives treated sewage effluent has been covered by *Eichhornia crassipes* on several occasions, despite chemical control measures that have reduced the cover to a small percentage of the whole lake each time they are applied. *Eichhornia crassipes* has also required vigorous chemical control measures to bring it under control on Lake McIlwaine, Zimbabwe. Here it was necessary to sustain a control programme for several years to eliminate germinating seedlings which otherwise would have reinfested the lake. Furthermore, as plants may occasionally enter the lake from the Makabusi River, a relatively small river flowing into its headwaters, it is necessary to maintain controls to remove any plants that are seen. The latest build-up of an excessive population of the weed occurred when these measures were suspended for several years (Jarvis *et al.*, 1982).

Hartebeespoort Dam near Pretoria, South Africa is another significant impoundment that has been severely infested by an invasion of *Eichhornia crassipes* (Scott *et al.*, 1979). The plant was first reported on the lake in 1959, and for a number of years it was successfully controlled by mechanical means. However, the dam wall was raised in 1971, and when the lake filled to its new capacity for the first time following heavy rains in 1974/75, the weed invaded areas where boat access was difficult. Here it proliferated rapidly (it was recorded to double in 30 days in September, 11.7 days in November, and 17 days in December). This resulted in the development of large troublesome populations that eventually covered 60% of the lake and were interfering with angling, sailing and boating so that there was a considerable public outcry against its presence. The decision was taken to spray the weed with the herbicide, terbutryn, which was licensed for use in aquatic environments in South Africa (2,4-D which is widely used against *Eichhornia* was not registered for use in water in South Africa). A detailed programme of physical, chemical and biological measurements were undertaken prior to, during and after the control programme. It is noteworthy that the presence of *Eichhornia crassipes* had reduced the numbers of bacteria, algae and zooplankton in the water. This could be regarded as an improvement in water quality, but any benefit in this regard was more than offset by the interference with recreation and possible accelerated loss of water by evapotranspiration. Following the spray programme there was reduction in oxygen concentration to between 2 and 3 mg L^{-1} which probably reduced zooplankton numbers but did not affect the fish fauna adversely.

Even though large quantities of plants were killed by the application of the herbicide over three to four months, extensive follow up operations with diquat, another herbicide, were necessary to eliminate rapidly proliferating survivors.

Thus, even though the fears being expressed at the early stages of the weed infestation in Lake Kariba did not materialise on that lake or on any of the other major man-made lakes in Africa, there can be no basis for complacency as the potential for serious aquatic weed problems to occur continues to exist particularly from the surface floating plants, *Salvinia molesta* and *Eichhornia crassipes*. For this reason it is important that pre-impoundment weed surveys be undertaken during the planning stages for a major new reservoir and that frequent surveys of lake basins be undertaken in existing lakes. These surveys should have the objective of looking for new growths of aquatic vegetation, particularly of aliens, which could potentially pose problems in the future. Due to the rapid vegetative growth of the plants a lot of time and money can be saved if the problems are tackled at an early stage, and early warning of the presence of these plants is very important.

7.2.6 *Weeds in wetlands and floodplains*

At the outset of this section it is important to distinguish between the plants in a wetland, which may interfere with the utilisation of that system by human society, and the vegetation characterising the wetland itself, which may be regarded as a potential threat to a human enterprise nearby. Also the wetland, as well as the vegetation in it, may be seen as undesirable in comparison with an alternative use for the area that is seen as more productive, such as drainage for agricultural land. The latter situation, though of critical importance to the future existence and management of wetlands in Africa, is a separate issue in the context of this chapter. However the wetland itself can markedly interfere with the utilisation of water resources and inhibit the development of whole regions on the continent. Perhaps the most outstanding example of this is the sudd of the Upper Nile. There can be no question that this vast area of wetland has proved to be an effective barrier to transport, travel and trade between communities on either side of it for many centuries. Such a situation underlies proposals to remove part of the system or to establish a by-pass canal around it such as the Jonglei Canal. In such cases the system rather than the plants is regarded as a nuisance, and the designation of individual plants as weeds is clearly inappropriate.

The situation would be different however if islands of vegetation break free, as explained in chapter 6, and interfere with human use of a water body away from the wetland. In some ways such plants could be regarded as weeds, but as control of the situation would require massive management of the wetland from which they originate with all the consequent environmental deterioration that would occur, it is more sensible to adjust to the situation rather than attempt to prevent it.

There are relatively few instances of weeds in natural wetlands in Africa. The presence of *Eichhornia crassipes* in the sudd, which is discussed in section 7.2.1, and the growth of submerged vegetation which interferes with the passage of boats between islands of emergent vegetation in Lake Chad, that is referred to in section 7.2.3, provide appropriate examples, and no further amplification is called for here. However paddy-rice constitutes a managed wetland with a well defined purpose. Any plants, other than rice, are clearly weeds. Most of these are emergent aquatic plants of which the most common infesting rice fields in Africa are grasses, such as *Echinochloa* spp., *Paspalum* spp. and *Panicum repens*, and sedges, such as *Cyperus* spp. Careful control of water supply in the paddy is necessary if terrestrial weeds are to be avoided. Aquatic weed problems in rice are best handled by manual weeding or the use of herbicides.

7.2.7 *Conclusions*

This review of water weeds in Africa reveals that the problems are basically due to two types of plants:

(a) plants native to Africa which grow in nuisance proportions because the environment has been disturbed or altered by man by the construction of artificial canals, drains and man-made lakes, or by anthropogenic enrichment of the water by plant nutrients;

(b) plants alien to Africa which are able to exploit aquatic habitats and build up large populations because of the absence of environmental controlling factors present in their native environments.

In both cases the plants can only be considered as weeds in terms of man's use of the water or water body, though it may be important to recognise potential weeds especially in the case of alien plants.

7.3 Techniques of management

More has been written on the techniques of controlling aquatic weeds than on the weeds themselves, and it is not necessary or appropriate to provide a detailed account of this aspect of aquatic weed management here. Instead, general principles will be enunciated and references made to the literature.

7.3.1 *Manual and mechanical methods*

The use of manual and mechanical methods was reviewed by Robson (1974) and Nichols (1974), while the use of complex machinery to deal with excessive growth of macrophyte vegetation in nutrient enriched lake systems was discussed by Livermore and Wunderlich (1969). There are two main forms of this method of weed control: cutting and dredging, the former carried out by hand and machinery and the latter by machinery only.

The cutting tools available for manual use are mostly modified forms of agricultural tools to cut terrestrial vegetation. They have the considerable benefit of precise selection of which vegetation to remove and at which level to cut it. However, the method is slow and tedious and in Africa brings the operators into direct contact with water which is highly likely to be infested with schistosoma carrying snails. However, if inexpensive labour is available in situations with high levels of unemployement, it is worth careful consideration, provided that precautions can be taken to minimise the risk of schistosomiasis infection.

Most mechanical devices for controlling aquatic vegetation are expensive and require the use of a trained operator. They are, therefore, usually only feasible for relatively large state-aided or managed projects, such as irrigation systems. Maintenance of complex equipment in remote situations also provides a problem. Both cutting and dredging equipment for use in irrigation systems have been employed in Africa. Most of this equipment consists of heavy machinery that is very effective when operated from the bank of a canal or drain. Such equipment includes drag-line excavators, hydraulic back-actors and back-hoes. Lighter machines which can be operated from boats have also been developed. Most of these are fitted with various forms of cutting blades that can be very effective in maintaining part of a lake free of weed to allow access of boats to open water and for other purposes. They can also be used effectively in channel systems.

There is a number of factors to take into account when cutting methods are employed to control aquatic vegetation, mainly the frequency and depth of cutting and the disposal of the cut material. For example, Nichols (1974) has demonstrated that the effectiveness of cutting in reducing the amount of *Myriophyllum spicatum* depends markedly on the season and frequency of cutting. Singh *et al.* (1976) found it was most effective to cut *Typha* plants at the flowering stage. If such plants were then submerged for 4 weeks there was almost no regeneration. Sale and Wetzel (1983) have shown that *Typha* growing with their rhizomes in anaerobic conditions are effectively drowned if they are repeatedly cut below the water surface. This is borne out by Husak's (1978) observation that two cuts of *Typha* below the water are necessary for long term control.

Plant material left in the water after cutting will decay and generate a considerable demand for oxygen. This can adversely affect the environment for other forms of life, and, for example, fish kills, may occur when a large amount of vegetation has been cut in a small water body. For this reason many of the more complex cutting machines are combined with a system for harvesting the cut plants. This not only reduces the risk of oxygen deficits but also makes it possible to make use of the harvested material (see section 7.5).

7.3.2 *Chemical methods*

During the 1950s and 1960s the use of herbicides to control aquatic weeds largely replaced mechanical and manual control in the economically wealthy countries of the world because of their increased effectiveness and rising labour costs. However, growing awareness of the risk of damage to the environment has brought chemical control under close scrutiny, and it is now used more sparingly and with more caution. The use of the many chemicals now available has been reviewed by Little (1968) and Blackburn (1974), but as new chemicals are regularly becoming available it is desirable to consult an experienced authority for the latest information. This approach should also be adopted for the selection of the most appropriate chemical to treat a particular plant as this is also under constant review and even recent compilations, such as that for *Salvinia molesta* by Harley and Mitchell (1981), are already out of date.

The dispersion and fate of chemicals in flowing systems are complex matters and have received especial study in the irrigation systems in Australia where they are widely used (O'Loughlin and Bowmer, 1975; Bowmer and Sainty, 1977; Bowmer, 1979; Bowmer *et al.*, 1979; Bowmer, 1982; Bowmer and Smith, 1984).

In spite of the difficulties and dangers of using herbicides, they are often the only means available for reducing a severe weed infestation to the level where it can be dealt with by other means. Also, they are often the most cost effective method. For all these reasons chemical control methods merit careful consideration.

7.3.3 *Biological control*

Of all methods of controlling water weeds, biological control has experienced the most far-reaching developments in the last twenty years. At the beginning of this period research into the biological control of *Alternanthera philoxeroides* in Florida, U.S.A., was in progress. The significant successes which were obtained against the plant after the release of the flea beetle, *Agascicles hygrophila*, in 1963 (Bennett, 1974; Gangstad, 1976) stimulated further investigations. Since then considerable research has been undertaken on the control of *Eichhornia crassipes*, principally involving two species of *Neochetina*, (a weevil) in the U.S.A. (Bennett, 1974) and Australia (Wright, 1981) and of *Salvinia molesta* in Zimbabwe (Bennett, 1974, Mitchell and Rose, 1979), Botswana (Edwards and Thomas, 1977) and Australia (Room *et al.*, 1981), while the use of the Chinese grass carp, *Ctenopharyngedon idella*, to control submerged weeds has also received attention (Van Zon, 1981). Spectacular success was achieved with the control of *Salvinia molesta* by the weevil, *Cyrtobagous singularis*, on Lake Moondarra, near Mount Isa, Australia (Room *et al.*, 1981).

Much of the research on biological control of aquatic weeds is now directed to finding a complex of phytophagous insects and plant pathogens which have an adverse effect on the pest organism and, in the case of the grass carp, of controlling stocking rates of the fish to obtain acceptable levels of control with the minimum of further management.

Originally, biological control was directed mainly against alien plants and was based on the principle that these plants were successful because, at least in part, they were not subject to attack from phytophagous animals and plant pathogens which had evolved with them in their native environment. Organisms were therefore sought in the plant's native distribution range which affected them adversely and which were host specific. Selected organisms were then evaluated against economically and ecologically important plants in the country where the pest plant was causing a problem. If their specificity was confirmed, the control organisms were bred in quarantine and 'cleaned' of any diseases or of animals which may predate on the control organism so that when the latter were released their numbers were only likely to be controlled by the availability of the pest plant. If the procedure is successful, an equilibrium is eventually reached where the pest is being controlled at fairly low numbers, though there may be seasonal fluctuations.

In the case of the grass carp, which shows little specificity (though it does exhibit distinct preferences) it is necessary to adjust stocking rates so that submerged vegetation is not completely eliminated. Thus research has largely been aimed at defining breeding requirements, so that these can be predicted and avoided, and at the production of non-breeding fish (Shelton *et al.*, 1982; Krasznai *et al.*, 1982). The former appear to be restrictive and only likely to be manifested in particular types of water body within fairly narrow climatic conditions. If breeding can be controlled, stocking rates can be adjusted by the introduction of known numbers of fish and replacement as required. Grass carps have been introduced into Africa for weed control in Egyptian irrigation systems (Dubbers *et al.*, 1981; El Gharably *et al.*, 1982; Siemelink *et al.*, 1982), and it is likely that, if these projects succeed, the fish will be introduced to other areas as has occurred in the United States of America.

7.3.4 *Environmental manipulation*

Many aquatic plants are sensitive to changes in water levels, particularly at certain times of the year. In some situations it is possible to draw down or raise water levels and thereby expose the weeds to environmental stress when they are most sensitive to it. This requires good knowledge of the ecology of the weed and of the hydrology of the infested system. Manning and Sanders (1975) describe the use of this approach almost to eliminate submerged plants, such as *Egeria densa* and *Cabomba caroliniana,* in a reservoir in Louisana, while Manning and Johnson (1975) combined drawdown of water level and the application of herbicides to control *Hydrilla verticillata* in another Louisiana reservoir. Malhotra (1972) has described how a procedure of draining canals in the irrigation scheme fed from the Bhakra Dam in India and leaving them dry for five days was successful in controlling weeds for the succeeding six months.

Other methods of manipulating environmental factors to put weeds under stress are the use of dyes to decrease light penetration, and of black plastic to blanket out submerged weeds from defined areas (Nichols, 1974; Mitchell, 1974; Dawson and Kern-Hansen, 1979). Substratum conditions can also be altered by dredging out nutrient rich muds or by covering them with sand or gravel.

7.3.5 *Integration of control measures*

As has been indicated in previous paragraphs, control programmes can often be made more effective by integrating more than one technique and using each to attack the pest plant at the most appropriate time so as to complement and/or support the effectiveness of another technique. In addition to those mentioned in section 7.3.4, such programmes have been formulated for *Alternanthera philoxeroides* control in the United States using both biological and chemical methods (Durden *et al.*, 1975; Gangstad *et al.*, 1975). This approach requires an understanding of the plant's ecology and of the operation and effectiveness of the control techniques. This approach is likely to become increasingly widely used throughout the world, and it will be important for the control of nuisance water plants in Africa.

7.4 The ecology of water weed management

From what has been written in preceding sections of this book it should not now be necessary to stress the importance of basing the management of nuisance populations of plants on the ecology of those plants and of the systems of which they are a component. Three factors have to be taken into account in this regard: (a) the role of macrophytes in aquatic ecosystems; (b) the ecological strategies and adaptations of the pest plants; (c) the environmental consequences of controlling the aquatic plants in terms of the effects of destroying the weed and of the use of particular control measures. These three factors will be considered in turn in the next three sections.

7.4.1 *The role of macrophytes in aquatic ecosystems*

The part played by vascular aquatic plants in aquatic ecosystems was reviewed by Gaudet (1974), Mitchell (1978) and Denny (Chapter 5 of this book). Briefly these plants are concerned with five important aspects which contribute to the stability of the system: primary production; nutrient uptake, release and retention; source of food; diversification of the habitat; and stabilisation of sediments. Details may be obtained from Chapters 5 and 6.

7.4.2 *Ecological strategies and adaptations of plants*

Some of the ecological strategies and adaptations of various growth forms of aquatic plants were examined in Chapters 2, 3 and 4. A knowledge of these strategies and adaptations is important to the success of any programme of control of these plants when they are weeds. For example, it is important to know how and when the plants reproduce. Similarly, knowledge of their substratum requirements and responses to chemical and physical factors in their environment is highly desirable. Control measures which do not take advantage of weaknesses in the plant's life cycle are less likely to be successful than those which continually provide conditions that the plants are best fitted to exploit.

7.4.3 *Environmental consequences of water weed control*

It should be clear that the sudden removal of large quantities of aquatic vegetation will have a severe effect on a number of other components of the aquatic ecosystem. The death and decay of other animals together with those of the destroyed plant material can often contribute to a marked degradation of water quality. Oxygen levels are likely to be lowered and nitrogen and phosphorus levels increased. As a consequence, the populations of phytoplankton and of bacteria may often be increased. In the same way, the use of certain control measures can contribute to undesirable changes in the ecosystem. This has been particularly investigated for herbicides (Brooker and Edwards, 1975; Newbold, 1975; Mitchell, 1978). The adverse effects of using a herbicide are variable and complex. Non-target organisms are often destroyed and this leads to changes in the species composition and population structure of the other plants and animals in the system. The destruction of the aquatic weed communities often leads to increased light penetration, and this might result in the replacement of the aquatic weeds by nuisance levels of algae. The loss of habitat and food for animals leads to the migration away from the system of some of the animals and the death of others. All of these events will have an effect on the quality of the water.

7.5 Utilisation of water weeds

The presence of a large quantity of unwanted plants provokes the obvious challenge of finding a use for them. Also the opportunity to change a nuisance into a benefit is attractive in countries where economic conditions favour low cost, labour-intensive developments. Possibilities for using aquatic weeds with particular reference for developing countries were canvassed in a publication released by the U.S. National Academy of Sciences (1976). Useful reviews have also been undertaken by Boyd (1974) and Little (1979), while Boyd (1972) provided a comprehensive bibliography on the use of aquatic weeds up to that time. Possible uses for *Eichhornia crassipes* were reviewed by Gopal and Sharma (1981) and the potential for *Typha* to be a crop rather than a weed was described by Morton (1975).

There are six main groups of use for aquatic weeds: as food for humans, livestock and aquatic animals, such as fish and crayfish; as compost, fertiliser, mulch or other forms of soil additives; for treatment of wastewaters; for pulp, paper and fibre production for building and weaving; and for energy either for burning directly or for generation of biogas and alcohol. Other uses for aquatic plants (not necessarily weeds) include production of medicinal products; in gardening, horticulture and freshwater aquaria (the international trade in these plants for these purposes has been responsible for the dispersal of many adventive weeds to new areas); and for stabilising sediments, lake shorelines and river banks.

Aquatic plants in the Nile Valley have been used from the times of ancient Egypt (Sculthorpe, 1967). *Cyperus papyrus* and *Nymphaea caerulea* were the emblems of Lower Egypt and Upper Egypt respectively. The former species was used for the construction of boats and making papyrus paper. For this, thin slices of culm pith were laid side by side so as to slightly overlap one another and a second layer arranged on top at right angles to the first. The strips were then moistened and stuck together with heavy pressure (Steenberg, 1968). *Nymphaea caerulea* and *N. lotus* were frequently illustrated on ancient Egyptian tombs and monuments. Both species were used for wreaths and floral decorations, and their seeds and rhizomes were used for food. The ancient Egyptians also used *Pistia stratiotes* to treat skin diseases and abrasions.

In modern times attention has turned more to the productive wetland plants and to acknowledged weed species, especially *Eichhornia crassipes*. The latter plant has been investigated for its nutritive value as feedstuff for ruminants in Sudan (Chalmers, 1968; Osman *et al.*, 1975). On its own the plant is not suitable as feed as it does not supply sufficient protein, is too high in water, is not particularly palatable and can induce diarrhoea. However, when partially dried, sprayed with molasses and supplemented with a concentrate of clover, sorghum grain and cotton seed cake, the plant could be used in ruminant diets. It is also interesting that in the Sudan and in Madagascar, the fresh plant is grazed by cattle during the dry season (Davies, 1959; Ramarokoto, 1965). The use of *E. crassipes* as a mulch and for fertiliser in the Sudan appears more promising (Kamal and Little, 1970). Suitable compost can be made from piles of partially dried plants mixed with cow dung and covered with earth, though more research is required to ascertain the best method. Also Abdalla and Abdel Hafeez (1969) showed that, when the plant is used as a mulch, it suppresses the growth of one of the major weeds in the Sudan, *Cyperus rotundus* (nutsedge), as well as conserving soil moisture.

Gopal and Sharma (1981) have reviewed previous experiences in the use of *Eichhornia crassipes*

Plate 7.4. Papyrus boats are still used on Lake Tana, Ethiopia. Thor Heyerdahl used Tana artisans to build his famous 'Ra' replica of Egyptian craft. Papyrus is also used for canoes by the River Bushmen in the Okavango Delta. (Photo J.J. Gaudet.)

to generate biogas, and the National Council for Research of Sudan and the National Academy of Sciences, U.S.A. (1975) have advocated this use for Africa. The plant is particularly suitable for anaerobic digestion which yields methane and other gases, such as hydrogen and carbon dioxide, because of its water content, favourable C:N ratio and readily digestible organic matter. The volume of gas and rate of production depends on the temperature of fermentation which appears to be optimal at 40°C when 330 dm^3 kg^{-1} are produced. Fermentation is improved when the plants are mixed with cow dung, and the cow dung-based gas plants that are successfully used in Indian and Chinese villages could be introduced to the African continent.

Salvinia molesta is another weed that has been investigated for possible forms of use. It is most suitable as a compost or mulch, and around Lake Kariba was used for the latter purpose when mixed with the dung of game animals (Boyd, 1974).

Typha spp., *Phragmites* spp. and *Cyperus papyrus* are probably the most productive aquatic plants in Africa, but these are seldom in large populations as weeds, and their utilisation in wetland systems has been considered in Chapter 6.

The potential to use herbivorous fish to control aquatic weeds and then to harvest the fish for protein has not been fully exploited in Africa. The Chinese grass carp, *Ctenopharyngedon idella*, has been introduced to Egypt to control submerged weeds in irrigation systems (section 7.3.3), but there is a number of herbivorous cichlid fish that are native to Africa, such as *Tilapia rendalli (T. melanopleura)*, which should also be investigated.

The use of water plants to treat wastewaters has been attracting increasing interest in the last 20 years. Three systems have been advocated: one using floating species on wastewater lagoons, another has emergent plants in trenches filled either with gravel or soil and the third employs existing natural swamp systems. Wolverton and his co-workers have been responsible for much of the early development of systems employing water hyacinth (*e.g.* Wolverton *et al.*, 1976), whilst Seidel and her co-workers have pioneered the use of emergent species such as *Schoenoplectus litoralis* and *Phragmites australis* in trench systems (*e.g.* Seidel *et al.*, 1978). All these systems have the attraction of being low cost and low maintenance systems which could be particularly applicable in rural situations.

While many of these uses appear to be attractive and some may be feasible in certain conditions, there are remarkably few recorded instances of the successful large-scale use of aquatic weeds. The main reason for this is the high percentage water content (often about 90%) of the plants. Mechanical harvesting and transport of fresh material is often rendered uneconomic so that the most feasible use is in situations where low cost labour is available and the plants can either be dried and compacted nearby, prior to transport, or utilised directly in the immediate vicinity. Any productive use of a weed, even if uneconomic in itself, will help offset the cost of control, but, if the harvesting and use becomes particularly successful, there is then a dilemma as to the weedy status of the plant. Consequently, proposals to utilise a weed should take account of its current weediness and the plant's capacity to produce sufficient material for the proposed use without adversely affecting the infested water body.

Another factor that cannot be stressed too strongly is that noxious aquatic weeds should not be introduced to areas where they are not present in order that they may provide a basis for the development of one of the forms of use that have been outlined. The potential adverse consequences of such an ill-advised act will certainly outweigh any possible benefits. The investigation of possible forms of utilisation of a known aquatic weed is only justified where and when the plant is a weed. This particularly applies to *Eichhornia crassipes*. Utilisation of native plants is an entirely different matter.

7.6 Principles and philosophy of water weed management

7.6.1 *Management of the whole system*

Since the causes of most water weed problems

Plate 7.5. The Istituto del Papiro in Siracusa (Italy) manufactures high quality papyrus parchment from imported Kenyan papyrus. Here, the second layer of bracing-strips of papyrus culm are being placed in position before pressing. (Photo K. Thompson.)

relate to the management of aquatic systems, and the control of the weed must take account of environmental factors, water weeds cannot be managed apart from the management of the ecosystem they are infesting. Thus water weed problems may be promoted by nutrient-rich run-off or stabilisation of water level and flow as a consequence of catchment management programmes. Similarly, weed plants are frequently an important component of the system by virtue of their biomass and the area they infest and a significant decrease in their population as a result of control measures could have a marked effect on the nature and function of other components of the system.

Therefore proposals for water weed management must be thoroughly integrated with the programme of management for the overall system. Where an overall plan is not expressly formulated, as may often be the case, proposals for weed management could include appropriate techniques of environmental manipulation and measures with potentially adverse ecological impacts, provided that the consequences of these proposals to the whole system are explicitly and clearly outlined for the decision maker.

7.6.2 *Objectives for management*

The objectives for the management of aquatic weeds are often so obvious that they are given no detailed consideration. However, when the problem is a major one with complex ecological ramifications, it is important to assess the nature of the problem, as far as possible in quantitative terms (see Mitchell, 1979*a*), and specify in detail the objectives of the programme. These, in turn, will

Plate 7.6. Before the Chinese invented paper, the Egyptians had a monopoly on 'writing parchment' made from papyrus. This is a particularly well-known scroll, but the illustration is actually of a superb reproduction by Corrado's Institute (Plate 7.5): even the wormholes have been 'built-in' (top left). (Photo K. Thompson.)

have to be related to the current and projected uses of the infested water body. It is important that objectives take account of socio-economic factors. This may involve consultation with local communities. Also objectives should incorporate a specific time span at the end of which they will be reassessed.

7.6.3 *Choice of alternative methods*

Different techniques for the management of water weeds were briefly referred to in section 7.3. The formulation of a programme of management from such techniques has been discussed by Mitchell (1978, 1979*b*). Simplified forms of cost-benefit analysis and cost-effectiveness analysis can be used to select the most suitable method or combination of methods. One of the alternatives that should always be considered is that of predicting the consequences of doing nothing and allowing the weed population to attain its maximum possible size. This often provides the most convincing reason for introducing control measures which otherwise may be unwelcome in terms of economic and/or environmental costs.

7.6.4 *Monitoring and re-evaluation*

Criteria by which a control programme will be judged should be determined at the outset and the means of objectively assessing the progress and impact of the programme incorporated in it. The collection of this information allows the programme to be re-evaluated and, if necessary, modified in the light of the objectives. The data can also be used to predict the eventual outcome of the programme and its applicability in other similar situations. The possible need for modification requires that the original programme be flexible. In initial stages this requirement acts against the expenditure of large sums of money on expensive equipment as this often commits all future control procedures to be based on the machine, even when it may be found to be less cost effective than another method.

7.6.5 *The steps in water weed management*

Mitchell (1974) and Soerjani (1977) recognised four phases in the management of water weed problems: problem identification, control selection,

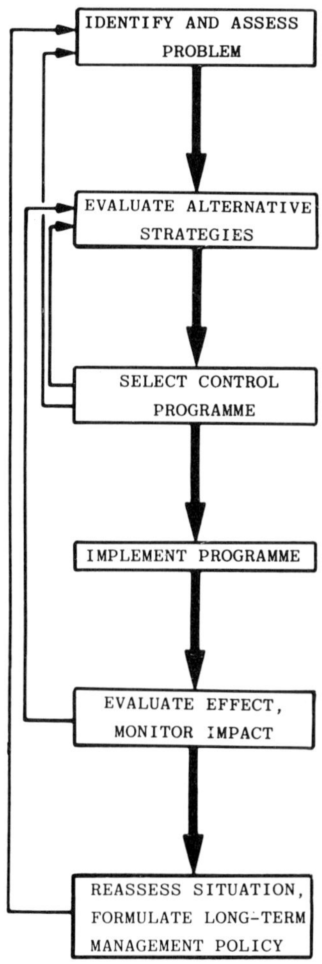

Fig. 7.1. The six stages embodied in water weed management.

control implementation and long term management. Later, Mitchell (1980) expanded this to six stages which can be embodied in a flow diagram as in Fig. 7.1. The management of long term, major or complex problems can be assisted by this systematic approach; minor, short term problems do not warrant it, though due care has to be paid to potential environmental hazards which could result from control measures.

7.6.6 *The role of responsible personnel*

Many of the worst aquatic weeds grow prolifically. For example, both *Eichhornia crassipes* and *Salvinia molesta* have the capacity to double in less than ten days in field conditions (Perkins, 1972; Mitchell and Tur, 1975). Infestations of these weeds are generally not noticed until the population is in the later stages of its exponential growth curve and the problem is then rightly perceived as being out of control. This situation has two implications for the responsible authority. First, rapid action must be taken at the earliest possible opportunity and obviously sensitive situations should be regularly and carefully surveyed for the presence of alien water plants. Prolonged bureaucratic procedures would of course be a waste of time and money. Second, there is likely to be little benefit from hasty over-reaction to a situation which is already out of control, though procedures aimed at preventing further spread should be instituted immediately. The systematic formulation of the most suitable programme can then occur.

7.7 References

Abdalla, A.A. & Abdel Hafeez, A.T., 1969. Some aspects of utilization of water hyacinth (*Eichhornia crassipes*). PANS, 15:204–207.

Beadle, L.C., 1981. The inland waters of tropical Africa: an introduction to tropical limnology. Second Edition. Longman, London.

Bennett, F.D., 1974. Biological control. In: D.S. Mitchell (ed.), Aquatic vegetation and its use and control. Unesco, Paris: 99–106.

Blackburn, R.D., 1974. Chemical control. In: D.S. Mitchell (ed.), Aquatic vegetation and its use and control. Unesco, Paris: 85–98.

Bogart, D.B., 1949. The effect of aquatic weeds on flow in Everglades canals. Soil Science Society of Florida, Proceedings, 9:32–52.

Bond, W.J. & Roberts, M.G., 1978. The colonization of Cabora Bassa, Moçambique, a new man-made lake, by floating aquatic macrophytes. Hydrobiologia, 60:243–259.

Boughey, A.S., 1963. The explosive development of a floating weed vegetation on Lake Kariba. Adansonia, 3:49–61.

Bowmer, K.H., 1979. Management of aquatic weeds in Australian irrigation systems. Proceedings of the Seventh Asian-Pacific Weed Science Society Conference, 1979: 219–222.

Bowmer, K.H., 1982. Residues of glyphosate in irrigation water. Pesticide Science, 13:623–628.

Bowmer, K.H. & Sainty, G.R., 1977. Management of aquatic plants with acrolein. Journal of Aquatic Plant Management, 15:40–46.

Bowmer, K.H., Sainty, G.R., Smith, G. & Shaw, K., 1979. Management of elodea in Australian irrigation systems. Journal of Aquatic Plant Management, 17:4–12.

Bowmer, K.H. & Smith, G.H., 1983. Herbicides for injection into flowing water: acrolein and endothal-amine. Weed Research, 24:201–211.

Boyd, C.E., 1972. A bibliography of interest in the utilization of vascular aquatic plants. Economic Botany, 26:74–84.

Boyd, C.E., 1974. Utilization of aquatic plants. In: D.S. Mitchell (ed.), Aquatic vegetation and its use and control. Unesco, Paris: 107–115.

Brezny, O., Mehta, I. & Sharma, R.K., 1973. Studies on evapotranspiration of some aquatic weeds. Weed Science, 21:197–204.

Brooker, M.P. & Edwards, R.W., 1975. Aquatic herbicides and the control of water weeds. Water Research, 9:1–15.

Bruwer, C.A., 1979. The economic impact of eutrophication in South Africa. South Africa Department of Water Affairs, Pretoria, Technical Report No. TR 94.

Burian, K., 1971. Primary production, carbon dioxide exchange and transpiration in *Phragmites communis* Trin. on the Lake Neusiedler See, Austria. Hidrobiologia, 12:203–218.

Carter, G.S., 1955. The papyrus swamps of Uganda. W. Heffer and Sons, Cambridge.

Chalmers, M.I., 1968. Report to World Food Programme on a visit to Sudan. In: Animal Production WFP. Mission report on animal nutrition in Sudan. Study on the use of water hyacinth in ruminant animal feeding and also as a means of weed control. FAO, Rome.

Cook, C.D.K., 1968. The vegetation of the Kainji Reservoir site in Northern Nigeria. Vegetatio, 15:225–243.

Davies, H.R.J., 1959. Effects of the water hyacinth (*Eichhornia crassipes*) in the Nile Valley. Nature, London, 184:1085–1086.

Dawood, K.E., Farooq, M., Dazo, B.C., Miguel, L.C. & Unrau, G.O., 1965. Herbicide trials in the snail habitats of the Egypt 49 project. Bulletin of the WHO, 32:269–287.

Dawson, F.H. & Kern-Hansen, U., 1979. The effect of natural and artificial shade on the macrophytes of lowland streams and the use of shade as a management technique. Internationale Revue der gesamten Hydrobiologie, 64:437–455.

Denny, P., 1984. Permanent swamp vegetation of the Upper Nile. Hydrobiologia, 110:79–90.

Dubbers, F.A.A., Ghoneim, S., Siemelink, M.E., El Gharably, Z., Pieterse, A.H. & Blom, J.E., 1981. Aquatic weed control in irrigation and drainage canals in Egypt by means of grass carp (*Ctenopharyngodon idella*). Proceedings of the Fifth International Symposium on Biological Control of Weeds, Brisbane, 1980: 261–271.

Durden, W.C., Blackburn, R.D. & Gangstad, E.O., 1975. Control program for alligatorweed in the south eastern States. Hyacinth Control Journal, 13:27–30.

Edwards, D. & Thomas, P.A., 1977. The *Salvinia molesta* problem in the northern Botswana and eastern Caprivi area. Proceedings of the Second National Weeds Conference of South Africa: 221–237.

El Gharably, Z., Khattab, A.F. & Dubbers, F.A.A., 1982. Experience with grass carps for the control of aquatic weeds in irrigation canals and drains in Egypt. Proceedings of the Second International Symposium on Herbivorous Fish, Novi Sad: 17–26.

El Hadidi, N.N., 1976. The riverain flora in Nubia. In: J. Rzoska (ed.), The Nile, biology of an ancient river. Junk, The Hague: 87–91.

Entz, B.A.G., 1976. Lake Nasser and Lake Nubia. In: J. Rzoska (ed.), The Nile, biology of an ancient river. Junk, The Hague: 271–291.

Entz, B.A.G., 1980. Ecological aspects of Lake Nasser – Nubia. The first decade of its existence, with special reference to the development of insect populations and the land and water vegetation. In: M. Kassas & S.I. Ghabbour (eds.), The Nile and its environment. Water Supply Management, 4(1–2):67–72.

Gangstad, E.O., 1976. Biological control operations on alligatorweed. Hyacinth Control Journal, 14:50–53.

Gangstad, E.O., Spencer, N.R. & Foret, J.A., 1975. Towards integrated control of alligatorweed. Hyacinth Control Journal, 13:30–33.

Gaudet, J.J., 1974. The normal role of vegetation in water. In: D.S. Mitchell (ed.), Aquatic vegetation and its use and control. Unesco, Paris: 24–37.

Gaudet, J.J., 1975. Mineral concentration in papyrus in various African swamps. Journal of Ecology, 63:483–491.

Gaudet, J.J., 1976*a*. Nutrient relationships in the detritus of a tropical swamp. Archiv fur Hydrobiologie, 78:213–239.

Gaudet, J.J., 1976*b*. *Salvinia* infestation on Lake Naivasha in East Africa. In: C.K. Varshney & J. Rzoska (eds.), Aquatic weeds in South East Asia. Junk, The Hague: 193–209.

Gaudet, J.J., 1977. Uptake, accumulation and loss of nutrients by papyrus in tropical swamps. Ecology, 58:415–422.

Gaudet, J.J., 1979*a*. Seasonal changes in nutrients in a tropical swamp: North Swamp, Lake Naivasha, Kenya. Journal of Ecology, 67:953–981.

Gaudet, J.J., 1979*b*. Weeds in African man-made lakes. PANS, 25:279–286.

Gopal, B. & Sharma, K.P., 1981. Water-Hyacinth (*Eichhornia crassipes*): the most troublesome weed of the world. Hindasia Publishers, Delhi.

Graham, W.A.E., 1973. Perennial submersed and floating aquatic weeds in South-eastern Australia. Proceedings of the Second Victorian Weeds Conference, Melbourne.

Guscio, F.J., Bartley, T.R. & Beck, A.N., 1965. Water resources problems generated by obnoxious plants. Journal of the Waterways Harbors Division. Proceedings of the American Society of Civil Engineers, 91:47–60.

Hall, J.B. & Okali, D.U.U., 1974. Phenology and productivity of *Pistia stratiotes* L. on the Volta Lake, Ghana. Journal of Applied Ecology, 11:709–725.

Hall, J.B., Laing, E., Hossain, M. & Lawson, G.W., 1969. Observations on aquatic weeds in the Volta Basin. In: L.E. Obeng (ed.), Man-made lakes. The Accra Symposium. Ghana Universities Press, Accra: 331–343.

Harley, K.L.S. & Mitchell, D.S., 1981. The biology of Australian weeds. 6: *Salvinia molesta*, D.S. Mitchell. Journal of the Australian Institute of Agricultural Science, 47:67–76.

Hattingh, E.R., 1962. Report on investigations into the control of *Salvinia auriculata* on Lake Kariba. Part I. Union Weedkiller Services, Johannesburg.

Hattingh, E.R., 1963. Report on investigations into the control of *Salvinia auriculata* on Lake Kariba. Part II: Field trials and recommendations. Union Weedkiller Services, Johannesburg.

Hira, P.R., 1969. Transmission of schistosomiasis in Lake Kariba, Zambia. Nature, London, 224:670–672.

Holm, L.G., Weldon, L.W. & Blackburn, R.D., 1969. Aquatic weeds. Science, New York, 166:699–709.

Husak, S., 1978. Control of reed and reed mace stands by cutting. In: D. Dykyjova & J. Kvet (eds.), Pond littoral ecosystems. Springer-Verlag, Berlin: 404–408.

Jacot Guillarmod, A., 1979. Water weeds in southern Africa. Aquatic Botany, 6:377–391.

Jarvis, M.J.F., Mitchell, D.S. & Thornton, J.A., 1982. Aquatic macrophytes and *Eichhornia crassipes*. In: J.A. Thornton (ed.), Lake McIlwaine: the eutrophication and recovery of a tropical African man-made lake. Junk, The Hague: 137–144.

Jordan, P. & Webbe, G., 1969. Human schistosomiasis. Heinemann Medical Books, London.

Kamal, I.A. & Little, E.C.S., 1970. The potential use of water hyacinth for horticulture in the Sudan. PANS, 16:488–496.

Krasznai, Z., Marian, T., Burizs, L. & Ditroi, F., 1982. Production of sterile hybrid grass carp (*Ctenopharyngodon idella* Val. × *Aristichthys nobilis* Rich.) for weed control. Proceedings of the Second International Symposium on Herbivorous Fish, Novi Sad: 55–61.

Lebrun, J., 1959. La lutte contre le developpement de l'*Eichhornia crassipes*. Bulletin Agricole du Congo Belge, 50:251–252.

Linacre, E.T., Hicks, B.B., Sainty, G.R. & Grauze, G., 1970. The evaporation from a swamp. Agricultural Meteorology, 7:375–386.

Little, E.C.S., 1967. Progress report on transpiration of some tropical water weeds. PANS, 13:127–132.

Little, E.C.S., 1968. The control of water weeds. Weed Research, 8:79–105.

Little, E.C.S., 1969. Weeds and man-made lakes. In: L.E. Obeng (ed.), Man-made lakes. The Accra Symposium. Ghana Universities Press, Accra: 284–291.

Little, E.C.S., 1979. Handbook of utilization of aquatic plants. FAO Fisheries Technical Paper No. 187.

Livermore, D.F. & Wunderlich, W.E., 1969. Mechanical removal of organic production from waterways. In: Eutrophication: causes, consequences, correctives. National Academy of Sciences, Washington, D.C.: 424–519.

Maembe, T.W., 1981. Final report on the improvement of fish processing and transportation on Lake Chad. FAO, Rome.

Malhotra, S.P., 1972. Remedy for aquatic weeds in Bhakra canals. Roxana Printers, Chandigarh, India.

Manning, J.H. & Johnson, R.E., 1975. Water level fluctuation and herbicide application: an integrated control method for hydrilla in a Louisiana reservoir. Hyacinth Control Journal, 13:11–17.

Manning, J.H. & Sanders, D.R., 1975. Effects of water fluctuation on vegetation in Black Lake, Louisiana. Hyacinth Control Journal, 13:17–21.

Marshall, B.E. & Junor, F.J.R., 1981. The decline of *Salvinia molesta* on Lake Kariba. Hydrobiologia, 83:477–484.

Mitchell, D.S., 1969. The ecology of vascular hydrophytes on Lake Kariba. Hydrobiologia, 34:448–464.

Mitchell, D.S., 1970. Autecological studies of *Salvinia auriculata* Aubl. Ph.D. Thesis, University of London.

Mitchell, D.S., 1973. Aquatic weeds in man-made lakes. In: W.C. Ackermann, G.F. White & E.B. Worthington (eds.), Man-Made lakes: their problems and environmental effects. American Geophysical Union, Washington, D.C.: 606–611.

Mitchell, D.S. (ed.), 1974. Aquatic vegetation and its use and control. Unesco, Paris.

Mitchell, D.S., 1976. Water weed problems in irrigation systems. In: E.B. Worthington (ed.), Arid land irrigation in developing countries: environmental problems and effects. Pergamon Press, Oxford: 317–328.

Mitchell, D.S., 1978. Aquatic weeds in Australian inland waters. Australian Government Publishing Service, Canberra.

Mitchell, D.S., 1979*a*. Assessment of aquatic weed problems. Journal of Aquatic Plant Management, 17:19–21.

Mitchell, D.S., 1979*b*. Formulating aquatic weed management programs. Journal of Aquatic Plant Management, 17:22–24.

Mitchell, D.S., 1980. Aquatic weeds. In: W.D. Williams (ed.), An ecological basis for water resource management. Australian National University Press, Canberra: 81–89.

Mitchell, D.S. & Rose, D.J.W., 1979. Factors affecting fluctuations in extent of Salvinia molesta on Lake Kariba. PANS, 25:171–177.

Mitchell, D.S. & Tur, N.M., 1975. The rate of growth of *Salvinia molesta* (*S. auriculata* Auct.) in laboratory and natural conditions. Journal of Applied Ecology, 12:213–225.

Mitchell, S.A., 1976. The marginal fish fauna of Lake Kariba. Kariba Studies, Paper No. 8:109–162.

Moghraby, A.I. el, 1982. The Jonglei Canal – needed development or potential ecodisaster? Environmental Conservation, 9:141–148.

Morton, J.F., 1975. Cattails (*Typha* spp.) – weed problem or potential crop? Economic Botany, 29:7–29.

National Council for Research of Sudan & National Academy of Sciences, USA, 1975. Aquatic weed management: some prospects for the Sudan and the Nile Basin. Aquatic Weed Management Workshop, 24–29 November, 1975. Khartoum.

Newbold, C., 1975. Herbicides in aquatic systems. Biological Conservation, 7:97–118.

Nichols, S.A., 1974. Mechanical and habitat manipulation for aquatic plant management. Technical Bulletin, Department of Natural Resources, Wisconsin, No. 77.

Okafor, L.I., 1980. Weeds of the Lake Chad and their potential problems in the South Chad Irrigation Project. Weed Science Society of Nigeria, Tenth Annual Conference, Benin City, Nigeria.

Okafor, L.I., 1982. A preliminary survey on the major aquatic weeds of Lake Chad, Nigeria. Proceedings of the Sixth International Symposium on Aquatic Weeds, Novi Sad: 10–19.

O'Loughlin, E.M. & Bowmer, K.H., 1975. Dilution and decay of aquatic herbicides in flowing channels. Journal of Hydrology, Amsterdam, 26:217–235.

Osman, H.E., El Hag, G.A. & Osman, M.M., 1975. Studies on the nutritive value of water hyacinth (*Eichhornia crassipes* Mart. Solms). In: M. Obeid (ed.), Aquatic weeds in the Sudan with special reference to water hyacinth. National Council for Research, Agricultural Research Council, Khartoum: 104–127.

Paperna, I., 1968. Studies on the transmission of schistosomiasis in Ghana. III: Notes on the ecology and distribution of *Bulinus truncatus rohlfsi* and *Biomphaearia pfeifferi* in the Lower Volta Basin. Ghana Medical Journal, 7:139–145.

Penfound, W.T. & Earle, T.T., 1948. The biology of the water hyacinth. Ecological Monographs, 18:447–472.

Penman, H.L., 1963. Vegetation and hydrology. Farnham Royal, Technical Communication, Commonwealth Bureau of Soils, No. 53.

Perkins, B.D., 1972. Potential for water hyacinth management with biological agents. Proceedings of the Fourth Tall Timbers Conference on Ecological Animal Control by Habitat Management: 53–64.

Pitlo, R.H., 1978. Regulation of aquatic vegetation by interception of daylight. Proceedings of the Fifth International Symposium on Aquatic Weeds, Amsterdam: 91–99.

Pitlo, R.H., 1982. Flow resistance of aquatic vegetation. Proceedings of the Sixth International Symposium on Aquatic Weeds, Novi Sad: 225–234.

Ramarokoto, M.R., 1965. Water hyacinth in Madagascar. Working Paper. Food and Agriculture Organization Conference on Quelea and Water Hyacinth Control, Douala.

Rijks, D.A., 1969. Evaporation from a papyrus swamp. Quarterly Journal of the Royal Meteorological Society, 95:643–649.

Robson, T.O., 1974. Mechanical control. In: D.S. Mitchell (ed.). Aquatic vegetation and its use and control. Unesco, Paris: 72–84.

Room, P.M., Harley, K.L.S., Forno, I.W. & Sands, D.P.A., 1981. Successful biological control of the floating weed salvinia. Nature, London, 294:78–80.

Rzoska, J., 1974. The Upper Nile swamps. Freshwater Biology, 4:1–30.

Sale, P.J.M. & Wetzel, R.G., 1983. Growth and metabolism of *Typha* species in relation to cutting treatments. Aquatic Botany, 15:321–334.

Schelpe, E.A.C., 1961. The ecology of *Salvinia auriculata* and associated vegetation on Kariba Lake. Journal of South African Botany, 27:181–187.

Scott, W.E., Ashton, P.J. & Steyn, D.J., 1979. Chemical control of the water hyacinth on Hartbeespoort Dam. Water Research Commission, National Institute for Water Research and Hydrological Research Institute, South Africa, Pretoria.

Sculthorpe, C.D., 1967. The biology of aquatic vascular plants. Edward Arnold, London.

Seidel, K., Happel, H. & Graue, G., 1978. Contributions to revitalisation of waters. Second Edition. Stiftung Limnologische Arbeitsgruppe, Dr Seidel e.V., Krefeld, F.R.G.

Shelton, W.L., Boney, S.P. & Rosenblatt, E.M., 1982. Monosexing grass carp by sex reversal and breeding (Abstract). Proceedings of the Second International Symposium on Herbivorous Fish, Novi Sad: 53–54.

Siemelink, M., Zonneveld, N. & El Enbaby, O.A.M., 1982. Controlled reproduction and cage culture of grass carp (*Ctenopharyngodon idella* Val.) in Egypt. Proceedings of the Second International Symposium on Herbivorous Fish, Novi Sad: 76–82.

Singh, S.P., Pahuja, S.S. & Moolani. M.K., 1976. Cultural control of *Typha angustata* at different stages of growth. In: C.K. Varshney & J. Rzoska (eds.), Aquatic weeds in South East Asia. Junk, The Hague: 245–247.

Soerjani, M., 1977. Integrated control of weeds in aquatic areas. In: J.D. Fryer & S. Matsunaka (eds.), Integrated control of weeds. University of Tokyo Press, Tokyo: 121–151.

Steenberg, B., 1968. Papyrus: problems in its utilisation for pulp and paper making. In: Pulp and paper development in Africa and the Near East. Vol. 2. FAO, Rome: 865–876.

Thompson, K., 1974. Ecology, management and utilization of aquatic and semi-aquatic vegetation in the Okavango Delta, Botswana: a review, with recommendations for further studies. Technical Report for UNDP Project BOT/506. FAO, Land and Water Development Division, Rome.

Timmer, C.E. & Weldon, L.W., 1967. Evapotranspiration and pollution of water by water hyacinth. Hyacinth Control Journal, 6:34–37.

United States National Academy of Sciences, 1976. Making aquatic weeds useful: some perspectives for developing countries. National Academy of Sciences, Washington, D.C.

Van der Bliek, A.M., El Gharably, Z., Pieterse, A.H. & Scheepens, M.H.M., 1982. Observations of the phenology of *Potamogeton pectinatus* L. and other submersed weeds in irrigation systems in Egypt. Proceedings of the Sixth International Symposium on Aquatic Weeds, Novi Sad: 37–44.

Van Zon, J.C.J., 1981. Status of the use of grass carp (*Ctenopharyngodon idella* Val.). Proceedings of the Fifth International Symposium on Biological Control of Weeds, Brisbane, 1980:249–260.

Wild, H., 1961. Harmful aquatic plants in Africa and Madagascar. Kirkia, 2:1–66.

Wolverton, B.C., Barlow, R.M. & McDonald, R.C., 1976. Application of vascular aquatic plants for pollution removal, energy, and food production in a biological system. In: J. Tourbier & R.W. Pierson (eds.), Biological control of water pollution. University of Pennsylvania Press, Philadelphia: 141–149.

Wright, A.D., 1981. Biological control of water hyacinth in Australia. Proceedings of the Fifth International Symposium on Biological Control of Weeds. Brisbane, 1980:529–535.

Author's address:
CSIRO Centre for Irrigation Research
Griffith, NSW 2680
Australia

CHAPTER 8

The conservation and management of African wetlands

CLIVE HOWARD-WILLIAMS and KEITH THOMPSON

Detailed contents

8.1 Introduction

The swamps associated with the complex fishpond ecosystems in Czechoslovakia, which have been under human management since about the 13th century, are now apparently threatened by modern hydrological and agricultural projects (Hejný, 1973). The floodplain soils of the lower Nile Valley and Delta have supported large human populations for 6000 years. Only late twentieth century technology has destroyed or damaged the natural system. The fact that such systems, which have been under man's active management for so long, are now threatened by his own activities makes one appreciate immediately similar dangers in other areas. This is particularly true for wetlands of the tropics, where man has until recently either ignored or 'reclaimed' them. For instance, Fosberg (1973) observes: 'The planner from a "developed" country usually finds it impossible to conceive of floods as anything but bad, so flood control usually accompanies "development" whether or not it is ecologically desirable'.

The implications of wetland conservation have been recognised internationally for the last ten years, essentially since the ratification of the so-called Ramsar Convention (sponsored by UNESCO in Ramsar, Iran, 2 Feb. 1971) on wetlands of international importance. At present the concepts of wetland conservation and management are being promoted by international agencies such as IUCN (International Union for Conservation of Nature and Natural Resources) and SCOPE (Scientific Committee on Problems of the Environment). In 1978 wetlands were included within the topics considered by the Aquatic Commission of the International Association for Ecology-INTECOL (INTECOL Newsletter B (6) 1978). The U.N. Environment Programme (UNEP) has recently decided to fund an international research programme on wetlands and shallow water bodies (SCOPE Newsletter 11, 1980) including African ones, and a special African wetlands project has now been set up. Following the publication of Project Aqua by Luther and Rzoska in 1971, the arguments for the conservation and wise management

Patrick Denny (ed.), The ecology and management of African wetland vegetation.
 ISBN 90 6193 509 1

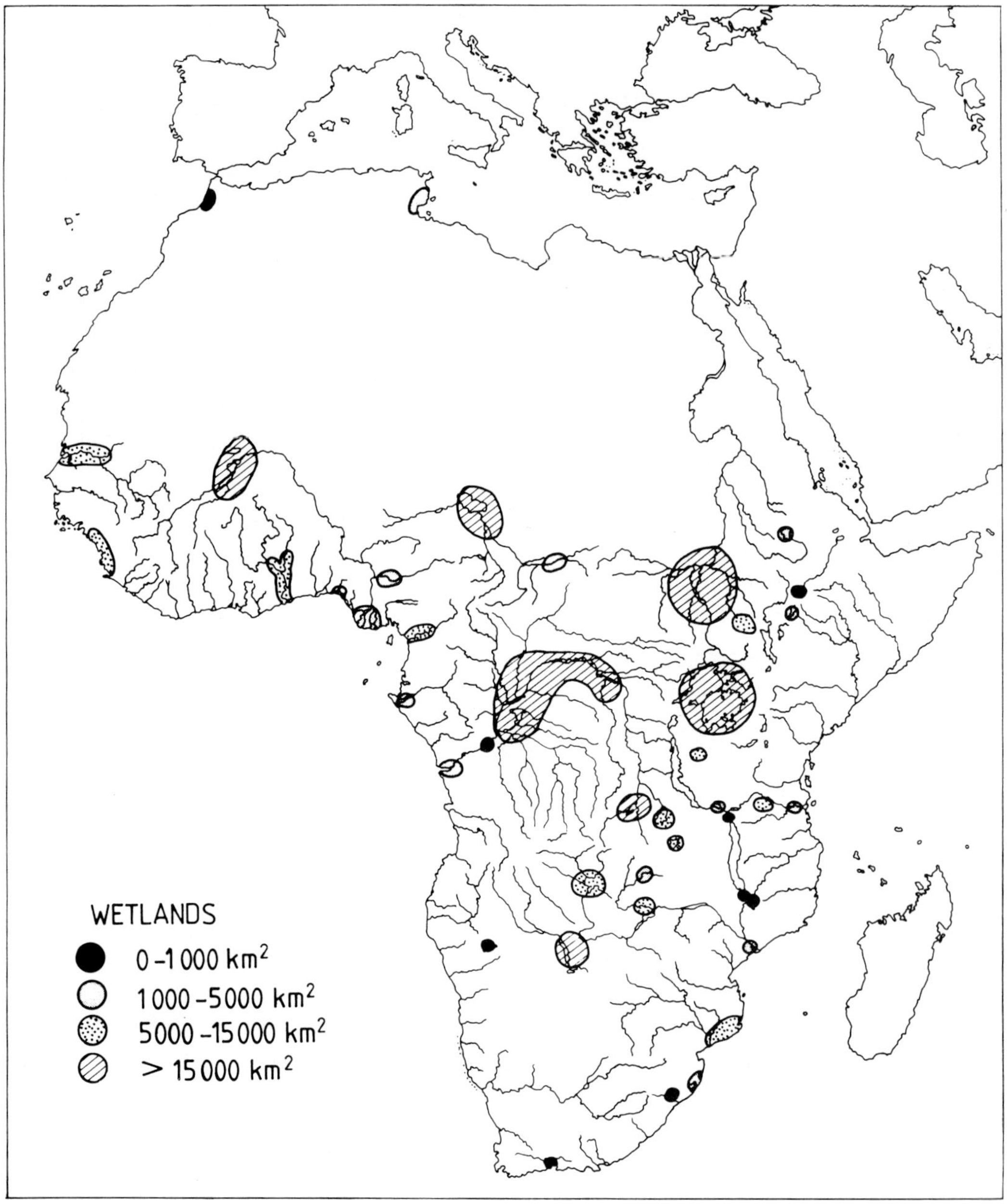

Fig. 8.1. Major areas of permanent and seasonal wetlands in Africa (refer Table 8.1. and Chapter 3 for fuller details). Most of the wetlands fall into three broad categories: (a) Endorheic systems (Okavango, Chad, *etc.*) (b) Upper floodplains of major rivers (Nile, Niger, Zaire, Zambezi). (c) Coastal riverine floodplains (Senegal, Niger, Zambezi, Incomati, *etc*). (d) Mountain wetlands (Drakensburg, Nyika, Rwanda/Burundi, Mt Kenya, Rwenzori, Mt Cameroun, Aberdares/Elgon).

Some of the areas delimited on Figure 8.1 (*e.g.* Morocco, coastal Moçambique) represent several separate wetlands, but in no case are the combined wetland areas less than about 250 km². Naturally, Africa also has very large numbers of small wetlands, particularly around lakes and in the lowland sections of western – and south-eastern – African rivers. Many of these would undoubtedly rate highly if they occurred in intensively researched countries of Europe or North America.

Many major edaphic grasslands are included in Figure 8.1. (*e.g.* Barotse Plains, Malagarasi Swamps, Chari-Logone floodplain, *etc*), but alkaline grasslands and dambo-type seasonal wetlands are still greatly under-represented (see Section 3.5.3).

of wetlands in Africa have been gaining ground (Beadle, 1974; Worthington, 1976; Gaudet, Mitchell and Denny, 1981; Thompson, 1974, 1976*a*). In the words of Worthington: 'General appreciation of the environment has reached a point today when it is hardly necessary to make the overall case for conservation and overall development of water and wetlands' (Worthington, 1976).

Beadle (1974) however, points out: 'The vast extent of the [African] swamplands poses some important problems of land use, to solve which we should know as much as possible of the peculiar conditions in swamps, and of the mechanisms by which plants and animals are adapted to them'.

Those interested enough to read this book (and to have progressed as far as this chapter) will almost certainly be familiar with the diverse arguments in favour of wetland conservation. This chapter is, therfore, not intended as merely a rehash of all the old arguments, even though many of these are very important. Rather, we would like to select a few of the many ways in which wetlands can be beneficial to man, and show with African examples that they are a valuable resource rather than just areas to be preserved.

8.2 Distribution, areas and conservation status

Considering that Africa is a dry continent (95% of the 30 million km^2 receives less than 500 mm of rain per year), it supports a surprisingly large area of wetland, some 345 000 km^2 or 1% of the total land area. This does not include the deep lakes and it also under-represents the dambo, pan and black day seasonal wetlands (*see* Section 3.4.2). The regional drainage systems can be seen in Fig. 8.1.

The most extensive swamp in the world, the Upper Nile Swamp in the Sudan, occupies an area of some 92 000 km^2 including floodplain of which 40 000 km^2 can be considered as permanent swamp (Beadle, 1974; Rzoska, 1974). Large swamp areas occur in Mali on the Middle Niger, in Chad, Zaire, Botswana and in Zambia. Lake Bangweulu has an associated swamp (6000 km^2) and floodplain (6000 km^2) of some 12 000 km^2 (Toews, 1975). To the west in northern Botswana, the Okavango Delta covers an area of 16 000 km^2 including both swamps and islands (Astle and Graham, 1976). As well as these large swamp areas, a network of swamps occupies about 6% (some 11 800 km^2) of the land surface of Uganda (Lind, 1956), 3% (some 24 000 km^2) of Zambia (Clarke, 1975) and an estimated 12 200 km^2 of small wetlands covers southern Africa south of the Limpopo River (D.M. Skead, unpublished). In addition, most of the African lakes, particularly the shallow ones, have associated fringing swamps.

Table 8.1 lists the larger African freshwater wetlands with estimated areas, although the difficulty of defining the area of a floodplain is obvious, and areas can vary considerably depending on who calculates them. Balek (1977) for instance, cites some swamp areas very much greater than those we believe to be reasonable ourselves (see Table 8.1). The list does not include the open water of deep lakes, reservoirs or coastal lagoons. Undoubtedly, many large wetlands are omitted. Morgan and Boy (1982) point out that wetland sites as large as 22 000 ha are not even represented on North African maps with a scale of 1:1 000 000–1:2 500 000.

The table highlights that the area of wetlands in Africa which is in some way protected is totally inadequate. Mavuti (1982) points out that less than 3% of the area of Kenya is occupied by swamp, but swamp conservation does not receive priority on a national scale. 'The tendency is to get rid of swamps because they harbour diseases or to drain and convert them into farmlands for food production.' From the data we have been able to assemble, only a small area of the larger wetlands have any conservation status – and much of this 'status' is of dubious value. In addition, none of the 'protected' areas is hydrologically independent. In spite of this overall low figure, some individual countries have a surprisingly high proportion of their wetlands under protection. In Zambia, for instance, where the national park system comprises 8.12% of the country, this system includes 30% of the country's wetlands (Clarke, 1975).

8.3 Importance and management potential

In modern Africa any conservation and manage-

Table 8.1. List of the larger freshwater wetland areas of the African continent and their known conservation status. Saline lakes and coastal lagoons are omitted. The wetlands are arranged in the drainage systems of Africa according to Pritchard, 1979. In addition to these listed here there are innumerable small wetlands across the continent. N.I. = no information available. (W) and (D) refers to wet and dry seasons respectively.

Drainage system	Name	Locality	Wetland type	Area km²	Conservation status	Special features	Reference
ZAIRE/ CONGO SYSTEM	Middle Congo swamps	Zaire, Congo	Riverine swamps and floodplain	40 550	N.I.	Permanent and seasonal	Balek (1977) Thompson (unpubl.)
	Kifukula depression	Zaire, Zambia	Swamps, floodplain	1 502	None	Swamps on the Luapula River. Variable area up to 10 m deep in places. Many small lakes and lagoons	Clarke (1975) Welcomme (1972)
	Mweru (Luapula River)	Zambia, Zaire	Shallow Lake (Mweru)	4 580	None	3–10 m deep, Major fishery	Clarke (1975) Welcomme (1972)
	Bangweulu	Zambia	Shallow lake, swamps and floodplain	8 800	Good – most of area in Game Management areas or in Isangano National Park	Black lechwe, waterfowl, Lake 4 m deep	Clarke (1975)
	Kamulondo depression	Zaire	Shallow lakes, swamps and floodplain	11 800 (W) 7 040 (D)	Approx. 20% protected in Parc National de L'Upemba	Lake Upemba (500 km²) and 50 other lakes on floodplain of Lualaba River	van Meel (1953) Welcomme (1972)
	Mweru Wantipa	Zambia	Shallow lake and swamps	300 1 204	Good – area lies in Mweru Wantipa National Park	Shallow lake, major fishery, abundant wildlife	Clarke (1975)
	Malebo Pool	Zaire, Congo	Shallow riverine lake	*ca.* 300			
	Tumba	Zaire	Shallow lake	767	N.I.	4 m deep	Welcomme (1972)
	Cheshi	Zaire	Shallow lake	41	N.I.	<3 m deep	Welcomme (1972)
	Maji Ndombe	Zaire	Shallow lake	2 294	N.I.	mean depth 5.2 m	Welcomme (1972)
	Rukwa	Tanzania	Shallow lake	2 300	Unprotected	variable size	Welcomme (1972) Lamprey (1975)
	Malagarasi	Tanzania	Floodplain and swamp	7 357	Unprotected	Important wildlife area	Balek (1977)
	Wembere	Tanzania	Floodplain and swamp	N.I.	Unprotected	Important wild bird area. Internal drainage	Lamprey (1975)
	Moyowosi	Tanzania	Swamp	N.I.	Unprotected	Wildlife abundant	Lamprey (1975)
NIGER-BENUE SYSTEM	Niger central delta	Mali	Floodplain, swamps and shallow lakes	20 000 (W) 3 900 (D)	Unprotected	Complex system of swamps and lakes. L. Niangaye is 402 km². Rice growing	Welcomme (1972)
	Niger river	Niger	Floodplain	907 (W) 270 (D)	Small section in Parc National du W		Welcomme (1972, 1975)
		Benin (Dahomey)	Floodplain	274 (W) 32 (D)			Bacalbasa-Dobrovici (1971)

	Niger river	Nigeria	Floodplain	4 800 1 800	(W) (D)			IUCN (1979)
	Toubouri depression	Cameroon, Chad	Shallow lakes and swamp	95	N.I.		Group of 4 lakes: Lére, Fianga, Tikem and Trené	Welcomme (1972)
	Sokoto river	Nigeria	Floodplain	*ca.* 400		Unprotected	Tributary of Niger R. Floodplain 8 km wide	Holden and Green (1960) Welcomme (1979)
	Benue river	Nigeria	Floodplain	3 100 1 240	(W) (D)	Unprotected		Motwani (1970) Welcomme (1975)
	Anambra river	Nigeria	Floodplain			Now affected by Kainji dam	50 lakes and ponds on floodplain. Important fishery	Awachie and Hare (1978)
NILE SYSTEM	Upper Nile swamps (Sudd)	Sudan	Swamp and floodplain	92 000 10 000	(W) (D)	4 small game reserves in the area: Zerat, Fanikang Is., Shambe and Rodigeru. Status uncertain. Part of area to be lost to Jonglei scheme	On Nile river, largest single swamp area in the world	Rzóska (1974) Lamprey (1975) Platenkamp (1978)
	Upper Kagera R.	Burundi, Rwanda, Tanzania	Shallow lakes and floodplain	340		Parc National de la Kagera covers much of the area	System of 15 shallow lakes including Lake Rugwero, and swamp	Welcomme (1972) IUCN (1980b)
	Lower Kagera R.	Rwanda, Tanzania	Shallow lakes and floodplain	180			Includes L. Hago, L. Ihema	Welcomme (1972)
	Ruzizi River	Burundi, Rwanda, Zaire	Floodplain and swamps	?		Proposal for 2 km^2 bird reserve in swamps	Flows from L. Kivu to L. Tanganyika. Large stands of papyrus	Welcomme (1972) Lamprey (1975)
	L. Victoria swamps	Uganda, Tanzania, Kenya	Swamp	?		Two small game reserves in swamp area near lake. Status now uncertain	Extensive papyrus swamps around lake	Beadle (1974) Lamprey (1975) Thompson (1976*c*)
	Ankole and Koki lakes	Uganda	Shallow lakes and floodplain	570		Part in Rwenzori National Park. Status now uncertain	Includes L. George (270 km^2)	Lamprey (1975) Welcomme (1972)
	Kenamuke swamp	Sudan	Swamp	13 955		N.I.		Balek (1977)
	Lotagipi	Sudan, Kenya	Swamp	12 900		Unprotected	Internal drainage basin	Balek (1977)
	Kyoga:							
	L. Kyoga	Uganda	Shallow lake	2 700			Average depth 6 m	Beadle (1982)
	L. Bisinia	Uganda	Shallow lake	160			Papyrus, *Miscanthidium* swamps	Welcomme (1972)
	L. Kwania and assoc. swamp	Uganda	Shallow lakes and swamps	465 1 700				
	Lake Tana	Ethiopia	Shallow lake	3 500		Unprotected	Average depth 8 m	Welcomme (1972)
	Semliki delta	Uganda, Zaire	Floodplain, delta	?		Status uncertain	Important wildlife area	Lamprey (1975)
	Lake Wamala	Uganda	Floodplain grassland	?		Unprotected		Lamprey (1975)

Table 8.1. Continued

Drainage system	Name	Locality	Wetland type	Area km^2	Conservation status	Special features	Reference
NILE SYSTEM (continued)	Yala swamp	Kenya	Swamp,delta	140	Unprotected	Currently water, development for agriculture	Mavuti (1982)
	Sondu-Ahero swamp	Kenya	Swamp, delta	140	Unprotected		Mavuti (1982)
ZAMBESI SYSTEM	Barotse plain (Zambesi floodplain)	Zambia	Floodplain	9 000 (W) 700 (D)	Part has Game Management area status	Heavily cultivated and grazed. Major fishery	Clarke (1975) Welcomme (1972)
	Liuwa plain	Zambia	Floodplain and swamp	3 500	Most of the area in Liuwa Plain National Park	Important wildlife area	Clarke (1975)
	Lukanga swamp	Zambia	Swamp	2 500	Unprotected	Papyrus, swamp fishery	Clarke (1975)
	Busanga plain	Zambia	Swamp and floodplain	300	80% in Kafue National Park, remainder in game management area	Papyrus. Wildlife abundant	Clarke (1975)
	Kafue flats	Zambia	Seasonal swamp, floodplain	6 000	Parts in Blue Lagoon and Lochinvar National Parks. Most lies in game management area	Major African wildlife area. Flood regime threatened by Itezhitezhi dam. Commercial fishery	Clarke (1975) Lamprey (1975)
	Chobe-Linyanti	Botswana, Namibia	Swamp	200	Small part in Chobe National Park	Mostly *Phragmites*	Lamprey (1975)
	Lake Liambezi	Namibia	Shallow lake	50	Unprotected		Seaman *et al.* (1978)
	Cameia	Angola	Floodplain	?	Cameia National Park	Floodplain of Luena R.	Huntley (1978)
	Lake Malombe	Malawi	Shallow lake and swamp	390	Part of swamp in Liwonde National Park	4 m deep. Receives outflow from L. Malawi	Welcomme (1972) Lamprey (1975)
	Shire River:						
	Elephant Marsh	Malawi	Swamp	520 (W) 140 (D)	Unprotected	Important swamp fishery	Welcomme (1972) Welcomme (1979)
	Ndindi Marsh	Malawi, Moçambique	Swamp	170 (W) 60 (D)			Hastings (1972) Welcomme (1979)
	Marromeu	Moçambique	Floodplain	?	Unprotected	Major wildlife area	Tinley (1975)
NORTH-WEST or MAGHREB SYSTEM only fresh-water wetlands included	Iriki	Morocco	Seasonal swamp	200 (W) 0 (D)	Unprotected	Important wild bird area	IUCN (1980a) Morgan & Boy (1982)
	Zima	Morocco	Shallow lake	6	Unprotected		Morgan & Boy (1982)
	Lacs du Moyen Atlas	Morocco	Shallow lakes	8	Some protected under Ramsar Convention	System of small lakes	Morgan & Boy (1982)

System	Wetland	Country	Type	Area (km²)		Status	Remarks	Reference
	Ichkeul	Tunisia	Shallow lake and swamp	120		Protected under Ramsar Convention	Important wild bird area. *Phragmites*, *Typha*	IUCN (1980a) Morgan & Boy (1982)
	Sebka Kelbia	Tunisia	Swamp	130		Unprotected	Permanent standing water. Important wild bird area	Morgan & Boy (1982)
	Chott el Fedjedj and Sebkhet el Hamma	Tunisia	Seasonal swamp	500 0	(W) (D)	Unprotected	Fresh water when full	Morgan & Boy (1982)
WESTERN SYSTEM	Ouémé	Benin, Nigeria	Floodplain and delta	896 52	(W) (D)	Unprotected	Includes lakes Azilli and Cele	Welcomme (1972) Welcomme (1979)
	Ogun River	Nigeria	Floodplain	63 25	(W) (D)	Partly protected by Ogun Forest Reserve		Welcomme (1979) IUCN (1979)
	Oshun River	Nigeria	Floodplain	37 20	(W) (D)	Unprotected		Welcomme (1979) Awachie and Hare (1978)
	Masilli River	Upper Volta	Floodplain	15 2	(W) (D)	Unprotected		Welcomme (1979)
	Volta River (White Volta)	Ghana	Floodplain	8532 1022	(W) (D)	N.I.		Welcomme (1979)
	Oti-Pendjari River	Upper Volta, Benin	Floodplain	4810	(W)	Protected by Arly reserve (Upper Volta) and Pendjari National Park (Benin)	Large flood prone lowlands – some forest. Rich wildlife areas	IUCN (1979)
	Gambia River	Gambia, Senegal	Floodplain			Part of floodplain in Niokolo Koba National Park (Senegal)	Heavily cultivated for rice	Welcomme (1975) IUCN (1979)
	Senegal River	Senegal, Mauritania	Floodplain, delta	4560 500 7970	(W) (D)	Djoudj National Park (120 km^2). Proposed extension (Toumbos Reserve) in Mauritania. Threatened by dams	Internationally important wild bird area. Important fishery	IUCN (1979) Welcomme (1975) Dorfman (1980)
	Lake Guiers	Senegal	Shallow lake	170		Unprotected		Welcomme (1972) IUCN (1979)
	Lake Rkiz	Mauritania	Shallow lake	120		Unprotected		Welcomme (1972) IUCN (1979)
	Lake Bam	Upper Volta	Shallow lake	13		Unprotected		Welcomme (1972)
WEST EQUA-TORIAL SYSTEM	Nyong River	Cameroon	Swamp forest	6688		Unprotected		Balek (1977) IUCN (1979)
	Dja River	Cameroon	Swamp forest	N.I.		Partly protected by Dja reserve		Balek (1977) IUCN (1979)
	Ogooué River	Gabon	Swamp, floodplain	N.I.		Unprotected	System of lakes, rivers and swamps. Abundant wildlife	IUCN (1979) Welcomme (1975)
	Lake Onangue	Gabon	Shallow lake	*ca.* 350		Unprotected		Welcomme (1972)
	Lake Mandje	Gabon	Shallow lake	*ca.* 50		N.I.		Welcomme (1972)

Table 8.1. Continued

Drainage system	Name	Locality	Wetland type	Area km^2	Conservation status	Special features	Reference
SOUTH EASTERN SYSTEM	Pongolo River	South Africa, Moçambique	Floodplain	100 (W) 26 (D)	Part in Ndumu Game Reserve	Sstem of shallow lakes on the floodplain. Heavily utilised for grazing and fishery	Heeg and Breen (1982)
	Pungué River	Moçambique	Floodplain	N.I.	Part in Gorongosa National Park	Rich wildlife area	Wild & Fernandes (1967)
	Incomati River	Moçambique	Floodplain, swamp	N.I.	Unprotected	Papyrus, *Typha* swamps	Wild & Fernandes (1967)
	Limpopo R. ?	Moçambique	Floodplain	N.I.	Part in Banhine National Park (7600 km^2 park area)	Rich wildlife area	Huntley (1978)
EASTERN SYSTEM	Lake Chilwa	Malawi, Moçambique	Shallow lake and swamp	600 800 (W) 700 (D)	Unprotected	Very important fishery. *Typha* Swamps	Kalk *et al.* (1979)
	Lake Chiuta	Malawi	Shallow lake and swamp	113	Unprotected	Drains to Lugenda River	Kalk *et al.* (1979) Lamprey (1975)
	Lake Amaramba	Malawi, Moçambique	Shallow lake and swamp	*ca.* 50	Unprotected	Drains to Lugenda River	Kalk *et al.* (1979) Lamprey (1975)
	Usangu	Tanzania	Floodplain and swamp (Utengule)	518	Unprotected	High agricultural potential High fishery potential	Bernascek (1981)
	Kilombero floodplain	Tanzania	Floodplain and swamp (Kibasira)	6647 89	Very small section in Selous Game Reserve	Highly productive fishery	Bernascek (1981)
	Rufiji River	Tanzania	Floodplain	1450	*ca.* 200 km^2 in Selous Game Reserve. Hydrology to be changed by Stieglers gorge dam	Active fishery. Internal delta area of small lakes	Bernascek (1981)
	Lorrian swamp	Kenya	Swamp	100	Unprotected	Iwaso Ngiro-Dera River	Mavuti (1982)
	Lake Naivasha	Kenya	Shallow lake and swamp	140	Some conservation status	Tourist area	Mavati (1982)
CHAD SYSTEM	Lake Chad	Chad, Niger, Nigeria, Cameroon	Shallow lake and swamps	13800	Small section in Lake Chad game sanctuary, Nigeria. Otherwise unprotected	Rich fishery. Lake of very variable size	Carmouze *et al.* (1972) Welcomme (1972)
	Yaérés	Chad, Cameroon	Floodplain	5957	Unprotected	On Chari and Logone rivers	Welcomme (1972) Welcomme (1979)
	Iro	Chad	Shallow lake, some swamp	195	Unprotected	On Salamat River	Welcomme (1972)

OKA-VANGO SYSTEM	Okavango delta	Botswana	Swamp	16 000 (W) 3 120 (D)	Small parts in Moremi Game Reserve and in Maun and Chobe National Parks	Abundant wildlife	Lamprey (1975) Astle and Graham (1978)
	Ngami	Botswana	Shallow lake	797	Unprotected	Rich fishery	Welcomme (1972) Welcomme (1979)
	Dow	Botswana	Shallow lake	> 100	Unprotected		Welcomme (1972) Lamprey (1975)
RUDOLPH SYSTEM	Abaya	Ethiopia	Shallow lake	1 161	Unprotected		Welcomme (1972)
Other lakes occur but these are saline	Awasa	Ethiopia	Shallow lake and swamp	202	Unprotected		Welcomme (1972)
	Shamo	Ethiopia	Shallow lake	551	Unprotected		Welcomme (1972)
	Ziway	Ethiopia	Shallow lake	437	Proposed Rift Valley lakes National Park	To be protected together with saline lakes Abiata, Shala and Langano	Lamprey (1975)
	Baringo	Kenya	Shallow lake	130	Part in a forest park		Lamprey (1975)
MADA-GASCAR	Alaotra	Madagascar	Shallow lake and swamp	200 1 000	N.I.	Rich fishery	Welcomme (1972) Moreau (1977)
	Itasy	Madagascar	Shallow lake	35	N.I.		Welcomme (1972)
	Kinkony and associated lakes	Madagascar	Shallow lakes	139–159	N.I.		Welcomme (1972)

ment issues must be viewed in the context of their place in the economies of the individual countries. As swamps are often associated with diseases such as schistosomiasis and malaria, and as swamp soils are often (but by no means always) relatively fertile and well-watered, there is thus a strong case for reclamation of swamps. However, there are equally strong arguments in favour of the maintenance of swamp areas, i.e. swamps:

(a) function as natural water storage and discharge areas;

(b) act as soil traps and prevent erosion in mountainous areas;

(c) are valuable wildlife refuges and hence tourist areas;

(d) can reduce nutrient loads in enriched waters and help decrease the salinity of influent water, especially during periods of low throughflow; and

(e) have a high productivity which can make significant contributions to human food supplies.

8.3.1 *Wetlands as natural water storage and discharge areas*

The role of swamps in African hydrology is partly reviewed in Balek (1977). He emphasised the fact that the presence of swamps reduced surface run-off. He demonstrated an inverse relationship between the annual run-off from central African catchments and the proportion of the catchment covered by swamps. Thus for a catchment with an annual rainfall of 750 mm, the annual run-off could be reduced from 56 mm to 3 mm if the swamp area increased from 1 to 10% of the catchment area. Thompson (1974) and Thompson and Hamilton (1983) pointed out the storage value of the swamps and lakes of the White Nile which provides 83% of the low flow (dry season) discharge to the lower Nile. Its two largest tributaries, the Blue Nile and Atbara, arise in sparsely vegetated areas and have almost no storage capacity. However, evaporative losses from swamps and other wetland types can be considerable. Fifty percent of the flow of the Nile is lost during its passage through the Sudd swamps (Hurst, 1933), and Balek (1977) estimated that 60% of the total inflow to Lake Bangweulu is lost by evapotranspiration. Wilson and Dincer (1976) estimate that evapotranspiration losses from the Okavango swamps account for 85% of the input. Russell (1981) points out that swamps in the upper reaches of rivers can hold considerable quantities of water releasing this slowly as inflows are reduced. However, many of these swampy areas can be drained and have soils suitable for agriculture. Drainage may result in a deterioration of the river flow regime although, as yet, there is little direct evidence of the extent of such damage.

In some arid areas swamps have been shown to reduce evaporative losses (Linacre *et al.*, 1970; Rijks, 1969 and see Chapter 6) due to shading, lower water temperatures and less air movement at the water surface. Such calculations, however, are based on a comparison with equal areas of open water. Wetlands have the property of spreading out the area of water, so no wetland can be as efficient at minimising evaporative losses as a deep reservoir or a deep canal. Natural wetlands do not cost money, however, and Balek (1977) pointed out that African swamps, 'acting as natural water reservoirs, form an important part of the African economy'.

Isotope enrichment studies (Wilson and Dincer, 1976) have demonstrated that the Ngamiland groundwater resources depend upon recharge from the Okavango swamps. About 2% of the Okavango input seeps into the regional aquifer which is now tapped at several locations in this arid area. Thibodeau and Ostro (1981) have quantified, in monetary terms, the value of a Massachusetts wetland as a flood control and water storage area. In addition these authors showed that wetlands can serve as groundwater recharge areas and are expected to be used as such to an increasing extent in Massachusetts.

For the optimum management of wetlands for water storage, the analysis of Thibodeau and Ostro (1981) showed that the best recourse was to leave a wetland in its natural state. A 20% loss in wetland storage capacity resulted in a greater than twofold financial loss to the surrounding area.

Thompson (1980) suggests; 'It is misleading to present a cost-benefit analysis in which a wetland has monetary value when under development, but no monetary value if it is left undeveloped'. A similar argument has recently been independently proposed by Pressey and Middleton (1982).

8.3.2 *Wetlands as soil and sediment traps*

The efficiency of wetlands as sediment traps has been discussed by Gaudet (1974). In South Africa *Phragmites australis* is planted in eroding water-courses to stabilise soils and riverbanks and reclaim eroded areas (Greyvenstein and de Villiers, 1975; Viljoen, 1976). Jacot-Guillarmod (1969, 1972) discussed the role of the bogs and sponges of the high altitude areas in Lesotho in the prevention of soil erosion, and points to the disastrous effects of burning and grazing these areas. Thompson (1975) discusses the changing patterns of the swamps of the Okavango Delta in Botswana as sediment builds up in localised areas, and in Chapter 5 Denny emphasises the role of euhydrophytes in the stabilisation of littoral shores.

Two of the largest rivers entering Lake Chilwa in Malawi, the Sombani and the Phalombe (see Kalk *et al.*, 1979) form a useful contrast. Both arise in the inselberg massif of Mt Mulanje in areas of forest reserve with similar geological features. The Sombani passes through dense swamps along its course before reaching Lake Chilwa, whilst the Phalombe meanders across a cultivated plain. The annual silt load for these rivers per km^2 of catchment area was estimated as 2 (Sombani) and 1693 (Phalombe) tons (Starmans, 1970).

In respect to sediment trapping and the prevention of erosion, it is interesting to note that small headwater wetlands can be just as important as large, lowland wetlands. As will be pointed out in later discussion, wetland size is not necessarily an indication of wetland value.

The long term stability of many wetlands depends on the silt which they trap. This is because silt provides nutrient replenishment and in some cases, such as the Nile and Mississippi Deltas, wetland subsidence occurs due to compaction when silt inputs are reduced (Stone *et al.*, 1978). Soil nutrient deficiency problems are widespread in Africa, but floodplains and other wetlands enjoy the benefits of water-borne nutrients. Floodplain and swamp dwellers in Africa do not have to practise shifting agriculture as land is 'refertilised' by trapped sediment during the annual or biannual floods. The Gezira area on the Nile floodplain (Barnett, 1977), the Rufiji Plains in Tanzania (Bernascek, 1981; Mwanyika, 1982) and the Pongolo floodplain in South Africa and Moçambique (Heeg and Breen, 1982) provide good illustrations of this sediment 'refertilisation'.

Thompson (1975) observes that the Kalahari sands of northern Botswana are deficient in every major plant nutrient so that subsistence agriculture can only be practised on the floodplains. Experience on the Nile Delta has shown that in the absence of annual floods there is an increasing need for fertiliser application, and Thompson (1976*c*) and Platenkamp (1978) point out that without prolonged annual flooding of African wetland soils the oxidation of soil organic matter would result in a dwindling fertility.

Developments such as dams and canals can seriously affect the pattern of sedimentation. Dams, of course, are most efficient sediment traps, and canals with relatively high water velocities can transport sediment past floodplains and swamps. Such developments seldom consider the downstream and adjacent impact of reduced sediment loads and flood control. Adeniji (1975) and Hilton and Kowu-Tsri (1970) draw attention to some of these side effects for, respectively, the Kainji and Volta dams. Work on the Pongolo floodplain has shown, however, how well-timed releases of water from a dam can 'simulate' to some extent annual floods, and the provision of discharge facilities from near the base of dams can, in some cases, scour sediment from the dam which would be of value to downstream wetland areas.

8.3.3 *Wetlands as valuable wildlife areas*

Wetlands throughout Africa are extensively utilised by wild game of all types. The enormous wildlife potential of African wetlands has been repeatedly described, and reviews are given by Huntley (1978), Clarke (1976), *etc.* (*see* Chapter 10).

Many of the rarer African vertebrates are dependent on floodplain and riverine systems. Large antelopes such as puku (*Kobus vardoni*), sitatunga

Plate 8.1. The Ramsar Convention (p. 203) acknowledges the need to preserve wetland habitat in order to retain bird diversity and numbers. Indeed, birds are excellent indicators of wetland 'condition'.
(a) Cormorants usually perch near a swamp fringe, where fish are particularly abundant.
(b) Goliath herons nest in swamp fringes where food is plentiful. (Photos K. Thompson.)

(*Tragelaphus spekei*) and the subspecies of lechwe (*Kobus leche*) as well as the hippopotamus (*H. amphibius*) and crocodile (*Crocodilus niloticus*) are all obligate wetland, shallow lake or river dwellers. Wetlands, with floodplains, usually provide good grazing and shelter during the dry months and are used by both game and domestic animals alike. In many parts of Africa (*e.g.* Ngorongoro, Kafue flats, Okavango, Upper Nile) game and domestic animals are found simultaneously on floodplains. The value of African wetlands as bird refuges and habitats, especially for migratory species is well known (see Moreau, 1972; Dowsett, 1980 and others). Batchelor (1978) points out that African waterfowl require two types of wetland, those which serve as a breeding habitat, and those which offer refuge and a feeding area during the moulting period (dry season). The former types are usually small isolated wetlands and backwaters, whilst the latter are larger stretches of water. From the point of view of conservation and management of waterfowl it is necessary to note that the preservation of small macrophyte covered ponds may be just as important as the protection of large water bodies. This argument can be equally applied to larger wildlife. East (1981) emphasises the great importance of small wildlife reserves where peculiar habitats are found. Examples are small reserves with suitable wetlands for rare species such as sitatunga and lechwe. In southern Africa many, if not most, of the small 'pans' of the Highveld have been destroyed by agricultural development (Milstein and Zaloumis, 1976), and overgrazing and cultivation have destroyed many of the small East African wetlands (Thompson and Hamilton, 1983).

In order to maintain the tourist potential of African wetlands as wildlife areas, perhaps the best management strategy would be to control water levels and inflows in order to guarantee an aquatic habitat, but still allow water levels to fluctuate. Some of the problems arising from stable water levels caused by the damming of wetland inflows or outflows are discussed later.

8.3.4 *Wetlands as regulators of salinity and nutrient enrichment*

The mechanisms by which salts, mineral ions and pollutants can be regulated by swamps are discussed in Chapter 6. Several studies have been published demonstrating that African wetlands can perform these functions for at least part of the year. Harrison *et al.*, (1960) in one of the earliest studies on nutrient filtering by swamps in Africa documented considerable water purification properties, including the removal of faecal bacteria, increasing water clarity and a summer removal of nitrate by a large *Phragmites* swamp. The removal of nutrients from the treated effluent of the Kampala sewage works by a papyrus swamp (*see* Plate 3.3) has also been documented (Anonymous, 1970).

Nutrient pathways in African swamps have been studied by Gaudet (1976, 1978, 1979) and Howard-Williams (1978), and some are discussed in Chapter 6 of this book. These studies indicate that the nutrient removal capacity depends on flow rates and the relative difference between uptake, elution and decomposition (Gaudet, Mitchell and Denny, 1981). In some cases nutrients can be added to flowing waters by swamps, as happens when water levels rise and standing dead vegetation is subjected to rapid aquatic leaching and decomposition (Howard-Williams and Lenton, 1975; McLachlan, 1971). In addition, flushes of nutrients from recently inundated floodplain soils are a fairly well documented phenomenon in Africa (Viner, 1974; McLachlan, 1970; Howard-Williams, 1972; Gaudet, 1977; Gaudet and Mithuri, 1981; Thompson and Hamilton, 1983). It must be stressed, therefore, that considerable caution needs to be applied to any scheme which aims to utilise African swamps for nutrient reduction, for if a system is operating as a nutrient trap it means that it is not operating in a 'steady state' and vegetative changes are, sooner or later, inevitable. So far no work has been done to establish the loading capacity of African wetlands to nutrient additions, although there is evidence of floristic changes to the Nakivubo swamp near Kampala as a consequence of sewage additions (Thompson, unpublished). Howard-Williams (1981) added varying levels of nitrate and phosphate fertiliser to a dense *Potamogeton pectinatus* community in a southern African coastal lake. The community was able to absorb the added nutrient for six months but the biological changes might be considered unfavourable as there was a massive development of epiphytic algae on the macrophytes. Stearns (1978), Stewart and Ornes (1975) and Valiela, Vince and Teal (1977) all suggest that any long term exploitation of natural wetlands by deliberate nutrient additions could be deleterious. The Norfolk Broads, England, are perhaps one of the best documented cases of wetland deterioration (in terms of plant and animal species diversity and fish productivity) following excessive unregulated nutrient inputs (see review in Moss, Leah and Clough, 1979).

Natural African wetlands should thus not be considered as nutrient effluent sinks, although, from the management standpoint, the establishment of 'artificial' wetlands for such a purpose should be evaluated.

There are two aspects to the role of African wetlands in ameliorating salinity problems. Firstly, wetland vegetation has been shown to alter the composition of major ions in some African waters (Talling, 1957; Viner, 1974; Gaudet, 1976 and Chapter 6) but Gaudet (1978) concluded that papyrus swamps would have little effect on the salt content of through-flow if discharge rates were high. Secondly, salt concentrations at soil surfaces occur in some wetlands due to evaporation during the drawdown phase. Rapid evapotranspiration in swamps where the water is relatively stagnant can also result in elevated salt concentrations in the swamp water (Gaudet 1976, Howard-Williams 1979*a*, Heeg *et al.*, 1978). The functioning of such systems in the long term relies on a periodic flushing of the salts from these soils and waters. Management of such floodplains and swamps for maximum productivity thus requires the maintenance of a flood regime. Drainage works in many wetland agriculture schemes in coastal Moçambique and south-western Uganda are constructed in such a way that salt accumulations can occasionally be flushed away by flooding the agricultural areas.

8.3.5 *Wetland production and human food supply*

It is in this context that we believe that African wetlands have both their greatest value and also face their greatest threat. Increased food and living space requirement is forcing more and more nations to adopt flood control systems, and yet it is the periodic flooding which many of the wetland ecosystems require in order to function properly.

Welcomme (1979) points out that floodplain ecosystems are comprised of two complementary phases; aquatic (during which they are submerged) and terrestrial, following drawdown. During the aquatic phase wetland plants can be harvested directly as fodder crops for livestock. Viljoen (1976) describes how *Phragmites australis* in southern Africa can be used for fodder, and analyses show that young plants up to 2 m in height have a good nutritional base. In the middle Niger swamps aquatic plants, principally *Echinochloa stagnina*, are extensively cut and sold as cattle food (Bacalbasa-Dobrovici, 1971).

During the dry season African wetlands are used extensively for livestock grazing areas or limited agriculture. Examples from Niger (Hurault, 1965), Zambia (Duerre, 1969), Mali (Bacalbasa-Dobrovici, 1971), Malawi (Kalk *et al.*, 1979) Sudan (Rzoska, 1974), Chad (Omerod, 1978), South Africa (Furness, 1982) and Botswana (Thompson, 1976*b*) amply illustrate this. Furness (1982) has demonstrated that the growth rate of *Cynodon dactylon* on the Pongolo floodplain is related to available soil moisture and decreases as the water level on the plain drops. The cattle follow the flush of grass growth on the newly exposed areas. However, as they spend much of the day off the floodplain, Furness (1982) concluded that there was a net export of nutrients in the form of urine and dung. This loss is replaced by nutrients in the silt deposited during the annual floods. Rzoska (1974, 1976) and Platenkamp (1978) describe the seasonal migrations of the Nuer and Dinka people of the Sudd on the Nile as they follow the good grazing on the edge of the falling and rising water levels. As peripheral areas around wetlands dry out cattle often wade into permanent swamps, (sometimes up to their necks) to graze species such as *Phragmites* spp., *Echinochloa* spp., *Vossia cuspidata, Vetiveria nigritana* and particularly *Panicum repens* and *Paspalidium geminatum* (Howard-Williams, 1973; Magadza, 1970; Thomas, 1941). The numbers of cattle supported by African wetland systems are enormous. Data for some of the larger floodplains of Africa are given in Table 8.2.

Grazing on wetlands can have very beneficial effects as nutrient cycling is speeded up considerably as a result. The first quantitative study of this was undertaken by McLachlan (1971) on the drawdown zone of Lake Kariba. She found that, in general, animal dung derived from grass released ions more rapidly than grass itself, and that the released nutrients were probably taken up by algae and submerged plants following reflooding of the drawdown zone. Shepherd (1976) estimated that some 500 kg y^{-1} of dung was dropped in every 100 m stretch around the Bangula Lagoon in Malawi. This is thought to contribute significantly to the exceptional fish production of this lagoon, and Fox (1976) presented data which indicated that fish standing stocks in the Okavango swamps increase in areas where cattle farming occurs. However, in some situations cattle graze for only part of the day

Table 8.2. Numbers of cattle associated with some of Africa's larger floodplain systems. (Original source: Welcomme, 1979).

Floodplain	Region	Numbers of cattle	Reference
SUDD	Southern Sudan	650 000	Rzoska 1974
BAROTSE	Western Zambia	310 000	Duerre 1969
KAFUE	Central Zambia	250 000	UN/FAO 1968
SHIRE	Malawi	148 000	Welcomme 1979
NIGER	Mali	200 000	Gallais 1967
GAMBIA	Gambia, Senegal	300 000	Welcomme 1979

on floodplain and wetlands. Hippopotamus feed on terrestrial or floodplain vegetation at night and defecate in the water by day. The data of Field (1970) and Heeg and Breen (1982) illustrate the considerable contribution these animals can make to the detrital inputs of swamps and shallow water bodies. In contrast Furness (1982) found that cattle grazing on the Pongolo floodplains for only part of the day resulted in a net export of nutrients from the area. Management of grazing practices to ensure the beneficial effects of nutrient recycling through dung and urine should not be too difficult to initiate in areas such as the Pongolo.

Maximal food production can be attained (in theory anyway) in African wetlands where during the dry season cattle are produced and, during the wet season, fish. Both aquatic and terrestrial phases rely on each other, but there is an upper limit to grazing pressure. Overgrazing, such as documented on the Ouémé floodplain by Hurault (1965) and the upper Zambesi floodplains by Duerre (1969), may cause breaching of banks of rivers and streams, siltation, and surface runoff. It also causes a reduction in the amount of aquatic vegetation, and consequently less food and shelter for fish, and lower fish stocks. Duerre (1969) recommended reduced grazing adjacent to tributaries and main river channels in the Upper Zambezi floodplain, and advised that burning should be better controlled.

Problems also occur where grazing activities conflict with cultivation. Ormerod (1978) describes how the Fulani people have, for about 100 years, organised grazing rights in the central delta of the Niger River. After harvesting of millet, cattle were encouraged to clear the harvested ground and deposit manure, which fertilised the next crop. However, new crops which take longer to mature, irrigation and intensive agricultural schemes have now closed areas to grazing, thus putting greater pressure on the remaining grazing land. A similar situation is beginning to develop in the Lake Chad basin.

A different conflict of interests has developed in the Okavango Delta, where huge areas are undepopulated due to trypanosomiasis. In recent years aerial spraying techniques against the tsetse fly vector have proved highly successful in eradicating this disease, and the Okavango swamps, being an isolated 'island' of infestation are an ideal candidate for such trial insecticide treatment. Thompson (1975, 1976*a*) and others, have suggested that such spraying operations may have ecological consequences beyond tsetse eradication, particularly due to the inevitable expansion of domestic livestock grazing in the absence of an enforceable land use policy able to regulate it. Davies and Bowles (1976) observe: 'Fears have been expressed that tsetse elimination would lead to drastic changes in the [Okavango] Delta, especially if invasion by cattle is unchecked. It has even been suggested that tsetse be allowed to remain to ensure the area remains unspoiled. However, it should be possible for future conservation to be carried out, and a land-use policy enforced, without relying on the help of a troublesome and disease carrying insect.' This, of course, is quite true, provided that an enforceable land use policy precedes tsetse eradication. As Omerod (1978) suggests for West Africa, the development of balanced land use policies does not seem to enjoy the same level of priority as does the opening up of new land for settlement. Indeed, he noted that virtually the only land in the Sudan and Northern Guinea zones of West Africa which hasn't been overgrazed is that which is either occupied by tsetse or has only recently been freed from the threat of trypanosomiasis. Many of the freed areas are seasonally flooded grasslands.

In some parts of the world (*e.g.* Rumania) reeds are used to manufacture paper, and there have been several proposals to utilise emergent aquatics such as papyrus in this way in Africa. However, although the technology is available, the economic viability of such projects is very uncertain. More specifically, Thompson (1975) points out that the need to regularly replant rhizomes, or to regulate harvesting frequency, presents problems with harvesting technology and the difficulties associated with short fibre length have held back interest in this type of industry. One of the more attractive prospects for floodplain and wetland development is 'biomass energy' production. There are three possibilites (Thompson, 1983):

(a) reed (*Phragmites*, papyrus) harvesting for

Plate 8.2. This 'scaffolding' on the Zaire River at Stanley Falls (near Kisangani) supports large numbers of funnel-type fish traps. The wide end of one such trap is partly visible just above centre. (Photo K. Thompson.)

fermentation of cellulose to ethanol;

(b) wood production for charcoal or fermentation to methanol;

(c) sugar production (sugar cane on floodplains, sugar palms such as *Nypa* in permanent swamp and *Borassus* in seasonal swamp), and subsequent fermentation to ethanol.

A further use of African wetlands for energy production involves the development of Africa's peatlands and peat deposits. Thompson (1973) estimated the peat reserves of Central and East Africa to be 4.3 km^3. The recent exploitation of sedge peats in Burundi (Thompson and Hamilton, 1983) presents an interesting case. This project has been conceived to help alleviate two acute local problems: food and fuel supplies for the region with Africa's highest population density. Some of the peatlands are being converted to agriculture, but some are being dug out to reduce oil imports and help minimise further deterioration of forests for firewood. Although wise management of these headwater and riverine peatlands is important to river hydrology at lower altitudes and to ensure that a useable land resource remains for future generations, it must be recognised that the short-term economic problems of these countries are acute. They need every possible help in managing this important wetland resource.

The fisheries potential of African wetlands is enormous and yet is under-exploited. Although swamp fishing techniques (largely trapping) are inefficient, 36% of the total African freshwater fish catch is from swamps and floodplains (Welcomme, 1972). Toews and Griffith (1979) found gillnet catches in the Bangweulu swamps to be three times higher than the open lake. The local fishermen find it easier to operate in the swamps, so that of the 12–15 000 tonnes yr^{-1} of fish taken in the Bangweulu system, approximately 50% was from the swamps (Bazigos *et al.*, 1976). Likewise over 30% of the fish production from Lake Chilwa originates from the surrounding swamps (Furse *et al.*, 1979). Twenty eight percent of the freshwater fish catch from Tanzania originates from small inland lakes, floodplains and ponds (Mapunda, 1981). Welcomme (1979) compared the areas of flooded land with total fish catches on African floodplain rivers. The data show that, in general, a mean annual yield of 40–60 kg ha^{-1} for exploited floodplain fisheries can be expected on a sustainable basis. Toews and Griffith (1979) point out that the expansion of fisheries in isolated lightly-exploited swamp areas seems to be an avenue for increased yields in African waters. Beadle (1981), for instance, is enthusiastic about the prospect for a modern commercial fishery in the Okavango Delta. Gilmore (1978) and Thompson (1980) draw attention to the problems inherent in disrupting social patterns by developing fisheries industries too quickly in such under-exploited regions (Plate 8.3).

A further possible avenue for exploitation is fish farming. Wetland areas subject to periodic flooding are not suitable for such ventures, but where developments such as irrigation schemes are proposed, fish ponds could be a viable source of protein (Krishnamurthy, 1981; Heeg and Breen, 1982). Maar (1965) proposed 'dual purpose' ponds for the Okavango, producing both fish and rice. Yields from well managed fish farms can be one to two orders of magnitude greater than those quoted above for floodplain fisheries.

An important aspect of fish productivity in African wetland systems is that of interface areas (ecotones). In areas where the emergent vegetation is dense, and the primary production: community respiration ratio is less than 1, deoxygenation is likely to occur. This, in fact, is a very common feature of African swamps (Carter, 1955; Beadle, 1981; Howard-Williams, 1979*a*, 1979*b*). However, at the swamp-lake, swamp-lagoon or river interface, both abundant oxygen and an abundant and varied food supply occurs. These interface areas are essential for providing feeding, breeding and sheltering areas for fish in all of the large shallow African fisheries (Ratcliffe, 1972; Bourn, 1972; Harding, 1960; Toews and Griffith, 1979; Gilmore, 1978; Furse, Morgan and Kalk, 1979; Furse *et al.*, 1979; Welcomme, 1972 *etc*). We will discuss further implications of this later in this chapter.

Of vital importance to the maintenance of wetland fisheries, particularly floodplain fisheries, is the annual cycle of flooding and drawdown. Agriculture on marginal floodplain soils and the use of river water for irrigation, both demand that floods be reduced in magnitude and number (Furness, 1982; Heeg, Breen and Rogers, 1980), a direct conflict with the needs of the fishery and livestock (Mwanyika, 1982). Welcomme (1979) observes that the majority of fishes inhabiting floodplains rivers are extremely sensitive to any restriction of the floodplain and modification of the flood cycle. In addition, maintenance of the maximal area of good grazing, where floodplains are used for cattle or game during the dry season, is totally dependent on a normal flooding regime. Attwell (1970) concluded that floodplain wildlife is apt to be as adversely affected by alterations of the flood regime as fish are.

Wetlands in general are described by Odum (1967) as being 'pulse stable' ecosystems. This means that their stability relies on alternating 'pulses' of environmental conditions (in this case wet and dry phases). The communities of pulse-stable systems are characteristically very resilient to change, but Welcomme (1979) points out that they are fragile in the face of stable conditions. Thus Arnold (1975) believes that the only effective long term management method is to keep the floodplain free from flood control.

The scheme to bypass the Upper Nile swamps by

Plate 8.3. Fishermen of the remote Upemba Swamps (Kamulondo Depression) in upper Zaire. The fishing is good, but the living hard.
(a) Dug-out canoes in the very shallow and highly productive Lake Kisale.
(b) The home of a fishing family on the swampy shores of the shallow Lake Upemba. Some of their catch is traded for floodplain-grown vegetables. (Photos K. Thompson.)

the Jonglei canal is estimated to reduce the area of permanent swamp by about 10%, but concern has been expressed over the ecological and sociological effects of this development (Mann, 1977; Platenkamp, 1978; Krishnamurthy, 1981). The area is inhabited by one million people, two million cattle, sheep and goats and about three million wild herbivores (Thompson and Hamilton, 1983). Although only a small area of permanent swamp will be lost there will be considerable changes on the floodplains. However, a major effort has been made to minimise sociological disruption (Anonymous, 1980; Moghraby, 1982). Dorfman (1980) analysed potential losses from the wetlands of the Senegal River if dams were to be built to control floods and supply water for irrigation. Fisheries losses alone in this system would amount to 22 000 tonnes y^{-1}, plus a complete disappearance of the commercial shrimp and crab fishery as has already happened in the Mediterranean near the Nile Delta. The Usangu floodplain in Tanzania was estimated by the UN/FAO (1961) to contain 2100 km^2 of soils suitable for irrigated agriculture – provided *total* flood control was achieved. Bernascek (1981) points out, however, that total flood control here is completely incompatible with floodplain fish production and fisheries.

Kok (1978) showed that the fishes of the Pongolo floodplain system depended completely on an annual cycle of floods of certain magnitudes and at certain times of the year. Man controlled floods from the upstream Jozini Dam produced catastrophic results when released at the wrong time of year. This affected particularly the wildlife of the area, destroying crocodile nests and inundating grazing areas utilised by game and wildfowl (Pooley, 1980). The extensive literature on the Pongolo floodplains and the effects of the development scheme on the plains is reviewed in Heeg and Breen (1982).

8.4 Appropriate management options

Clearly in the light of the arguments put forward in sections 8.3.1–8.3.5 above, the best management strategy to maintain African wetlands at their natural (*i.e.* high) level of productivity and to retain their services as erosion barriers and water storage sites is to leave them alone.

However, because of the considerable economic value of wetlands to most African governments, a moratorium on wetland development would be quite unacceptable. The key to retaining the many values and benefits of Africa's remaining wetland resources is wise multiple use management. Thompson(1983) discusses the benefits of developing swamps as swamps. Some of the available management options appear to be:

(a) River floodplains can be used for intensive agriculture, such as is practised in the Gezira on the Nile or the middle Niger Delta. However, there is now enough accumulated experience to show clearly that intensive single purpose development on a large scale almost invariably has serious and expensive long-term side effects. Future intensive developments must employ sound 'ecological engineering' techniques based on the much greater knowledge of wetland functioning which we have in the 1980s. We suggest some of these techniques below.

(b) River waters can be used for the irrigation of surrounding land by dam construction, but the downstream wetlands need not be completely sacrificed. The recent Nile by-pass of the Jonglei Canal in the Sudan (Anonymous, 1980), where only 10% of the permanent swamp is affected with minimal sociological disruption, is a good example. Several others are described in Kapetsky (1981). In the early 1970s the lower few kilometres of the Boro outlet of the Okavango swamps were dredged and bunded to improve water supplies to the Orapa diamond mine. Although Lubke *et al.*, (1981) have expressed concern over the damage to the vegetation of the floodplain, in fact the damage, thus far, is relatively slight considering the total extent of the Okavango. This water abstraction project has provided tangible benefits to one of Botswana's major exporting industries (Ernest, 1977). However, yet further schemes for use of this water are proposed which are, as Beadle (1981) states, 'aimed at exploiting the water more effectively without destroying the ecosystem which, it is generally agreed, should be preserved.' Beadle

(1981) suggests clearing and deepening channels would draw more water from the swamps during flood periods and so reduce evapotranspiration losses. However, if this is done, a considerable proportion of the Delta will cease to be a wetland! We suggest that in such cases a decision should be made first as to what is an acceptable wetland loss, *then* plan waterworks which will meet but not exceed the specifications. Work on the functioning of the wetlands on the Pongolo River floodplain (Furness, 1982; Heeg and Breen, 1982; and others) shows that they can still function as natural units with only a small proportion of the original river flow. However, in order to achieve this, artificial 'floods' of the right magnitude and at the right time have to be released down the river. In the case of the Rufiji Basin floodplains in Tanzania, Bernacsek (1981) suggested that instead of total flood control for irrigated agriculture a better alternative would be to regulate natural floods to levels compatible with a wet season crop (*i.e.* rice) and employ irrigation only for dry season crops. This would allow a substantial floodplain fishery to exist. Further examples of this type of argument are found in other parts of Africa (Welcomme, 1978; Kapetsky, 1981).

(c) Better use can be made of the primary production of wetland aquatic plants. If marginal grazing land around wetlands is converted to cultivation, the aquatic plants in deeper water or permanent waters could be deliberately harvested for cattle to replace some of the lost grazing pasture. As pointed out earlier, many swamp plants, particularly large, fast-growing grasses such as *Vossia cuspidata, Echinochloa* spp. and *Panicum repens* are very valuable fodder plants.

Although plant harvesting for paper production has not been particularly successful, it may become a viable proposition if pulp and paper costs continue to rise. Furthermore, we believe there is real potential in utilising wetland areas for 'biomass energy' as described earlier (section 8.3.5).

(d) Better use can be made of the highly productive swamp/open-water interface areas.

The relatively high fish productivity of these areas in Africa has already been referred to. The most productive wetlands from the wildlife and fishery point of view are those with a complex structure and zonation, with areas of open water interspersed with vegetation (Weller and Frederickson, 1974; Weller, 1978). In other words, they have a long and complex interface zone where euhydrophyte vegetation can develop fully and provide feeding areas and refuge for the fauna (see Chapter 5). However, a characteristic of the lake edges of many African swamps is that they are fairly straight because of the growth characteristics of swamp plants and their response to changes in water depth and wind action. This usually results in a minimal swamp/open-water interface zone compared with the total areas of swamp and open water separately.

There is little doubt that artificially increasing the length of the interface areas will increase wetland secondary production. Artificial manipulation of water levels can produce a complex wetland system (Weller, 1978) but this would seldom be possible in Africa. However, channels are already a natural feature of many floodplain swamps. Levée spillways of major water courses in the Upper Nile swamps (Sutcliffe, 1976) and the Okavango (Thompson, 1975) provide access for boats to the backwaters. These spillways are often enlarged and maintained by hippo, and it would be relatively easy to increase the network still further. In certain areas a cheap way of doing this or of increasing the interface would be to cut (or promote fishermen to cut) a series of channels into swamps from the open water. Such a man-made channel system is occasionally found in the Lake Chilwa swamps, for instance. These channels, and other small man-made clearings in the swamp have high daily dissolved oxygen values, extensive periphyton development and growth of submerged aquatic plants, *i.e.* they provide ideal conditions for high secondary production. This is in contrast to the low oxygen-containing, low pH waters under the adjacent swamp vegetation (Howard-Williams, 1979*a, b*). Fish traps and nets are placed all around the clearings and channels and the cut vegetation is often used to build floating houses (Kalk *et al.*, 1979). Elaborate trap systems are used to catch fish in both artificial and natural channels in the swamps of central Zaire (Thompson, un-

Plate 8.4. Channels cut through floating, or semi-floating, swamps not only provide access for fishermen, but also raise the fish production by increasing the area of swamp/open-water interface. Many channels originate 'naturally' as hippo tracks. (Photo K. Thompson.)

published). More recognition should be given to the existence of such channels and clearings where they occur. Such recognition should go hand in hand with protection of these areas as advocated by Duerre (1969) for the central Barotse floodplain.

The attraction of such an artificial channel system is that it should improve a natural food resource without damaging the swamp system. It is labour-intensive, but not likely to be socially disruptive in the same way as large-scale commercial operations. Nevertheless, we draw attention to the apparently unrealised potential of swamp fisheries together with a word of caution: that feasibility depends on practical pilot studies in each swamp. Small changes in channel profiles and swamp flow patterns may in some cases bring about major shifts

Plate 8.5. A valley swamp in the heavily-populated upland Kigezi Region of south-western Uganda. The papyrus is cut for thatch, and would normally be left to regenerate (see Plate 6.6). Nowadays, increasingly, agriculture is encroaching on these highly productive wetland soils. (Photo K. Thompson.)

in sedimentation patterns.

(e) Finally, it is clear from the wetland functions described above, that management should not just be restricted to the wetland itself, but should include upstream and downstream areas. A floodplain fishery can be ruined by even a low dam downstream which prevents fish migrations. Upstream damming, pollution or siltation will obviously influence wetland ecology down below. In modern Africa we cannot recommend that every hectare of wetland be declared a protected or managed area in perpetuity. Two African studies, for instance, show that dams have not seriously affected the productivity of the downstream floodplain fishery (Otobo, 1981; Awachie and Hare, 1981). Thus we should ask the question 'how much or how many of our wetlands can we afford to lose in the interests of electricity or agriculture without serious long term implications for the catchments concerned?' In Africa we don't even know the extent of the wetland areas. Table 8.1 is very crude and does not include the myriad of small wetlands so important for local winter grazing, wildlife, fish breeding, soaks and sponges for water retention

Plate 8.6. The climate in the Kigezi Region of Uganda is ideal for intensive agriculture, and many of the valley swamps have now been drained and converted to cash crop production. (Photo K. Thompson.)

and so on. Clearly the first management priority is to delineate and classify the wetlands of Africa before they disappear. The approach used on the recent survey of North African wetlands by Morgan and Boy (1982) should be extended to other parts of Africa. Management and development plans for African wetlands should centre firstly on the large ones, as these have both the greatest potential for development and the greatest value for wildlife conservation. In particular we suggest that the large wetlands in those countries with the most serious water limitations are those deserving priority attention. Wetlands, including floodplains, are the only areas where subsistence farming is feasible in many parts of Africa. 'Western' concepts of wetland conservation are often inappropriate in the African context. For instance in 1970 a concerted effort by several eminent scientists and conservationists successfully blocked the funding for a proposal, approved at the national level, to generate hydro-electricity at the famous Kabalega (then Murchison) Falls in Uganda. The revenue from tourism generated by these falls can scarcely be compared with the loss of potential electricity. The international community does not usually provide compensation for such losses. As a second example, Kerfoot (1975*a*, 1975*b*) has expressed in very strong terms, concern over the conservation status of the Okavango Delta, and over the environmental implications of development proposals and activities there. He has received a great deal of support from others in the international community who have similar views. We unquestionably need a strong international conservation movement, but we should question what does it, or should it, achieve at the local level where the protection measures would be felt. Tourist and 'world heritage' values simply do not have much meaning at a local level in a continent which has many of the poorest countries in the world. We believe that multiple *use* is the key to the maintenance of those of Africa's wetlands which remain, but unfortunately many development aid projects are single purpose. The integration of different individual projects in dry land areas is difficult enough (see Pereira, 1973), but almost impossible to achieve in wetlands which respond to, say, a single flood régime. Many development aid projects are short term agreements, a serious problem in wetland schemes where environmental difficulties (*e.g.* soil acidification) do not arise until a number of years after wetland 'reclamation'. Hansen and Koesoebiono (1980) suggest that development agencies give a 10–20 year commitment to wetland schemes in Indonesia to ensure proper resolution of the possible problems. Heeg and Breen (1982) point out that many of the assets of the Pongolo floodplain have not been considered in development plans because none of the planners has ever visited it, and Bernacsek (1981) suggests that the major reason for the failure of some development projects is the exclusion of impact area residents from the decision making and planning processes.

Surely it is not too much to ask that African wetland (including floodplain) development projects be integrated at the planning stage, have a long term financial commitment and be co-ordinated by someone, not only familiar with the area, but familiar with the ideals and aspirations of the local people.

8.5 References

Adeniji, E.O., 1975. Observations on agriculture and fisheries downstream of the Kainji Dam. In: A.M.A. Imevbore & O.S. Adegoke (eds.). The Ecology of Lake Kainji, the transition from River to Lake. University of Ife Press, Ife-Ife, Nigeria: 181–186.

Anonymous, 1970. Master plans for water supply and sewage for the Greater Kampala/Jinja areas until the year 2000. UNDP/SF/WHO Project 28. Terminal Report UNDP, Kampala, Uganda.

Anonymous, 1980. Jonglei Canal. A development project in the Sudan. Published by the Executive Organ, National Council for the Development of the Jonglei Canal Area, Khartoum.

Arnold, M.D., 1975. Floods as man-made disasters. Environmental Conservation, 2:257–263.

Astle, W.G. & Graham, A., 1976. Ecological investigations of the UNDP in the Okavango Delta. Proceedings of the Symposium on the Okavango Delta and its Future Utilisation. Published by the Botswana Society, Gaborone: 81–91.

Attwell, R.I., 1970. Some effects of Lake Kariba on the ecology of the floodplain of the Mid-Zambesi River of Rhodesia. Biological Conservation, 2:189–196.

Awachie, J.B.E. & Hare, L., 1978. The fisheries of the Anambra, Ogun and Oshun river systems in Southern Nigeria. In: R.L. Welcomme (ed.), Symposium on River and Floodplain Fisheries in Africa. CIFA Technical Papers 5:170–184.

Bacalbasa-Dobrovici, N., 1971. The economic importance of the burgu (*Echinochloa stagnina* P. Beauv.) on the middle Niger in the Niger Republic. Hidrobiologia (Bucuresti) 12:41–46.

Balek, J., 1977. Hydrology and water resources in tropical Africa. Developments in Water Science 8. Elsevier Scientific Publishing Company, Amsterdam.

Barnett, T., 1977. The Gezira Scheme. Frank Cass Ltd., London.

Batchelor, G., 1978. Why conserve wetlands. Fauna and Flora, 33:8–9.

Bazigos, G.P., Grant, C.D. & Williams, C., 1976. Revised fisheries time series of the natural waters of Zambia (1966–1974). UN/FAO report ZAM/73/009, St. 5/3, Rome.

Beadle, L.C., 1974. The inland waters of tropical Africa (1st edition) Longmans, London.

Beadle, L.C., 1981. The inland waters of tropical Africa (2nd edition) Longmans, London.

Bernascek, G.M., 1981. Freshwater fisheries and industry in the Rufiji River Basin, Tanzania: The prospects for co-existence. In: J.M. Kapetsky (ed.), Seminar on River Basin Management and Development. CIFA Technical Papers 8, 69–88.

Bourn, D.M., 1972. The feeding, diet and ecological relations of three species of fish of economic importance in southern Malawi. M.Sc. Thesis, University of Edinburgh.

Carter, G.S., 1955. The papyrus swamps of Uganda. W. Heffer and Sons, Cambridge.

Carmouze, J.P., Dejoux, C., Durand, J.R., Grass, R., Iltis , A., Lauzanne, G., Lemoalle, J., Lévêque, C., Loubens, G.

& Sant-Jean, L., 1972. Grandes zones ecologiques du lac Tchad. Cahiers ORSTOM, serie Hydrobiologie, 7:17–23.

Clarke, J.E., 1975. Zambia's wetlands. Black Lechwe 12:14–17.

Davies, J.E. & Bowles, J., 1976. Ecology of the tse-tse fly in the Okavango Delta, Proceedings of the Symposium on the Okavango Delta and its Future Utilisation. Published by the Botswana Society, Gaborone: 141–152.

Dorfman, D., 1980. Senegal River, Africa: Fishes or Dams? Underwater Naturalist 12:4–6.

Dowsett, R.J., 1980. The migration of coastal waders from the palaeartic across Africa. Le Gerfaut, 70:3–35.

Duerre, D.C., 1969. Report to the Government of Zambia on Fisheries Development in the Central Barotse Floodplain. UN/FAO Report No. TA 2638, Rome.

East, R., 1981. Species-area curves and populations of large mammals in African savanna reserves. Biological Conservation, 21:111–126.

Ernest, H.M., 1976. Manipulation of the Okavango Delta for increased outflow; its impacts and constraints. Proceedings of the Symposium on the Okavango Delta and its Future Utilisation. Published by the Botswana Society, Gaborone: 337–346.

Field, C.R., 1970. A study of the feeding habits of the hippopotamus (*Hippopotamus amphibius* Linn.) in the Queen Elizabeth Park, Uganda, with some management implications. Zoologica Africana, 5:71–86.

Fosberg, F.R., 1973. Temperate zone influence on tropical forest land use: a plea for sanity. In: B.J. Meggers, E.S. Ayensu & W.D. Duckworth (eds.), Tropical Forest Ecosystems in Africa and South America, a Review. Smithsonian Institution, Washington, D.C.: 345–350.

Fox, P.J., 1976. Preliminary observations on fish communities of the Okavango Delta. Proceedings of the Symposium on the Okavango Delta and its Future Utilisation. Published by the Botswana Society, Gaborone: 125–130.

Furness, H.D., 1981. The plant ecology of seasonally flooded areas of the Pongolo River floodplain, with particular reference to *Cynodon dactylon* (L.) Pers. Ph.D. Thesis, University of Natal, Pietermaritzburg, South Africa.

Furse, M.T., Kirk, R.C., Morgan, P.R. & Tweddle, D., 1979. Fishes: Distribution and biology in relation to change. In: M. Kalk, A.J. McLachlan & C. Howard-Williams (eds.), Lake Chilwa. W. Junk, The Hague: 175–208.

Furse, M.T., Morgan, P.R. & Kalk, M., 1979. The fisheries of Lake Chilwa. In: M. Kalk, A.J. McLachlan & C. Howard-Williams (eds.), Lake Chilwa. W. Junk, The Hague: 209–230.

Gallais, J., 1967. Le Delta interieur du Niger; étude de geographie regionale. Memoirs de l'Institut Fondamental d'Afrique Noire, 79:1–619.

Gaudet, J.J., 1974. The normal role of vegetation in water. In: D.S. Mitchell (ed.), Aquatic Vegetation and its Use and Control. UNESCO, Paris: 24–37.

Gaudet, J.J., 1975. Mineral concentrations in papyrus in various African swamps. Journal of Ecology, 63:483–491.

Gaudet, J.J., 1976. Nutrient relationships in the detritus of a tropical swamp. Archiv für Hydrobiologie, 78:213–239.

Gaudet, J.J., 1977. Uptake and loss of mineral nutrients by papyrus, Ecology, 58:415–422.

Gaudet, J.J., 1978. Effect of a tropical swamp on water quality. Verhandlungen der internationale Vereinigung für theoretische und angewandte Limnologie, 20:2202–2206.

Gaudet, J.J., 1979. Seasonal changes in nutrients in a tropical swamp: North swamp, Lake Naivasha, Kenya. Journal of Ecology, 67:953–981.

Gaudet, J.J., Mitchell, D.S. & Denny, P., 1981. Macrophytes (Aquatic vegetation) In: J.J. Symoens, M. Burgis & J.J. Gaudet (eds.), The Ecology and Utilisation of African Inland Waters. UNEP, Nairobi: 27–36.

Gaudet, J.J. & Mithuri, F.M., 1981. Nutrient relationships in shallow water in an African lake, Lake Naivasha. Oecologia, 49:109–118.

Gilmore, K.S., 1978. The river and floodplain fishery of the Okavango Delta, Botswana. In: R.L. Welcomme (ed.), Symposium on River and Floodplain Fisheries in Africa. CIFA Technical Papers 5:165–169.

Greyvenstein, P.S. & de Villiers, C.P.M., 1975. Common reed as a reclamation plant. Farming in South Africa, Engineering Series No. E1/1975. Department of Agricultural Technical Services, Pretoria.

Hansen, A.J. & Koesoebiono, A., 1980. Settling coastal swamplands in Sumatra: a case study for integrated resource management. In: L. MacAndrews & Chia Lin Sien (eds.), Developing Economies and the Environment. McGraw Hill Publishing Company: 121–195.

Harding, D., 1960. Lake Kariba: The hydrology and development of the fisheries. In: R.H. McConnell (ed.), Man-made Lakes. Academic Press, London: 7–18.

Harrison, A.F., Kellar, P. & Dimovic, D., 1960. Ecological studies on Olifantsvlei, near Johannesburg. Hydrobiologia, 15:89–124.

Hastings, R.E., 1972. Fisheries Research Unit, Lower Shire. Interim Report 1970–1972. Fisheries Department, Zomba, Malawi. Unpaginated.

Heeg, J., Breen, C.H., Colvin, P.M., Furness, W.D. & Musil, C.F., 1978. On the dissolved solids of the Pongolo Floodplain pans. Journal of the Limnological Society of Southern Africa, 4:59–64.

Heeg, J. & Breen, C.M., 1982. Man and the Pongolo Floodplain. South African National Scientific Programmes Report No. 56, CSIR, Pretoria, S. Africa.

Heeg, J., Breen, C.M. & Rogers, K.H., 1980. The Pongolo Floodplain: A unique ecosystem threatened. In: M.N. Bruton & K.H. Cooper (eds.), Studies on the Ecology of Maputaland. Rhodes University Press, Grahamstown, South Africa: 374–381.

Hejný, S., 1973, (ed.), Ecosystem study on wetland biome in Czechoslovakia. Czechoslovak IBP/PT-PP report No. 3. Czechoslovak Academy of Sciences, Trebon.

Hilton, T.E. & Kowu-Tsri, J.Y., 1970. The impact of the Volta scheme on the lower Volta floodplains. Journal of Tropical Geography, 30:24–37.

Holden, M.J. & Green, J., 1960. The hydrology and plankton of the River Sokoto. Journal of Animal Ecology, 29:65–84.

Howard-Williams, C., 1972. Limnological studies in an African swamp: seasonal and spatial changes in the swamps of Lake Chilwa, Malawi. Archiv für Hydrobiologie, 70:379–391.

Howard-Williams, C., 1973. Vegetation and environment in the marginal areas of a tropical African lake (L. Chilwa, Malawi). Ph.D. Thesis, University of London.

Howard-Williams, C., 1979*a*. The aquatic environment II. Chemical and physical characteristics of the Lake Chilwa swamps. In: M. Kalk, A.J. McLachlan & C. Howard-Williams (eds.), Lake Chilwa. W. Junk Publishers, The Hague: 79–90.

Howard-Williams, C., 1979*b*. Interactions between swamp and lake. In: M. Kalk, A.J. McLachlan & C. Howard-Williams (eds.), Lake Chilwa. W. Junk Publishers, The Hague: 231–245.

Howard-Williams, C., 1981. Studies on the ability of a *Potamogeton pectinatus* community to remove nitrogen and phosphorus compounds from lake water. Journal of Applied Ecology, 18:619–637.

Howard-Williams, C. & Howard-Williams, W.A., 1978. Leaching of nutrients from the swamp vegetation of a shallow tropical lake. Aquatic Botany, 4:257–267.

Howard-Williams, C. & Lenton, G.M., 1975. The role of the littoral zone in the functioning of a shallow tropical lake ecosystem. Freshwater Biology, 5:445–459.

Huntley, B.J., 1978. Ecosystem conservation in Southern Africa. In: M.J.A. Werger (ed.), Biogeography and Ecology of Southern Africa. W. Junk Publishers, The Hague: 1333–1384.

Hurault, J., 1965. Les principaux types de peuplement du sud-est Dahomey et leur representation cartographique. Etud. Photo-Interpretation Inst. Geogr. National. Paris. 2.

Hurst, H.E., 1933. The Sudd region of the Nile. Journal of the Royal Society of Arts, 81:721–736.

IUCN, 1979. The Distribution of Protected Areas in Relation to the Needs of Biotic Community Conservation in West and Central Africa . IUCN,Gland, Switzerland.

IUCN, 1980a. Directory of Wetlands of International Importance in the Western Palaearctic. Compiled by E. Carp. IUCN, Gland, Switzerland.

IUCN, 1980b. 1980 United Nations List of National Parks and Equivalent Reserves. IUCN , Gland, Swizerland.

Jacot-Guillarmod, A., 1969. The effect of land usage on aquatic and semi-aquatic vegetation at high altitudes in Southern Africa. Hydrobiologia, 34:3–13.

Jacot-Guillarmod, A., 1972. The bogs and sponges of the Orange River catchment within Lesotho. Civil Engineering in Southern Africa, 14:84–85.

Kalk, M., McLachlan, A.J. & Howard-Williams, C. (eds.), 1979. Lake Chilwa. W. Junk Publishers, The Hague.

Kapetski, J.M. (ed.), 1981. Seminar on River Basin Management and Development. CIFA Technical Papers 8.

Kerfoot, O., 1975a. The Okavango controversy: a real cause for alarm. South African Journal of Science, 71:67.

Kerfoot, O., 1975b. The Okavango controversy – a follow up. South African Journal of Science, 71:196–197.

Kok, H.M., 1978. Ecological studies on some important fish species of the Pongolo floodplain, Kwazulu. Proceedings of the Workshop on Man and the Pongolo floodplain. Document 10. University of Natal, Pietermaritzburg, South Africa. Cyclostyled.

Krishnamurthy, V.G., 1981. The impact forseen of the Jonglei Canal Scheme on the Fisheries of the Sudd region: The problems and solutions. In: J.M. Kapetsky, (ed.), Seminar on River Basin Management and Development. CIFA Technical Papers 8, 105–119.

Lamprey, H.F., 1975. The distribution of protected areas in relation to the needs of biotic community conservation in Eastern Africa. IUCN Occasional Paper, 16:1–83.

Linacre, E.T., Hicks, B.B., Sainty, G.R. & Grauze, G., 1970. The evaporation from a swamp. Agricultural Meteorology, 7:375–386.

Lind, E.H., 1956. Studies in Uganda swamps. Uganda Journal 20:166–176.

Lubke, R.A., Raynham, G.L. & Reavell, P.E., 1981. Reassessment of plant succession on spoil heaps along the Boro River, Okavango Delta, Botswana. South African Journal of Science, 77.

Luther, H. & Rzoska, J., 1971. A Source Book of Inland Waters Proposed for Conservation. IBP Handbook 21. Blackwell Scientific Publication, Oxford.

Maar, A., 1965. Fisheries survey in Bechuanaland Protectorate during the years 1963–1964. OXFAM publications. Cyclostyled.

Magadza, C.H.D., 1970. A preliminary survey of the vegetation of the shore of Lake Kariba. Kirkia, 7:253–267.

Mapunda, X.E. The economic impact of the fisheries of the Kilombero River Basin on the Morogoro region of Tanzania. In: J.M. Kapetsky (ed.), Seminar on River Basin Management and Development. CIFA Technical Papers 8, 89–95.

Mann, O., 1977. The Jonglei Canal. Environmental and social aspects. Report for the Environment Liaison Centre, Nairobi, Kenya.

Mavuti, K.M., 1982. Wetlands and shallow water bodies in Kenya: characteristics and general status of knowledge. Proceedings of the International Scientific Workshop on Ecosystem Dynamics in Freshwater Wetlands and Shallow Water Bodies, 1. SCOPE/UNEP Publication, Moscow: 188–204.

McLachlan, S.M., 1970. The influence of lake level fluctuations and the thermocline on water chemistry in two gradually shelving areas in Lake Kariba, Central Africa. Archiv für Hydrobiologie, 66:499–510.

McLachlan, S.M., 1971. The rate of nutrient release from grass and dung following immersion in lake water. Hydrobiologia, 37:521–530.

Milstein, P. le S. & Zaloumis, E.A., 1976. Threatened wetland habitat and associated species. Proceedings of a Symposium on Endangered Wildlife in southern Africa. Endangered Wildlife Trust, Johannesburg: 105–115.

Moghraby, A.I. el, 1982. The Jonglei Canal – Needed development or potential ecodisaster. Environmental Conservation, 9:141–148.

Moreau, R.E., 1972. The Palearctic-African bird migration systems. Academic Press, New York.

Moreau, J., 1977. Le lac Alaotra a Madagascar. Evolution geographique passée et actuelle. Annales de Limnologie, 13:261–274.

Morgan, N.C. & Boy, V., 1982. An ecological survey of the standing waters in North West Africa: 1. Rapid survey and classification. Biological Conservation, 24:5–44.

Moss, B., Leah, R.T. & Clough, B., 1979. Problems of the Norfolk Broads and their impact on freshwater fisheries. Proceeding of the First British Freshwater Fisheries Conference: 67–85.

Motwani, M.P., 1970. Fishery investigations on the Niger and Benue rivers in the northern region and development of a programme of riverine fishery management and training. UNDP/FAO Technical Report to the Government of Nigeria (TA 2771).

Mwanyika, N.A., 1982. Human use of wetlands in Tanzania. The case of lower Rufiji flood plain. Proceedings of the International Scientific Workshop on Ecosystem Dynamics in Freshwater Wetlands and Shallow Water Bodies, 2. SCOPE/UNEP Publication. Moscow: 357–379.

Odum, E.P., 1971. Fundamentals of Ecology. W.B. Saunders, Philadelphia.

Omerod, W.E., 1978. The relationship between economic development and ecological degradation: how degradation has occurred in West Africa and how its progress might be halted. Journal of Arid Environments, 1:357–379.

Otobo, F.O., 1978. The commercial fishery of the Middle River Niger, Nigeria. In: R.L. Welcomme (ed.), Symposium on River and Floodplain Fisheries in Africa. CIFA Technical Papers, 5:185–208.

Pereira, H.C., 1973. Land use and water resources. Cambridge University Press, Cambridge.

Platenkamp, J.D.M., 1978. The Jonglei Canal, its impact on an integrated system in the southern Sudan. Inst. of Cultural Anthropology and Sociobiology of Non-western Peoples, University of Leiden. Publication No. 26.

Pooley, A.C., 1980. Crocodile research in Maputaland. In: M. Bruton & K.H. Cooper (eds.), Studies on the Ecology of Maputaland. Rhodes University Press, Grahamstown, S. Africa: 293–299.

Pressey, R.L. & Middleton, M.J., 1982. Impacts of flood mitigation works on coastal wetlands in New South Wales. Wetlands, 2:27–44.

Pritchard, J.M., 1979. Landform and Landscape in Africa. Edward Arnold, London.

Ratcliffe, C., 1972. The fishery of the lower Shire river area. Malawi Fisheries Bulletin No. 3. Fisheries Department. Lilongwe, Malawi.

Rijks, D.A., 1969. Evaporation from a papyrus swamp. Quarterly Journal of the Royal Meteorological Society, 95:643–649.

Russell, E.W., 1981. Role of watershed management for arable land use in the tropics. In: R. Laland & E.W. Russell (eds.), Tropical Agriculture and Hydrology. John Wiley & Sons, New York: 11–16.

Rzóska, J., 1974. The Upper Nile Swamps, a tropical wetland study. Freshwater Biology, 4:1–30.

Rzóska, J., (ed.), 1976. The Nile: Biology of an Ancient River. W. Junk Publishers, The Hague.

Seaman, M.T., Scott, W.E., Walmsley, R.C., van der Waal, B.C.W. & Toerien, D.F., 1978. A limnological investigation of Lake Liambezi, Caprivi. Journal of the Limnological Society of Southern Africa, 4:129–144.

Shephard, C.J. (ed.), 1976. Investigation into fish productivity in a shallow freshwater lagoon in Malawi. Ministry of Overseas Development Report. London.

Starmans, G.A.N., 1970. Soil erosion of selected African and Asian catchments. (Cited in Balek, 1977, p. 118.)

Stearns, F., 1978. Management potential: summary and recommendations. In: R.E. Good, D.F. Whigham & R.L. Simpson (eds.), Freshwater Wetlands. Academic Press, New York: 357–364.

Steward, K.K. & Ornes, W.H., 1975. Assessing a marsh environment for wastewater renovation. Journal of the Water Pollution Control Federation, 47:1880–1891.

Stone, J.M., Bahr, L.M. & Day, J.W., 1978. Effects of canals on freshwater marshes in coastal Louisiana and implications for management. In: R.E. Good, D.F. Whigham & R.L. Simpson (eds.), Freshwater Wetlands. Academic Press, New York: 299–320.

Sutcliffe, J.V., 1976. Hydrological study of the southern Sudd region. Proceedings of the Symposium on the Okavango Delta and its Future Utilisation. Botswana Society, Gaborone: 277–298.

Talling, J., 1957. The longitudinal succession of water characteristics in the White Nile. Hydrobiologia, 9:73–89.

Thibodeau, F.R. & Ostro, B.B., 1981. An economic analysis of wetland protection. Journal of Environmental Management, 12:19–30.

Thomas, A.S., 1941. The vegetation of the Sese Islands, Uganda. Journal of Ecology, 29:330–352.

Thompson, K., 1973. The ecology of peatlands in East and Central Africa and their classification for agriculture. In: R.A. Robertson (ed.), Classification of Peat and Peatlands. Proceedings of the International Peat Society, Glasgow: 60–73.

Thompson, K., 1974. Ecological effects of wetland utilisation in Africa. Paper presented at 1st International Congress of Ecology, The Hague.

Thompson, K., 1975. Ecology, management and utilisation of aquatic and semi-aquatic vegetation in the Okavango Delta, Botswana. FAO/UNDP Terminal Report of Project BOT/71/506. UN/FAO. Rome.

Thompson, K., 1976*a*. The Okavango Delta and its utilisation: An attempt at a synthesis of the proceedings. Proceedings of the Symposium on the Okavango Delta and its Future Utilisation. Botswana Society, Gaborone: 67–79.

Thompson, K., 1976*b*. Primary productivity of African wetlands, with particular reference to the Okavango delta. Proceedings of the Symposium on the Okavango Delta and its Future Utilisation. Botswana Society, Gaborone: 67–79.

Thompson, K., 1976*c*. Swamp development in the headwaters of the White Nile. In: J. Rzoska (ed.), The Nile. Biology of an Ancient River. W. Junk Publishers, The Hague: 177–196.

Thompson, K., 1980. Stress environments and economic development. INTECOL Bulletin, 7/8:53–69.

Thompson, K., 1983. The rise and fall of the floodplain: a study of resource use and arisuse. Journal of the Commonwealth Human Ecology Council, 6:13–16.

Thompson, K. & Hamilton, A.C., 1983. Peatlands and swamps of the African continent. In: D.W. Goodall (ed.), Ecosystems of the world, 48:325–365.

Tinley, K.L., 1975. Marromeu – wrecked by the big dams. African Wildlife. 24:22–25.

Toews, D., 1975. Limnology of Lake Bangweulu. UN/FAO Report No. FI:DP/SAM/681511/7/REV:1. Rome.

Toews, D. & Griffith, J.S., 1979. Empirical estimates of potential fish yield for the Lake Bangweulu System, Zambia, Central Africa. Transactions of the American Fisheries Society, 108:241–252.

UN/FAO, 1961. The Rufiji Basin, Tanganyika. Report to the Government of Tanganyika, FAO, Rome.

UN/FAO, 1968. Multipurpose survey of the Kafue River Basin, Zambia. Final Report FAO/SF:35/ZAM. FAO, Rome.

Valiela, I., Vince, S. & Teal, J.M., 1977. Assimilation of sewage by wetlands. In: M. Wiley (ed.), Estuarine Processes. Vol. 1. Academic Press, New York: 234–253.

van Meel, L., 1953. L'étude du lac Upemba. In: Exploration du Parc National de l'Upemba, Mission G.G. de Witte. Institute des Parcs Nationaux du Congo Belge, Bruxelles, 9:1–190.

Viljoen, J., 1976. Uses of *Phragmites australis*. Hand, Weidingsveren, Sud Afrika, 11:19–22.

Viner, A.B., 1974. The supply of minerals to tropical rivers and lakes (Uganda). In: G. Olson (ed.), An Introduction to Land-water Relationships. Academic Press, New York: 227–261.

Welcomme, R.L., 1972. A brief review of the floodplain fisheries of Africa. FAO Committee for Inland Fisheries of Africa. Report CIFA/72/S. 17.

Welcomme, R.L., 1975. The fisheries ecology of African floodplains. CIFA Technical Papers, 3:1–51.

Welcomme, R.L. (ed.), 1978. Symposium on River and Floodplain Fisheries in Africa. CIFA Technical Papers, 5.

Welcomme, R.L., 1979. Fisheries Ecology of Floodplain Rivers. Longmans, London.

Weller, M.W., 1978. Management of freshwater marshes for wildlife. In: R.E. Good, D.F. Whigham & R.L. Simpson, Freshwater Wetlands. Academic Press, New York: 267–284.

Weller, M.W. & Fredrickson, L.H., 1974. Avian ecology of a managed glacial marsh. Living Bird, 12:269–291.

Werger, M.J.A. & Ellenbroek, G.A., 1980. Water resource management and floodplain ecology: an example from Zambia. In: J.I. Furtado (ed.), Tropical Ecology and Development. Int. Soc. Trop. Ecol., Kuala Lumpur: 693–702.

Wild, H. & Fernandes, A., 1967. Vegetation map of the Flora Zambesiaca Area. Supplement to Flora Zambesiaca. M.O. Collins, Salisbury.

Wilson, V.H. & Dincer, T., 1976. An introduction to the hydrology and hydrography of the Okavango Delta. Proceedings of the Symposium on the Okavango Delta and its Future Utilisation. Botswana Society, Gaborone: 33–47.

Worthington, E.B., 1976. The conservation of wetlands in Africa. Proceedings of the Symposium on the Okavango Delta and its Future Utilisation. Botswana Society, Gaborone: 61–66.

Authors' addresses:
C. Howard-Williams
Department of Scientific and Industrial Research
Taupo Research Laboratory
P.O. Box 415
Taupo, New Zealand

K. Thompson
Department of Biological Sciences
University of Waikato
Hamilton, New Zealand

CHAPTER 9

African wetland vegetation – concluding perspective

DAVID S. MITCHELL, PATRICK DENNY and CLIVE HOWARD-WILLIAMS

African wetlands include some of the most remote and beautiful areas of the continent but, with a kind of morbid inevitability, their solitude is being impinged upon by the excesses of the modern way of life. The last two chapters have considered means by which their natural attributes may be exploited whilst their unique qualities are preserved. Exploitation need not be the antithesis of conservation, although the ecological equilibria and interdependent nature of organisms within wetland habitats are often very finely poised so that small imbalances, brought about through either natural or man-induced phenomena, may have far-reaching, unpredictable and irreversible consequences. It is the very delicacy of the ecosystem which should alert conservationists, politicians and those in the pursuit of leisure to the dangers of interfering with the system. However, management of wetland vegetation can only be really comprehensive and satisfy the needs of all parties if the plants, and their interrelationships in the natural environment, are fully investigated. The aquatic macrophytes are the cornerstones of the wetland habitat, but our lack of knowledge of so many aspects of their biology is alarming. To take an example: one of the largest areas of wetland in Africa that is potentially threatened by development is the Sudd in southern Sudan. Recently, considerable interest in the area has been shown by biologists, sociologists and conservationists because of the construction of the Jonglei Canal which will divert some of the swamp water. One of us (P.D.) travelled through the Sudd in 1979 with a team of biologists from the Hydrobiological Research Unit, The University of the Sudan, and in the accompanying survey a plant new to Science was found (Denny, 1984). The surprise was not that it had been found (it has now been described and named *Suddia sagittifolia* (Renvoize, Lock and Denny, 1984) but that it is large (up to 3.0 m tall), of considerable taxonomic interest, and appears to be not uncommon in the Sudd. Nothing is known about its biology. If such obvious specimens have maintained their anonymity until now, and so little is known about other plants and animals, clearly, there is much more work to be done. It is most important that biologists receive the necessary support from governments to investigate the real potential of the wetland ecosystems; management may then be based on the fullest information.

There are far too many occasions when major agricultural schemes or engineering constructions in Africa have had undesirable and unforeseen ecological side effects. Consequently, if tragic blunders are to be avoided, future management decisions should be taken in the light of the most complete information and understanding possible. The advice of experts with specialised knowledge and experience is essential. However, there are far too few occasions when the contribution from 'experts' has been linked to the aspirations of the inhabitants of the area where the management decisions are to be implemented. Thus it is highly desirable to involve members of local communities

Patrick Denny (ed.), The ecology and management of African wetland vegetation.
 ISBN 90 6193 509 1

Plate 9.1. A plant new to science, *Suddia sagittifolia,* up to 3 m tall, found poking its leaves through a papyrus canopy in the Sudd of the Upper Nile. This clearly demonstrates that there is still much to be learned about African wetlands and fundamental research requires support. (Photo P. Denny.)

affected by these decisions. Unfortunately, many forms of environmental manipulation are little understood by members of these communities who, nevertheless, are intimately affected by management decisions and procedures. Such people are seldom consulted and, if they are, they are made to feel inadequate by the unintelligible nature of the explanation given. Hence, they usually play little or no part in the decision making process, and so bear no responsibility for the success or failure that follows. They experience no sense of reward in the former and are not motivated to avoid the latter.

Cairns (1975) points out that, in general, in river-basin management, technological developments are further advanced than the sociopolitical changes necessary for their successful implementation. There is thus a need for ecologists to ensure that the necessary information reaches both decision makers and the general public. The problem of involving the community in environmental management issues which require fully informed decisions is most acute amongst populations where there is a wide range in levels of education and economic wealth and where the least favoured educationally and economically are most dependent on environmental conditions for their day to day welfare. This situation is manifest throughout Africa, and proposals for the management of any component of the environment must, therefore, be approached with caution and with due regard to the difficulties described above. There are too many examples in Africa of carefully planned and well executed schemes of development where the full benefits have not been realised or that have foundered because they were foreign to the aims and aspirations of the local population.

Even a brief exposition such as this reveals clearly the difficulties of making, implementing, and profiting from major management decisions. Managers and decision makers frequently resort to time-honoured trial and error procedures in a world that can no longer afford the errors in this approach. The way forward is obscure. Much of the required information is not readily available, and the means for consulting all the people involved is seldom present. Nevertheless, there are at least seven clearly identifiable procedural steps that can be carried out which cumulatively provide a systematic alternative to *ad hoc* trial and error decision making. These are:

(a) establish broad but relevant guiding principles, which set the goals and explain the need for development and/or conservation;

(b) determine criteria to be used in selecting between options;

(c) elucidate specific objectives and list them in priority order;

(d) examine options, and arrange them in order of suitability;

(e) determine criteria and means for evaluating the extent to which objectives are achieved following implementation;

(f) implement the most suitable option, (or options in an integrated pattern);

(g) monitor progress.

Different steps involve different sectors of the community. The first is primarily the concern of the politicians and their advisors on behalf of the wider community and should provide both framework and justification for management. The second is derived from the first but is best carried out by a multidisciplinary group of experts with a collective background in engineering, ecology and socio-economics.

Consideration must be given to the relative importance of such potentially conflicting issues as: economic efficiency (cost-effectiveness or favourable benefit: cost ratios) and long term ecological resilience; local or regional needs and national benefits; national development and international concern for environmental issues; present, short term gains and sustained, long term returns. The third step of setting objectives should involve the maximum consultation within the community. No matter how complex the details of a particular matter, most people can express desires for goals and agreement or disagreement with designated objectives. Steps four to seven are again best carried out by selected groups of experts assigned to the task.

Management decisions of a minor nature obviously do not call for such a detailed procedure, but appropriate components of the same systematic approach can be employed by the person or persons responsible, in an effort to avoid subjective, inadequately considered decisions.

The management of water resources and catchment run-off is of particular importance to the welfare of the African continent and its inhabitants. All the above difficulties are present, and yet the future development of the vast natural resources in Africa for the benefit of its peoples are dependent on the reliable provision of water for purposes such as irrigation and other agriculture, hydro-electricity, secondary and tertiary industry, and day to day domestic requirements in both rural and urban situations. The construction of irrigation systems, by-pass canals and man-made lakes; the drainage of wetlands; and the excessive cultivation, or grazing, of catchments, thereby leading to erosion, all have profound effects on the nature, extent and distribution of Africa's water resources and of the plants growing in or adjacent to them. Indeed, the management of aquatic vegetation cannot be separated from that of water resources which, in turn, cannot be separated from the management of catchments.

Goals for the management of aquatic vegetation, thus, have to be set in the context of overall management of the environment and, depending on their magnitude, according to appropriate steps in the framework set out earlier. For example, if there is a weed problem, the extent and detrimental cost of the weed plant population will determine the magnitude of the problem; while the ecological nature of the plant and its environment will determine the level to which control of the plant can be achieved within reasonably acceptable costs. Most weed problems occur in aquatic systems that have been constructed by man or are being subjected to major alterations, both physical and chemical, of the natural regime in which the system evolved and developed. Purposes for the water body are generally explicit and consequent goals for the management of the aquatic vegetation are obvious. However, occasionally the criteria for successful achievement of the management programme may not be reached within reasonable economic limits. If so, it is necessary to accept a modification of the original purposes of the water body or some suitable compromise, at least on a temporary basis.

In some weed problems there is potential for the excessive plant material to be productively used (Chapter 7), but conservation of the weed plants or the affected system is seldom, if ever, an issue. Conservation matters are only likely to be raised in relation to other biotic components of the system which may be adversely affected by the presence of the weeds or by the measures being employed to control them.

The management of natural wetlands is very different from the management of water weed problems and provides a contrasting example. Overall goals frequently do not take account of the wetlands; and conservation issues may need to be emphasised. The principal reason for this is that most decision makers in Africa place little value on wetlands and see only benefit in limiting the fluctuation in water flows and levels on which many of these systems depend. Indeed, the need for compromise between development and conservation of such systems is generally not apparent until ecological responses interfere with the purposes set for development.

Even when consideration is given to the value of natural systems and conservation objectives form part of a programme for development and management, these are generally secondary to the main thrust of obtaining greater economic return from the resources present on the continent. Our understanding of wetland systems has increased considerably in the last decade, and it is probably fair to say that we now know the basic roles of wetlands in providing ecological and hydrological stability in catchments. Because of the likely conflict between the conservation of wetland areas and the requirements of agriculture for their often fertile land, both economic studies and ecological research on wetlands urgently need emphasis. The economic (rather than only the ecological) value of wetlands must be clearly demonstrated in an effort to increase general awareness of the need for wetland protection.

The minimum area of wetland that will nevertheless ensure all the benefits of natural wetlands in a catchment should be established. For instance, it may be that a few small, strategically located wetlands can confer the same hydrological or ecological (and therefore economic) value to catchments as large wetland areas. This would provide a suitable compromise between the necessary conservation of wetlands and the pressure for development of catchments to make maximum use of land and other scarce resources.

There are two ways applied ecologists might approach these questions in their research activities: by deferred action studies involving further basic research if the current state of knowledge is recognised by the manager(s) to be inadequate, and/or by adaptive management studies (*c.f.* Mentis 1980). This latter process, once clear objectives have been set, involves the careful elucidation, testing and subsequent modification or refinement of actual management strategies which are applied in the field to the system itself.

While we believe that the setting of management goals should involve the aspirations of the local population, the carrying out of research or adaptive management to fulfil these goals must be left to experts who have been specially trained in these fields. In most African countries these experts have been largely expatriates. While training for doctors, engineers, economists and sociologists is recognised as being of immediate benefit to developing countries, similar emphasis is not placed on the training of professionals for the management of water resources. The cycle of droughts in the Sahel and southern parts of the continent, which have resulted in famine and enormous losses of life, should surely have highlighted the importance of water as Africa's most precious resource. As aquatic vegetation has such important effects on the availability of water and influences its utilisation for agriculture and fisheries, we are constantly surprised that there are so few aquatic botanists with African experience, and even today these live mostly in continents other than Africa.

Africa is no longer the sleeping giant or enigmatic continent that it was a century ago: it is vibrant, dynamic and developing. As such it is more vulnerable to ecological change than it has ever been and, consequently, most liable to experience the management errors which may cause irreversible damage. It is entering a future where man will increasingly manipulate his environment on a large scale in anticipation of deriving considerable benefit from it and greater independence of it. The extent to which success meets his efforts will depend on the care with which he plans and the level of understanding he employs in his decision making. Our role as aquatic plant biologists is to provide the most complete and clearest possible understanding of the biology, ecology and economic value of these plants and their systems.

This has been the major purpose of the book, and we all hope it goes some way to discharge the debt we owe to the continent that has nurtured, sustained and excited us, and will always live in us wherever we may be in the world.

References

Cairns, J., 1975. Quality control systems. In: B.A. Whitton (ed.), River Ecology. Blackwell Scientific Publications, Oxford.

Denny, P., (1984). Permanent swamp vegetation of the Upper Nile. Hydrobiologia, 110:79–90

Mentis, M.T., 1980. Towards a scientific management of terrestrial ecosystems. South African Journal of Science, 76:536–540.

Renvoize, S.A., Lock, J.M. & Denny, P., (1984). A remarkable new grass genus from the southern Sudan. Kew Bulletin, 39:455–61

Authors' addresses:
David Mitchell
CSIRO Centre for Irrigation Research
Griffith, NSW 2680
Australia

Patrick Denny
School of Biological Sciences
Queen Mary College
The University of London
London E1 4NS England

Clive Howard-Williams
Department of Scientific and Industrial Research
Taupo Research Laboratory
P.O. Box 415
Taupo, New Zealand

CHAPTER 10

A cross-indexed bibliography of African wetland plants and vegetation

KEITH THOMPSON, CLIVE HOWARD-WILLIAMS and DAVID MITCHELL

The African continent has many very large rivers with extensive floodplains, large numbers of both natural and man-made lakes with well-developed littoral zones, some of the largest swamp complexes in the world, and vast areas of land which are seasonally flooded or waterlogged but hydrologically isolated from the main drainage systems. The sheer number and variety of these wetlands, ranging from sea-level to over 5000 metres altitude, means that the African wetland flora is very extensive, and the range of wetland plant communities considerable.

Moreover, since Africa is, overall, a dry continent, the economic importance of its wetlands and wetland plants is enormous: the impact of aquatic weeds, the role of littoral and riverine vegetation and swamps in maintaining highly productive fisheries, the influence of permanent and seasonal wetlands on catchment water balance, aquatic plants and plant communities as reservoirs for human parasites and disease vectors, the critical role played by edaphic grasslands in maintaining the most varied populations of wild grazing mammals in the world, as well as huge numbers of domestic livestock.

African wetlands and wetland plants are indisputably very valuable, and yet it is all too frequently observed that useful literature is difficult, or impossible, to locate. The bibliography which comprises this chapter is not merely a supplement to the other chapters in this book: it is designed as a reference work in its own right – a guide to the most useful literature available on all aspects of African aquatic plants and wetlands. It is aimed at all advanced undergraduates, research workers and administrators with interests in African wetlands.

In compiling the bibliography, every effort has been made to ensure that all citations are complete and understandable: the poor quality of citations is a problem which has frustrated many students of African literature. Thus, since most of the bibliographic entries are located in less well known literature sources, abbreviations, which so often confuse, have been entirely dispensed with. It has, however, been necessary to include several restricted publications, such as reports of government departments, where they contain important data not available in any other form. We realise that many of these obscure publications may be difficult to obtain outside the countries concerned, but at least a knowledge of their existence will enable the determined researcher to track them down.

The only major exclusions from this bibliography are accounts of 'general interest' and works of purely taxonomic value. Thus, regional floras and publications dealing with vegetation in a very general way, such as national park guides, have been omitted, except where they have substantial sections dealing with the ecology of aquatic plants. We have also confined the bibliography to strictly African material, so that general accounts of, for instance, aquatic weeds, do not feature, unless they have specific sections dealing with Africa, or were

Patrick Denny (ed.), The ecology and management of African wetland vegetation.
 ISBN 90 6193 509 1

written with an African perspective.

However, the categories TAXONOMY, CHECKLIST and CLASSIFICATION have been included in the index to draw together certain particularly useful species lists and phytosociological réléves. VEGETATION DESCRIPTION, with over 25% of the total entries, is a category which quite clearly demonstrates that most of the references to African aquatic plants and wetlands are either 'surveys' or other types of purely descriptive accounts. On the other hand, less than 15% of the citations quantitatively relate the distribution, abundance and growth of aquatic plants to environmental factors, and the really important ecosystem processes, such as PRODUCTIVITY, PHYSIOLOGY and DECOMPOSITION attract less than 80 entries continent-wide.

There are three general categories: ECOLOGY, ECOLOGICAL PROBLEMS and ENVIRONMENTAL FACTORS. Whilst these are not likely to be useful in tracking down specific reference material, they may be helpful to the student unfamiliar with Africa who wishes to establish fairly quickly a 'feel' for the African literature. Most of this literature has been written with 'a problem' in mind, and many of these problems refer to AQUATIC WEEDS. Thus, 150 citations are accounted for by EICHHORNIA, PISTIA and SALVINIA alone, and most of the 23 MYRIOPHYLLUM entries also refer to the plant's impact as a weed.

The other area in which government and international agency funding is at a high level is DEVELOPMENT (*ca.* 100 entries) and CONTROL/MANAGEMENT (*ca.* 120). Many of these accounts occur in restricted circulation reports and are of a fairly superficial nature. Note that the category UTILISATION refers specifically to plant species, and not to wetland types. Wetland CONSERVATION, on the other hand, has a mere 24 entries.

The uneven distribution by country – SOUTH AFRICA (110), SUDAN (77), ZIMBABWE (61), UGANDA (57), ZAMBIA (44), ZAIRE (43), but KENYA (18), TANZANIA (15), NIGER (6), ANGOLA (5), CONGO (3) and MALI (2) – partly reflects land (particularly wetland) area and partly the level of a country's economy and the availability of suitably trained personnel. Furthermore, the disparity is actually increasing: literature deficiencies are not being corrected. Thus a 1984 issue of the Journal of the Limnological Society of Southern Africa is devoted entirely to wetland material, whilst Uganda, for instance, with the highest proportion of permanent wetland of any country in Africa, has virtually no active current research in this field. The man-made lakes, such as KARIBA and VOLTA feature more often than those without major development projects, and of the 41 HYDROLOGY entries, only one or two are not connected with development projects.

The bibliography provides separate listings of different wetland types under ESTUARY, FLOODPLAIN, LAKE, RIVER and SWAMP, and these may be cross-referenced to species. Note that SWAMP FOREST and SWAMP SAVANNA (= edaphic grassland) are listed separately, as also are DAMBO and ESOBE.

Authors' addresses:
Keith Thompson
Department of Biological Sciences
University of Waikato
Private Bag
Hamilton, New Zealand

Clive Howard-Williams
Department of Scientific and Industrial Research
Taupo Research Laboratory
P.O. Box 415
Taupo, New Zealand

David Mitchell
CSIRO Centre for Irrigation Research
Griffith NSW 2680
Australia

1 ABDALLA, A.A. and ABDEL HAFEEZ, A.T., 1969. Some aspects of utilisation of water hyacinth (*Eichhornia crassipes*). PANS, 15:204–207.
KEYWORDS: Egypt, *Eichhoria*, utilisation (man), aquatic weeds

2 ABDELMALIK, W.E.Y., EL-SHINAWY, R.M.K., ISHAK, M.M. and MAHMOUD, K.A., 1943. Uptake of radionuclides by some aquatic macrophytes of Ismailia Canal, Egypt. Hydrobiologia, 42:3–12.
KEYWORDS: Egypt, *Potamogeton*, *Charophytes*, chemistry, river (Ismailia Canal)

3 ABIODUN, A.A. and IMEVBORE, A.M.A., 1977. Monitoring of aquatic weeds in Lake Kainji using LANDSAT data. In: SIL Congress, Copenhagen, 1977. DIS Congress Service, Copenhagen.
KEYWORDS: Nigeria, aquatic weeds, lake (Kainji), control/management (*Eichhornia*)

4 ABU-GIDEIRI, Y.B. and YOUSIF, A.M., 1974. The influence of *Eichhornia crassipes* on planktonic development in the White Nile. Archiv für Hydrobiologie, 74:463–467.
KEYWORDS: Sudan, utilisation (animals), chemistry, plankton, *Eichhornia*, ecological problems, river (Nile)

5 ACOCKS, J.P.H., 1976. Riverine vegetation of the semi-arid and arid regions of South Africa. Journal of the South African Biological Society, 17:21–35.
KEYWORDS: South Africa, river (South Africa), vegetation description (South Africa), riverine

6 ADAM, J.G., 1958. Flore et végétation de la réserve botanique de Noflaye (environs de Dakar Sénégal). Bulletin de l'Institut Françaïse d'Afrique Noire, Serie A, 20(3):809–868.
KEYWORDS: Sénégal, vegetation description
(In French)

7 ADAM, J.G., 1964. Contribution a l'étude de la végétation du lac Guiers (Senegal). Bulletin de l'Institut Françaïse d'Afrique Noire, Serie A, 26:1–72.
KEYWORDS: Sénégal, vegetation description (Sénégal), lake (Guiers)
(in French)

8 ADEJUWON, J.O., 1971. The ecological status of fresh water swamp savannas in the forest-zone of Nigeria. Journal of the West African Science Association, 16(2):133–154.
KEYWORDS: Nigeria, swamp (savanna), ecology

9 ADLER, E., 1977. Hartebeespoort Dam – asset or disaster (In Afrikaans). Veldtrust (Pretoria), March 1977:14–15.
KEYWORDS: South Africa, development (water), *Eichhornia*, ecological problems, lake (Hartebeespoort)

10 AHN, P.M., 1958. Regrowth and swamp vegetation in the western forest areas of Ghana. Journal of the West African Science Association, 4:163–173.
KEYWORDS: Ghana, vegetation description (Ghana), swamp (Ghana), development (land/swamp), swamp (forest)

11 AHTI, T., HAMET-AHTI, L. and PETTERSSON, B., 1973. Flora of the inundated Wadi Halfa reach of the Nile, Sudanese Nubia, with notes on adjacent areas. Annales Botanici Fennici, 10:131–162.
KEYWORDS: Sudan, vegetation description (Sudan), swamp (Wadi Halfa), river (Nile)

12 ALEEM, A.A. and SAMAAN, A.A., 1969. Productivity of Lake Mariut, Egypt. Part II. Primary production. Internationale Revue der gesamten Hydrobiologie und Hydrographie, 54:491–527.
KEYWORDS: Egypt, productivity, lake (Mariut), plankton

13 ALLANSON, B.R., 1959. Investigations into the ecology of polluted rivers in the Transvaal. A report on the physical, chemical and biological conditions in the Jukskei – Crocodile system, C.S.I.R.–National Institute for Water Research, Pretoria.
KEYWORDS: South Africa, chemistry, river (Jukskei/Crocodile), ecology, ecological problems

14 ALLANSON, B.R., 1960. Investigations into the ecology of polluted inland waters in the Transvaal. Part I, The physical, chemical and biological conditions in the Jukskei – Crocodile river system. D. Phil. Thesis. University of Cape Town.
KEYWORDS: South Africa, chemistry, river (Jukskei/Crocodile), ecology, ecological problems

15 ALLANSON, B.R., 1961. Investigation into the ecology of polluted inland waters in the Transvaal. Part 1. Hydrobiologia, 18:1–76.
KEYWORDS: chemistry, South Africa, ecological problems

16 ALLANSON, B.R. (Editor), 1979. Lake Sibaya. W. Junk b.v., Monographiae Biologicae 36, The Hague.
KEYWORDS: South Africa, lake (Sibaya), vegetation description (South Africa), aquatic plants, chemistry

17 ALLANSON, B.R., BRUTON, M.N. and HART, R.C., 1974. The plants and animals of lake Sibaya, Kwazulu, South Africa: a checklist. Revue de Zoologie et de Botanique Africaines, 88:507–532.
KEYWORDS: South Africa, lake (Sibaya), checklist, aquatic plants

18 ALMEIDA, F.S., 1971. Terceiro simposio de ervas aquaticas. Instituto da Investigaçao de Agricole de Moçambique, Informacoes Technicoes, 37, 37 pp.
KEYWORDS: Moçambique, development (land/swamp), swamp (Moçambique) (in Portuguese)

19 ALMEIDA, F.S., 1972. Problema de vegetaçao aquatica do Lago Kariba. Instituto da Investigaçao de Agricole de Moçambique, Informacoes Technicos, 57, 57 pp.
KEYWORDS: Zimbabwe, lake (Kariba), *Salvinia*, ecological problems (in Portuguese)

20 ALSTON, A.H.G., 1959. The Ferns and Fern Allies of West Tropical Africa. Crown Agents, London.
KEYWORDS: Africa, checklist, ferns

21 ANDREWS, F.W., 1945. Water Plants in the Gezira canals. Annals of Applied Biology, 32:1–14.
KEYWORDS: Sudan, development (water), aquatic weeds, *Myriophyllum*, river (Nile)

22 ANDREWS, F.W., 1956. The flowering plants of the Sudan I–III. Sudan Government Publication.
KEYWORDS: checklist, Sudan

23 ANDREWS, P., GRAVES, C.P. and HORNE, J.F.M., 1975. Ecology of the lower Tana River floodplain (Kenya). Bulletin of the East African Natural History Society, 151:1–31.
KEYWORDS: Kenya, floodplain (Tana), vegetation description (Kenya), ecology, river (Tana)

24 ANONYMOUS, 1956. Bestrijding van de water hyacinth in Belgisch Congo. Africa Institution Bulletin, 10:87–89.
KEYWORDS: Zaïre, *Eichhornia*
(in Dutch)

25 ANONYMOUS, 1957. O jacinto da agua. Gazeta Agricole, 9:99–101.
KEYWORDS: Moçambique, *Eichhornia*
(in Portuguese)

26 ANONYMOUS, 1957. Symposium on *Eichhornia crassipes*, Leopoldville. Conseil Scientifique pour l'Afrique au Sud du Sahara (C.S.A.), Publication 27, 27 pp.
KEYWORDS: Zaïre, *Eichhornia*, ecological problems

27 ANONYMOUS, 1957. The water hyacinth. Now a "prohibited immigrant". East Africa Agricultural Journal, 22:129.
KEYWORDS: Africa, *Eichhornia*, ecological problems

28 ANONYMOUS, 1960. *Myriophyllum* eradication. Cape Provincial Administration Department of Nature Conservation, Report, 25:104.
KEYWORDS: South Africa, *Myriophyllum*, control/management (*Eichhornia*)

29 ANONYMOUS, 1961. A fern makes a take-over bid for Kariba Lake. New Scientist, 9:731.
KEYWORDS: Zimbabwe, *Azolla*, lake (Kariba), ecological problems

30 ANONYMOUS, 1966. Water hyacinth eradication programme stepped up. South African Sugar Journal, 50:379.
KEYWORDS: South Africa, *Eichhornia*, control/management (*Eichhornia*)

31 ANONYMOUS, 1968. Multipurpose survey of the Kafue River Basin, Zambia, I. United Nations/FAO Report SF.35/Zambia, Rome.
KEYWORDS: Zambia, vegetation description (Zambia), ecology, floodplain (Kafue), development (water), river (Kafue)

32 ANONYMOUS, 1972. The extent and degree of occurrence of *Salvinia molesta* (Kariba weed) in the Chobe–Linyanti–Kwando River System. Report of the Joint Botswana/South African Survey, CSIR Pretoria/Botswana Ministry of Agriclture, Gaborone, Botswana.
KEYWORDS: Botswana, Zimbabwe, Namibia, *Salvinia*, river (Chobe/Linyanti/Kwando)

33 ANONYMOUS, 1975. Aquatic weeds in the Sudan, with special reference to water hyacinth, National Research Council, Report PN–AAC–696, Washington, D.C, 165pp.
KEYWORDS: Sudan, *Eichhornia*, control/management (*Eichhornia*)

34 ANONYMOUS, 1975. Workshop on Management and Utilisation of Aquatic Weeds in the Sudan and Nile Basin. Khartoum, Sudan, S.A.S., Report PN–AAC–641, Washington, D.C, 63pp.
KEYWORDS: Egypt, Sudan, aquatic weeds, control/management (*Eichhornia*), river (Nile)

35 ANONYMOUS, 1976. Kariba-weed spreading in Botswana. Agricultural News, 8(40):4.
KEYWORDS: Botswana, *Salvinia*, ecological problems

36 ANONYMOUS, 1976. Weed harvester – the Aqua-Trio weed harvesting system in use at Sandvlei. Municipal Engineer, 7(3):21.
KEYWORDS: South Africa, aquatic weeds, control/management, lake (Sandvlei)

37 ANONYMOUS, 1977. Control of water hyacinths in South Africa. Munisipale Ingenieur, 8(6):39–43.
KEYWORDS: South Africa, *Eichhornia*, control/management (*Eichhornia*)

38 ANONYMOUS, 1980. Destroying the devil's flower, (*Eichhornia crassipes*). African Wildlife, 34(4):35–36.
KEYWORDS: Africa, *Eichhornia*, control/management (*Eichhornia*)

39 APPLETON, C.C., 1974. A check-list of the flora and fauna of the Gladdespruit, Nelspruit District, Eastern Transvaal. Limnological Society of Southern Africa Newsletter, 22:49–51.
KEYWORDS: South Africa, checklist

40 ASHTON, P.J., 1974. The effect of some environmental factors on the growth of *Azolla filiculoides* Lam. In: E.M. Van Zinderen Bakker (Editor) The Orange River Project, Progress Report, The University of the Orange Free State, Bloemfontein, 123–138.
KEYWORDS: South Africa, environmental factors, *Azolla*

41 ASHTON, P.J., 1977. Factors affecting the growth and development of *Azolla filiculoides* Lam. Proceedings of the Second National Weeds Conference of South Africa. pp 249–268.
KEYWORDS: South Africa, environmental factors, *Azolla*

42 ASHTON, P.J. and WALMSLEY, R.D., 1976. The aquatic fern *Azolla* and its *Anabaena* symbiont. Endeavour, 35:39–43.
KEYWORDS: South Africa, physiology, *Azolla*

43 ASHTON, P.J., SCOTT, W.E. and STEYN, D.J., 1981. The chemical control of the water hyacinth *Eichhornia crassipes* (Mart.) Solms. Progress in Water Technology, 13:865–882.
KEYWORDS: South Africa, *Eichhornia*, control/management (*Eichhornia*)

44 ASHTON, P.J., SCOTT, W.E., STEYN, D.J. and WELLS, R.J., 1979. The chemical control programme against water hyacinth *Eichhornia crassipes* (Mart.) Solms on Hartebeespoort Dam: historical and practical aspects. South African Journal of Science, 75:303–306.
KEYWORDS: South Africa, *Eichhornia*, control/management (*Eichhornia*), lake (Hartebeespoort)

45 ASTLE, W.G. and GRAHAM, A., 1976. Ecological investigations of the UNDP in the Okavango Delta. Proceedings of the Symposium on the Okavango Delta and its Future Utilisation. The Botswana Society, Gaborone, pp 81–91.
KEYWORDS: Botswana, swamp (Okavango), floodplain (Okavango), wildlife, vegetation description (Botswana)

46 ASTLE, W.L., 1965. The edaphic grasslands of Zambia. Proceedings of 9th International Grassland Congress. pp 363–373.
KEYWORDS: Zambia, swamp (savanna), vegetation description (Zambia), dambos

47 ASTLE, W.L., 1965. The grass cover of the Chambeshi Flats, Northern Province, Zambia. Kirkia, 5:37–50.
KEYWORDS: Zambia, floodplain (Chambeshi), swamp (savanna), vegetation description (Zambia)

48 ATTIONU, R.H., 1970. Eco-physiological studies on *Pistia stratiotes* L. with special reference to its occurrence on the Volta Lake in Ghana. M.Sc. Thesis. University of Ghana, Legon.
KEYWORDS: Ghana, chemistry, *Pistia*, lake (Volta), physiology

49 ATTIONU, R.H., 1976. Some effects of water lettuce (*Pistia stratiotes* L.) on its habitat. Hydrobiologia, 50:245–254.
KEYWORDS: Ghana, *Pistia*, ecology

50 ATTWELL, R.I.G., 1970. Some effects of Lake Kariba on the ecology of a flood-plain of the mid-Zambezi Valley of Rhodesia. Biological Conservation, 2:189–195.
KEYWORDS: Zimbabwe, swamp (savanna), lake (Kariba), floodplain (Zambezi), development (water), ecological problems

51 BACALBASA-DOBROVICI, N., 1971. The economic importance of the burgu (*Echinochloa stagnina* P. Beauv.) on the middle Niger in the Niger Republic. Hidrobiologia, Bucurest, 12:41–46.
KEYWORDS: Niger, utilisation (animals), *Echinochloa*, utilisation (man), floodplain (Niger), river (Niger)

52 BAKSHI, T.S., 1964. Control of *Vossia cuspidata*. Span, 7:37–39.
KEYWORDS: Africa, *Vossia*, control/management (*Vossia*)

53 BALEK, J., Swamps. In: Hydrology and Water Resources in Tropical Africa. Elsevier, Amsterdam, pp 154–168.
KEYWORDS: Africa, swamp, hydrology, development (water)

54 BALON, E.K. and COCHE, A.G. (Editors), 1974. Lake Kariba: A Man-Made Tropical Ecosystem in Central Africa. W. Junk b.v., The Hague.
KEYWORDS: Zimbabwe, Zambia, lake (Kariba), chemistry, vegetation description (Zimbabwe), aquatic plants, swamp (Zimbabwe/Zambia), development

55 BARD, J., 1960. La lutte contre la végétation aquatique. Proceedings of the Third Symposium on Hydrobiology and Inland Fisheries: Major Lakes, Lusaka, 1960. Conseil Scientifique pour l'Afrique au Sud du Sahara, CCTA/CSA Technical and Scientific Publication, 63:22pp.
KEYWORDS: Africa, vegetation description, lake, aquatic plants
(In French)

56 BARD, J., 1965. Végétaux aquatiques tropicaux. Bois Forêts Tropicales, 99:3–11.
KEYWORDS: Zaïre, Cameroun, vegetation description (Zaïre), *Eichhornia*, *Echinochloa*

57 BARI, E.A., 1968. Sudan. In: I and O Hedberg, (Editors), Conservation of Vegetation in Africa South of the Sahara. Acta Phytogeographica Suecica, 54:58–64.
KEYWORDS: Sudan, vegetation description (Sudan), conservation, swamp (Upper Nile/Sudd)

58 BARRADAS, L.A., 1944. As possibiladades agricolas dos terrenos de machongo. Boletim da Sociedade de Estudos da Colonia de Moçambique, 46:1–14.
KEYWORDS: Moçambique, swamp (Moçambique), development (land/swamp)

59 BARRADAS, L.A., 1951. Machongos reclamation. In: Proceedings of the Conference on Hydrology and Water Resources. East African High Commision, Nairobi, 2 pp.
KEYWORDS: Moçambique, swamp (Moçambique), development (land/swamp), peatland

60 BATANOUNY, K. and EL-FIKY, 1975. The water hyacinth (*Eichhornia crassipes*) in the Nile , Eygpt. Aquatic Botany, 1:243–252.
KEYWORDS: Sudan, *Eichhornia*, ecological problems, river (Nile)

61 BATCHELOR, G.R., 1978. Why conserve wetlands?. Fauna et Flora, 33:8–9.
KEYWORDS: South Africa, conservation, birds, wildlife

62 BATES, G.R. and PHIPPS, J.B., 1958. Water hyacinth and its control in Southern Rhodesia. Proceedings of the African Weed Control Conference, Victoria Falls, 1958. Fisons Pest Control, Saffron Walden, U.K, pp 135–139.
KEYWORDS: Zimbabwe, *Eichhornia*, control/management (*Eichhornia*)

63 BAUER, S., 1966. Helicopter against water hyacinth in the Sudan. Agricultural Aviation, 8:116–118.
KEYWORDS: Sudan, *Eichhornia*, control/management (*Eichhornia*), river (Nile)

64 BAUER, S.W., 1975. A mathematical swamp model. University of the Witwatersrand, Johannesburg, Report No 5:p75.
KEYWORDS: South Africa, swamp, hydrology, environmental factors

65 BEADLE, L.C., 1930. Scientific results of the Cambridge expedition to the East African lakes 1930. 3. Observations on the bionomics of some East African swamps. Zoological Journal of the Linnaean Society, 38:135–155.
KEYWORDS: Kenya, swamp (East Africa), ecology, fauna

66 BEADLE, L.C., 1930. Scientific results of the Cambridge expedition to the East African lakes 1930. 4. The waters of some East African lakes in relation to their flora and fauna. Zoological Journal of the Linnaean Society, 38:156–211.
KEYWORDS: Kenya, chemistry, lake (East Africa), vegetation description (East Africa), fauna

67 BEADLE, L.C., 1958. Hydrobiological investigations on tropical swamps. Verhandlungen der Internationalen Vereinigung fúr theoretische und angewandte Limnologie (Verh. Internat. Verein. Limnol.), 13:855–857.
KEYWORDS: Uganda, chemistry, swamp (Uganda)

68 BEADLE, L.C., 1974 and 1981. The Inland Waters of Tropical Africa: An Introduction to Tropical Limnology. (1st and 2nd editions). Longmans, London and New York, 365 and 475 pp.
KEYWORDS: Africa, review, chemistry, aquatic plants, hydrology, plankton, lake, swamp, river, fish, invertebrates, development (water)

69 BEADLE, L.C. and LIND, E.M., 1960. Research on the swamps of Uganda. Uganda Journal, 24:84–87.
KEYWORDS: Uganda, swamp (Uganda), chemistry, invertebrates, vegetation description (Uganda)

70 BEAM, W., 1914. Papyrus and paper manufacture. Cairo Scientific Journal, 8:283–285.
KEYWORDS: Egypt, *Cyperus*, utilisation (man)

71 BEAUCHAMP, R.S.A., 1958. Utilising the natural resources of Lake Victoria for the benefit of fisheries and agriculture. Nature, London, 181:1634–1636.
KEYWORDS: East Africa, lake (Victoria), development (land/swamp), fish

72 BEBAWI, E.F., 1972. Studies on the ecology of *Eichhornia crassipes* (Mart) Solms in the Sudan. Msc thesis, University of Khartoum.
KEYWORDS: Sudan, *Eichhornia*, ecology

73 BEGG, G.W., 1971. Preliminary tests using formalin as a possible means of chemically controlling *Salvinia auriculata* and other undesirable floating plants in freshwater impoundments. Limnological Society of Southern Africa Newsletter, 17:22–26.

KEYWORDS: Zimbabwe, *Salvinia*, control/management (*Salvinia*), aquatic weeds

74 BEGG, G.W., 1973. The biological consequences of discharge above and below Kariba Dam. Proceedings of the Onzième Congrès des Grands Barrages 1973. Commission Internationale des Grands Barrages, Madrid, pp 421–430.
KEYWORDS: Zimbabwe, hydrology, development (water), ecological problems, lake (Kariba)

75 BEGG, G.W., 1978. The estuaries of Natal. Natal Town and Regional Planning Commission, Pietermaritzburg, Report No.41, 41:1–657.
KEYWORDS: estuary (South Africa), South Africa

76 BELL, R.H.V. and GRIMSDELL, J.J.R., 1973. The persecuted black lechwe of Zambia. Oryx, 12(1):77–92.
KEYWORDS: Zambia, wildlife, lechwe, conservation, swamp (savanna), floodplain (Bangweulu)

77 BENECH, V., LEMOALLE, J. and QUENSIERE, J., 1976. Mortalités de poissons et conditions de milieur dans le Lac Tchad au cours d'une période de sècheresse. Cahiers d'Office de la Récherche Scientifique et Technique Outre-Mer, Serie Hydrobiologique, 10(2):119–130.
KEYWORDS: Tchad, chemistry, lake (Tchad), fish, ecological problems

78 BENNETT, F.D., 1972. Some aspects of the biological control of aquatic weeds. Proceedings of the Second International Symposium on the Biological Control of Weeds, Rome 1971.
KEYWORDS: Africa, *Eichhornia*, *Salvinia*, *Myriophyllum*, control/management (weeds)

79 BERG, A., 1959. Analyse des conditions improprès au développement de la jacinthe d'eau, *Eichhornia crassipes* (Mart.) Solms dans certaines rivières de la cuvette Congolaise. Bulletin Agricole du Congo Belge, 50:365–401.
KEYWORDS: Zaïre, *Eichhornia*, ecological problems, river (Zaïre)
(in French)

80 BERG, A., 1960. Determination of conditions unsuitable for the growth of the water hyacinth *Eichhornia crassipes* (Mart.) Solms in certain rivers of the Congo basin. .Bulletin du Centre Belge d'Etudes et de Documentation des Eaux, 47:28–42.
KEYWORDS: Zaïre, *Eichhornia*, chemistry, ecological problems, river (Zaïre)

81 BERG, A., 1961. Role écologique des eaux de la cuvette congolaise sur la croissance de la jacinthe d'eau, *Eichhornia crassipes* (Mart.) Solms. Mémoires de l'Academie royale des sciences d'Outre-Mer: nat. med, 8(12):1–120.
KEYWORDS: Zaïre, *Eichhornia*, chemistry, ecological problems, river (Zaïre)
(in French)

82 BERG, C.L., 1950. Swamp reclamation and irrigation. Annual Report of the Department of Hydrological Survey, Uganda Government (1949), pp 6–9.
KEYWORDS: Uganda, swamp (Uganda), development (land/swamp), control/management (swamp), irrigation

83 BERG, C.L., 1951. Swamp reclamation. Proceedings of the Conference on Hydrology and Water Resources. East African High Commission.
KEYWORDS: Uganda, swamp (Uganda), development (land/swamp), irrigation, control/management (swamp)

84 BERRY, L. and WHITEMAN, A.J., 1968. The Nile in the Sudan. Geographical Journal, 134:1–37.
KEYWORDS: Sudan, hydrology, vegetation description (Nile), river (Nile), development

85 BESHIR, M.E., 1978. Gravity-flow irrigation and the spread of aquatic weeds in the Sudan. Environmental Conservation, 5(2):143–146.
KEYWORDS: Sudan, development (land/swamp), aquatic weeds, control/management, irrigation

86 BEWS, J.W., 1917. The plant ecology of the Drakensberg Range. Annals of the Natal Museum, 3(3):511–561.
KEYWORDS: South Africa, vegetation description, peatland

87 BIGGS, R.C., 1976. The effects of the seasonal flood regime on the ecology of Chief's Island and the adjacent floodplain system. Proceedings of the Symposium on the Okavango Delta and its future Utilisation. The Botswana Society, Gaborone, pp 113–120.
KEYWORDS: Botswana, hydrology, floodplain (Okavango), swamp (Okavango), vegetation description (Okavango)

88 BISHAI, H., 1961. The effect of the water hyacinth on the fisheries of the Sudan. Report of the Hydrobiological Research Unit, University of Khartoum, 8:29–36.
KEYWORDS: Sudan, fish, *Eichhornia*, ecological problems, river (Nile)

89 BISHAI, H.M., 1962. The water characteristics of the Nile in the Sudan with a note on the effect of *Eichhornia crassipes* on the hydrobiology of the Nile. Hydrobiologia, 19:357–382.
KEYWORDS: Sudan, chemistry, hydrology, *Eichhornia*, ecological problems, river (Nile)

90 BISHOP, W.W., 1958. Raised swamps of Lake Victoria. Records of the Geological Survey of Uganda 1955–1956, Entebbe, pp 1–10.
KEYWORDS: Uganda, *Cyperus*, swamp (Uganda), hydrology, lake (Victoria)

91 BOLTT, R.E., HILL, B.J. and FORBES, A.T., 1969. The benthos of some Southern African lakes. I. The distribution of aquatic macrophytes and fish in Lake Sibayi. Transactions of the Royal Society of South Africa, 38:241–248.
KEYWORDS: South Africa, lake (Sibayi), fish, aquatic plants

92 BOND, W.J. and ROBERTS, M.G., 1978. The colonization of Cabora Bassa, Moçambique, a new man-made lake, by floating African macrophytes. Hydrobiologia, 60:243–259.
KEYWORDS: Moçambique, development (water), lake (Cabora Bassa), aquatic weeds

93 BOUCHER, C. and JARMAN, M.L., 1977. The vegetation of the Langebaan area, South Africa. Transactions of the Royal Society of South Africa, 42:241–271.
KEYWORDS: South Africa, vegetation description, swamp (Langebaan), development (land/swamp)

94 BOUGHEY, A.S., 1957. The physiognomic delimitation of West African vegetation types. Journal of the West African Science Association, 3(2):148–165.
KEYWORDS: classification, West Africa

95 BOUGHEY, A.S., 1963. The explosive development of a floating weed vegetation on Lake Kariba. Adansonia, 3:49–61.
KEYWORDS: Zambia, Zimbabwe, *Salvinia*, lake (Kariba), ecological problems

96 BOULOS, L., 1967. On the weed flora of Aswan, Egypt. Botaniska Notiser, 120:368–372.
KEYWORDS: aquatic weeds, Egypt, lake (Nasser)

97 BOULOS, L. and EL–HADIDI, N.M., 1967. Common Weeds in Egypt. Dar el–Maaref. Press, Cairo.
KEYWORDS: Egypt, *Salvinia*, aquatic weeds

98 BOURN, D.M., 1972. The feeding, diet and ecological relations of three species of fish of economic importance in southern Malawi. M.Sc. Thesis. University of Edinburgh.
KEYWORDS: Malawi, fish, utilisation (animals)

99 BOUSSIOUD–CORBIERS, F. and SAVOURE, B., 1978. Contribution à l'étude des ceintures de végétation d'une lagune continentale du nord de la Tunisie (Lac Ischkeul). Hydrobiologia, 58:25–32.
KEYWORDS: Tunisia, lake (Ischkeul), vegetation description (Tunisia)

100 BOUXIN, G., 1976. Ordination and classification in the upland Rugege Forest (Rwanda, Central Africa). Vegetatio, 32:97–115.
KEYWORDS: classification, Rwanda, swamp (forest), peatland, swamp (Rwanda/Burundi)

101 BOWMAKER, A.P., 1960. Seasonal hydrological changes affecting the ecology of Luaka Lagoon, Lake Bangweulu: some preliminary observations. Proceedings of the Third Symposium on Hydrobiology and Inland Fisheries: Major Lakes. Lusaka, 1960. Conseil Scientifique pour l'Afrique au Sud du Sahara, CCTA/CSA Technical and Scientific Publication, 63:22pp.
KEYWORDS: Zambia, lake (Bangweulu), swamp (Bangweulu), *Cyperus*, fish, hydrology

102 BOWMAKER, A.P., 1969. A study of the hydrology of the Sinamwenda River and "Estuary", Lake Kariba. The Nuffield Lake Kariba Research Station, Sinamwenda, Report 1962–1968, 15–17.
KEYWORDS: Zimbabwe, river (Sinamwenda), lake (Kariba), hydrology

103 BOWMAKER, A.P., 1973. Hydrophyte dynamics in Mwenda Bay, Lake Kariba. Kariba Studies, Paper No. 3:42–59.
KEYWORDS: Zimbabwe, fish, *Salvinia*, lake (Kariba)

104 BRAIN, C.K., 1937. Weeds of Southern Rhodesia. Part II: Water weeds. Rhodesia Agricultural Journal, 34:612–615.
KEYWORDS: Zimbabwe, aquatic weeds

105 BREEN, C.M. and JONES, I.D, 1970. A preliminary list of angiosperms collected in the vicinity of Lake Sibayi. Transactions of the Royal Society of South Africa, 39:235–245.
KEYWORDS: South Africa, checklist, swamp (Lake Sibayi), lake (Sibayi)

106 BREEN, C.M., EVERSON, C. and ROGERS, K., 1977. Ecological studies on *Sporobolus virginicus* (L.) Kunth. with particular reference to salinity and inundation. Hydrobiologia, 54:135–140.
KEYWORDS: South Africa, hydrology, floodplain, swamp (savanna), chemistry, *Sporobolus*

107 BREEN, C.M., LILLIG, C., OLIVER, J. and FURNESS, H.D., 1976. A quantitative sampler for aquatic macrophytes. Journal of the Limnological Society of Southern Africa, 2:59–62.
KEYWORDS: South Africa, aquatic plants, productivity

108 BRELSFORD, V., 1961. The story of a swamp: Lake Bangueulu in Northern Rhodesia. Horizon, 3(7).
KEYWORDS: Zambia, vegetation description, swamp (Bangweulu), floodplain, lake (Bangweulu)

109 BRIND, W.G., 1954. The Okavango delta. Report of the 1951–53 Field Surveys, Bechuanaland Government Department of Geological Survey, 58pp.
KEYWORDS: Botswana, development (land/swamp), swamp (Okavango), floodplain (Okavango)

110 BRITTON, P.L., 1979. Seasonality, density and diversity of birds in a papyrus swamp in western Kenya. Ibis, 120(4):450–466.
KEYWORDS: Kenya, birds, *Cyperus*, swamp (Kenya), wildlife

111 BROEKHUYSEN, G.J. and TAYLOR, H., 1959. The ecology of South African estuaries. VIII. Kosi Bay estuary system. Annals of the South African Museum, 44:279–296.
KEYWORDS: South Africa, vegetation description (South Africa), estuary (Kosi Bay), ecology

112 BROWN, A.F., 1905. Some notes on the Sudd formation of the upper Nile. Botanical Journal of the Linnaean Society, 37:51–58.
KEYWORDS: Sudan, hydrology, *Cyperus*, swamp (Upper Nile/Sudd), river (Nile)

113 BUDDENHAGEN, I.W. and PERSLEY, G.J., 1978. Rice in Africa. Academic Press.
KEYWORDS: *Oryza*, taxonomy

114 BULMAN, R.B., 1951. Swamp reclamation. Annual Report of the Department of Hydrological Survey, Uganda Government (1950), 44–47.
KEYWORDS: Uganda, swamp (Uganda), development (land/swamp)

115 BULMAN, R.B., 1952. Some problems in tropical swamp reclamation. Annual Report of the Department of Hydrological Survey, Uganda Government (1951), pp 9–18.
KEYWORDS: Uganda, swamp (Uganda), development (land/swamp), *Cyperus*, ecological problems

116 BUNTING, A.M. and LEA, J.D., 1962. The soils and vegetation of the Fung, east central Sudan. Journal of Ecology, 50:529–558.
KEYWORDS: Sudan, swamp (savanna)

117 BURTON, M.N., WILSON, G. and MUSIL, C.F., 1978. Environmental impact assessment: proposed Umfolozi dams on Lake Tiza. Unpublished Report to Natal Parks Board, 10pp.
KEYWORDS: South Africa, lake (Tiza), river (Umfolozi), ecological problems, development (water)

118 CALDER, E.A., 1959. Nitrogen fixation in a Uganda swamp soil. Nature, London, 184:746.
KEYWORDS: Uganda, chemistry, swamp (Uganda)

119 CAMUS, C., 1957. Problèmes posés par les papyrus au Lualaba Supérieur. Académie Royale de Sciences Coloniales, Bruxelles, Bulletin des Séances, 3:1164–1185.
KEYWORDS: Zaïre, *Cyperus*, swamp (Upemba), river (Lualaba), ecological problems

120 CANTRELL, M.A., 1981. Bilharzia snails and water level fluctuations in a tropical swamp. Oikos, 36:226–232.
KEYWORDS: utilisation (animals), disease, invertebrates, swamp, hydrology, control/management (bilharzia)

121 CARMOUZE, J-P. (*et al*), 1972. Grandes zones écologiques du Lac Tchad. Cahiers d'Office de la Récherche Scientifique et Technique Outre-Mer, Série Hydrobiologique, 6:103-169.
KEYWORDS: Tchad, lake (Tchad), swamp (Tchad), vegetation description (Tchad), floodplain (Tchad)

122 CARMOUZE, J-P., DURAND, J.R. and LEVEQUE, C., 1983. The lacustrine ecosystem during the "Normal Chad" period and the drying phase. In: J.P. Carmouze, J.R. Durand and C. Lévêque (Editors), Lake Chad. W. Junk b.v., The Hague, Monographiae Biologicae 53, pp 527-560.
KEYWORDS: Tchad, lake (Tchad), swamp (Tchad), river (Chari), littoral

123 CARMOUZE, J.P., DEJOUX, L., DURAND, J.R., GRAS, R., ILTIS, A., LAUZANNE, L., LEMOALLE, J., LEVEQUE, C., LOUBENS, G. and SAINT-JEAN, L., 1972. Grandes zones écologiques du Lac Tchad. Cahiers d'Office de la Récherche Scientifique et Technique Outre-Mer, Série Hydrobiologique, 6:103-169.
KEYWORDS: Tchad, chemistry, hydrology, swamp (Tchad), aquatic plants, lake (Tchad), vegetation description (Tchad)

124 CARMOUZE, J.P., FOTIUS, G. and LEVEQUE, C., 1978. Influence qualitative des macrophytes sur la regulation hydrochimique du lac Tchad. Cahiers d'Office de la Récherche Scientifique et Technique Outre-Mer, Série Hydrobiologique, 12:65-69.
KEYWORDS: Tchad, lake (Tchad), chemistry, aquatic plants

125 CARNEGIE, A.J.M., 1978. The borer *Eldana saccharina* Walker. ISSCT Entomology Newsletter, 4:7-8.
KEYWORDS: South Africa, invertebrates, *Cyperus*, disease

126 CARTER, G.S., 1955. The papyrus swamps of Uganda. W. Heffer and Sons, Cambridge, 25pp.
KEYWORDS: Uganda, utilisation (animals) , *Cyperus*, swamp (Uganda), chemistry

127 CHACHU, R.E.O., 1979. The vascular flora of Lake Kainji. In: Proceedings of the International Conference on Kainji Lake and River Basins Development in Africa, Ibadan, 11-17 December 1977. KLRI, New Busa, Nigeria, 2:479-487.
KEYWORDS: Nigeria, lake (Kainji), aquatic plants, vegetation description (Nigeria)

128 CHADWICK, M.J., 1961. Some observations on the ecology of *Eichhornia crassipes* Solms. Annual Report of the Hydrobiological Research Unit, University of Khartoum, 8:23-28.
KEYWORDS: Sudan, *Eichhornia*, *Phragmites*, ecology

129 CHADWICK, M.J. and OBEID, M., 1966. A comparative study of the growth of *Eichhornia crassipes* (Mart.) Solms and *Pistia stratiotes* L. in water culture. Journal of Ecology, 54:563-576.
KEYWORDS: Sudan, productivity, *Eichhornia*, *Pistia*

130 CHALMERS, M.I., 1968. Report to World Food Programme on a visit to Sudan. In: Animal Production WFP. Mission report on animal nutrition in Sudan. Study on the use of water hyacinth in ruminant animal feeding and also as a means of weed control. Food and Agriculture Organisation of the United Nations, Rome.
KEYWORDS: control/management, utilisation (animals), Sudan, aquatic weeds

131 CHATEAU, R., 1957. La flore adventive des rivières. Bulletin Agricole du Congo Belge, 48:561–580.
KEYWORDS: Zaïre, *Myriophyllum*, river, vegetation description (Zaïre)

132 CHATEAU, R., 1957. Le désherbage en riziculture. Journal d'Afrique tropique et Botanique Appliquée, 4:490–526.
KEYWORDS: development (land/swamp)

133 CHAUDHRY, A.B. and CHACHU, R.E.O., 1979. Preliminary study of some indigenous vascular flora in the drawdown area of Kainji Lake and its probable significance to lakeshore farming. In: Proceedings of the International Conference on Kainji Lake and River Basins Development in Africa, Ibadan, 11–17 December 1977. KLRI, New Busa, Nigeria, 2:529–539.
KEYWORDS: Nigeria, aquatic plants, lake (Kainji), vegetation description (Nigeria), development (land/swamp)

134 CHENERY, E.M., 1953. A digest of the chemistry of dead swamps in Kigezi District, S.W. Uganda. Annual Report of the Department of Hydrological Survey, Uganda Government (1952), pp 9–14.
KEYWORDS: Uganda, chemistry, swamp (Uganda), control/management (swamp)

135 CHENERY, E.M., 1955. Acid sulphate soils in Central Africa. Fifth International Congress of Soil Science (1954). 4:195–198.
KEYWORDS: Uganda, chemistry, swamp (Uganda), control/management (swamp)

136 CHEVALIER, A., 1908. L'Afrique centrale française. Mission Chari–lac Tchad 1902–1904. Paris.
KEYWORDS: Tchad, vegetation description (Tchad), lake (Tchad)
(In French)

137 CHISHOLM, I.F., 1979. A laboratory investigation of *Paulinia acuminata* (De Beer) (Orthoptera: Acididae) as a biological agent for *Salvinia molesta*. Bulletin of Entomological Research, 69:111–114.
KEYWORDS: Zimbabwe, *Salvinia*, control/management (*Salvinia*)

138 CLARK, J.E., 1976. Problems of managing African wetlands. IUCN Publications, New Series, Supplementary Paper No. 45, Morgues, Switzerland, 172–180.
KEYWORDS: Eastern Africa, floodplain (Eastern Africa), swamp (savanna), conservation, control/management (swamp)

139 CLARKE, J.E., 1975. Zambia's Wetlands. Black Lechwe, 12:14–17.
KEYWORDS: Zambia, swamp (Bangweulu), floodplain (Bangweulu), lake (Bangweulu), conservation, swamp (savanna), wildlife

140 CLAYTON, W.D., 1957. The swamps and sand dunes of Hadejia. Nigeria Geographical Journal, 1:31–37.
KEYWORDS: Nigeria, swamp (Hadejia), vegetation description (Nigeria), swamp (forest)

141 CLAYTON, W.D., 1958. A tropical moor forest in Nigeria. Journal of the West African Science Association, 4:1–3.
KEYWORDS: Nigeria, swamp (forest)

142 COATES, D., and REDDING–COATES T.A., 1981. Ecological problems associated with irrigation canals in the Sudan with particular reference to the spread of bilharziasis, malaria and aquatic weeds and the ameliorative role of fishes. International Journal of Environmental Studies, 16:207–212.
KEYWORDS: Sudan, ecological problems, irrigation, disease, aquatic weeds, fish, development (water)

143 COCHE, A.G., 1974. Semi aquatic and aquatic vegetation. In: E.K. Balon and A.G. Coche (Editors), Lake Kariba: A Man-Made Tropical Ecosystem in Central Africa. Junk, The Hague, pp 43–46.
KEYWORDS: Zimbabwe, lake (Kariba), *Pistia*, *Salvinia*, *Potamogeton*, vegetation description (Zimbabwe)

144 COLE, N.H.A., 1968. Review of classification of vegetation in Sierra Leone. Journal of the West African Science Association, 13:81–92.
KEYWORDS: Sierra Leone, swamp, vegetation description (Sierra Leone)

145 COLE, N.H.A., 1973. Soil conditions, zonation and species diversity in a seasonally flooded tropical grass-herb swamp in Sierra Leone. Journal of Ecology, 61:831–847.
KEYWORDS: Sierra Leone, swamp (savanna), vegetation description (Sierra Leone), ecology

146 COOK, C.D.K., 1968. The vegetation of the Kainji Reservoir site in Northern Nigeria. Vegetatio, 15:225–243.
KEYWORDS: Nigeria, lake (Kainji), vegetation description (Nigeria), development (water)

147 CORILLON, R. and GUERLESQUIN, M., 1972. Récherches sur les Charophycées d'Afrique Occidentale, systematique, phytogéographie et écologie, cytolgie. Travail et Memoire Laboratoire Biologie Végétation et de Phytogéographie, Angers, 25 pp.
KEYWORDS: Central Africa, *Charophytes*, ecology
(in French)

148 CORREMAN, J., 1959. *Tetranychus telarius* (Linne.) (Acari: Tetranychidae) parasite de la jacinthe d'eau, *Eichhornia crassipes*, Mart. Bulletin Agricole du Congo Belge, 50:395–401.
KEYWORDS: Zaïre, control/management (*Eichhornia*), *Eichhornia*, invertebrates
(in French)

149 COUTINHO, L.P., 1948. A brief report on the peatlands of Moçambique: a preliminary study of their properties and their treatment for crop production. Proceedings of the Conference Africaine des Sols, Goma, Congo Belge. 8–15 Novembre 1948, 21pp.
KEYWORDS: Moçambique, swamp (Moçambique), development (land/swamp), peatland

150 CURSON, H.H., 1947. Notes on Eastern Caprivi Strip. South African Journal of Science, 43:124–157.
KEYWORDS: Namibia, vegetation description, swamp (Caprivi)

151 DAMAS, H., 1963. Les populations animales et végétales du lac de barrage de la Lufira (Mwadingusha). Publications de l'Université Elizabethville, 6:88–94.
KEYWORDS: Zaïre, river (Lufira), vegetation description (Zaïre), fish
(in French)

152 DANDY, J.E., 1937. The genus *Potamogeton* in tropical Africa. Botanical Journal of the Linnaean Society, 1:507–540.
KEYWORDS: Africa, *Potamogeton*, ecology

153 DATTA, S.C. and BISWAS, K.K., 1969. Physiology of germination in *Pistia stratiotes* seeds. Hydrobiologia, 24:70–79.
KEYWORDS: Egypt, *Pistia*, physiology

154 DAVIES, B.R., 1975. Cabora Bassa hazards. Nature, London, 254:477–478.
KEYWORDS: Moçambique, development (water), ecological problems, lake (Cabora Bassa)

155 DAVIES, B.R., 1978. The secondary productivity of the mussel ***Musculus vivigiliae*** Barnard, in the littoral aquatic macrophyte beds of Svartvlei. Institute of Freshwater studies, Rhodes University, Annual Report, No 10.
KEYWORDS: South Africa, aquatic plants, invertebrates, utilisation (animals), lake (Swartvlei)

156 DAVIES, B.R., 1982. Studies of the zoobenthos of some Southern Cape coastal lakes. Spatial and temporal changes in the benthos of Svartvlei, South Africa, in relation to changes in the submerged littoral macrophyte community. Journal of the Limnological Society of Southern Africa, 8:33–45.
KEYWORDS: South Africa, aquatic plants, lake (Swartvlei), invertebrates, utilisation (animals)

157 DAVIES, B.R., HALL, A. and JACKSON, P.B.N., 1975. Some ecological aspects of the Cabora Bassa dam. Biological Conservation, 8:189–201.
KEYWORDS: South Africa, development (water), ecological problems, lake (Cabora Bassa)

158 DAVIES, H.R.J., 1959. Effects of water hyacinth (*Eichhornia crassipes*) in the Nile Valley. Nature, London, 184:1085–1086.
KEYWORDS: Sudan, *Eichhornia*, ecological problems, river (Nile)

159 DAVIES, H.R.J., 1959. The effect of *Eichhornia crassipes* on the peoples of the Sobat, and the White Nile between Sobat mouth and Kosti. Annual Report of the Hydrobiological Research Unit, University of Khartoum, 6: pp 26–28.
KEYWORDS: Sudan, *Eichhornia*, ecological problems, river (Nile)

160 DAWKINS, H.C. and PHILIP, M.S., 1962. Working Plan for the Mpanga Forest for the period 7/7/60–30/6/65, Uganda Forest Department, Entebbe, 31 pp.
KEYWORDS: Uganda, classification, swamp (forest)

161 DAWOOD, K.I., FAROOQ, M., DAZO, B.C. MIGUEL, I.C., and UNRAU, G.O., 1965. Herbicide trials in the snail habitats of the Egypt 49 Project. Bulletin of the World Health Organisation, 32:269–287.
KEYWORDS: Egypt, invertebrates, control/management (bilharzia), disease

162 DAY, J.H., 1951. The ecology of South African estuaries. Part 1. Transactions of the Royal Society of South Africa, 33:53–89.
KEYWORDS: South Africa, vegetation description (South Africa), estuary (South Africa)

163 DAY, J.H., 1974. The ecology of the Morrumbene estuary, Moçambique. Transactions of the Royal Society of South Africa, 41:43–97.
KEYWORDS: Moçambique, ecology, estuary (Morrumbene), vegetation description (Moçambique)

164 DAY, J.H., MILLARD, N.A.H. and BROEKHUYSEN, G.J., 1953. The ecology of South African estuaries. IV. The St. Lucia system. Transactions of the Royal Society of South Africa, 34:129–156.
KEYWORDS: South Africa, vegetation description (South Africa), estuary (St Lucia)

165 DEAN, 1967. Grasslands of the Rukwa Valley. Journal of Applied Ecology, 4:45–57.
KEYWORDS: Tanzania, swamp (savanna), lake (Rukwa)

166 DEAN, G.J.W., 1967. Grasslands of the Rukwa Valley. Journal of Applied Ecology, 4:45–57.
KEYWORDS: Tanzania, vegetation description (Tanzania), swamp (savanna), floodplain (Rukwa)

167 DEAN, W.R.J., 1979. The use of *Potamogeton pectinatus* rafts as foraging substances by birds. Ostrich, 50(3):186.
KEYWORDS: Africa, *Potamogeton*, utilisation (animals), birds, wildlife

168 DEBENHAM, F., 1952. Studies of an African swamp. Northern Rhodesia Journal, 88pp.
KEYWORDS: Zambia, swamp (Bangweulu), floodplain (Bangweulu), vegetation description, lake (Bangweulu)

169 DEBENHAM, F., 1952. Study of an African swamp. Report of the Cambridge University expedition to the Bangweulu Swamps, Northern Rhodesia, 1949. Her Majesty's Stationery Office, London.
KEYWORDS: Zambia, vegetation description (Zambia), swamp (Bangweulu), floodplain (Bangweulu), lake (Bangweulu)

170 DEJOUX, C. AND SAINT-JEAN, L., 1972. L'Etude des communautés d'herbiers du lac Tchad: récherches préliminaires. Cahiers d'Office de la Récherche Scientifique et Technique Outre-Mer, Série Hydrobiologique, 6(1):67–83.
KEYWORDS: Tchad, vegetation description (Tchad), lake (Tchad)
(in French)

171 DENNY, P., 1966. A study of some factors affecting the distribution of aquatic macrophytes. Ph.D. Thesis. St. Andrews University.
KEYWORDS: Uganda, chemistry, *Potamogeton*, lake (Bunyoni), aquatic plants

172 DENNY, P., 1971. Zonation of aquatic macrophytes around Habukara Island, Lake Bunyoni, S.W. Uganda. Hidrobiologia, Bucurest, 12:249–257.
KEYWORDS: Uganda, lake, *Potamogeton*, vegetation description (Uganda), aquatic plants, lake (Bunyoni)

173 DENNY, P., 1972. Lakes of south-western Uganda. I. Physical and chemical studies on Lake Bunyonyi. Freshwater Biology, 2:143–158.
KEYWORDS: Uganda, lake (Bunyoni), chemistry

174 DENNY, P., 1972. Sites of nutrient absorption in aquatic macrophytes. Journal of Ecology, 60:819–829.
KEYWORDS: Uganda, chemistry, physiology, *Potamogeton*

175 DENNY, P., 1973. Lakes of south-western Uganda. II: Vegetation studies on Lake Bunyoni. Freshwater Biology, 3:123–135.
KEYWORDS: Uganda, *Myriophyllum*, aquatic plants, ecology, lake (Bunyoni), vegetation description (Uganda)

176 DENNY, P., 1984. Permanant swamp vegetation of the Upper Nile. Hydrobiologia, 110:79–90.
KEYWORDS: Sudan, swamp (Upper Nile/Sudd), vegetation description (Sudan), river (Nile)

177 DENNY, P. and LYE, K.A., 1973. The *Potamogeton schweinfurthii* complex in Uganda. Kew Bulletin, 28(1):117–120.
KEYWORDS: Uganda, *Potamogeton*

178 DENNY, P., and BAILEY, R.G. (Editors), 1976. A Biological Survey of Nyumba Ya Mungu Reservoir, Tanzania, July-Sept. 1974. 253pp.
KEYWORDS: Tanzania, fish, vegetation description, utilisation (man), lake (Nyumba ya Mungu)

179 DENNY, P., BOWKER, D.W. and BAILEY, R.G., 1978. The importance of the littoral epiphyton as food for commercial fish in the recent African man-made lake, Nyumba ya Mungu Reservoir, Tanzania. Biological Journal of the Linnaean Society, 10:139–150.
KEYWORDS: Tanzania, utilisation (animals), littoral, lake (Nyumba ya Mungu), utilisation (man), aquatic plants, fish

180 DEUSE, B., 1960. Etude écologique et phytosociologique de la végétation des ésobes de la région est du lac Tumba (Congo Belge). Academie Royale des Sciences Outre-Mer, Sciences Naturelles et Médicales, Nouveau Série, 11(3).
KEYWORDS: Zaïre, ésobes, swamp (savanna), ecology, vegetation description (Zaïre)

181 DEUSE, P., 1959. Esquisse de la végétation des tourbiéres et marais tourbeaux du Ruanda-Urundi. Bulletin de la Société Royale Sciences de Liége, 1-2:47-53.
KEYWORDS: Rwanda, Burundi, swamp (Rwanda/Burundi), vegetation description, utilisation (man), peatland
(in French)

182 DEUSE, P., 1966. Contribution a l'étude des tourbières du Rwanda et du Burundi. Treizième Rapport d'Institut pour la Récherche Scientifique en Afrique Centrale (I.R.S.A.C.), 1960–1964, pp 53–115.
KEYWORDS: Rwanda, Burundi, swamp (Rwanda/Burundi), vegetation description, utilisation (man), peatland
(in French)

183 DODGE, J.S. and LADAN, A., 1966. A note on *Pistia* control. Mosquito News, 26:443–445.
KEYWORDS: Africa, *Pistia*, control/management, aquatic weeds

184 DOUTHWAITE, R.J. and VAN LAVIEREN, L.P., 1977. A description of the vegetation of Lochinvar National Park, Zambia. Technical Report 34, National Council for Scientific Research, Lusaka, 66 pp.
KEYWORDS: Zambia, floodplain (Kafue), swamp (savanna), swamp (Kafue), vegetation description (Zambia)

185 DOWNING, B.H., 1968. Notes on the ecology of Natal Highveld-Lowveld vleis. Proceedings of the Grassland Society of South Africa. 3:131–134.
KEYWORDS: South Africa, vegetation description (South Africa), swamp (South Africa), swamp (savanna), ecology

186 DRAR, M., 1951. The problem of sudd in relation to stablising and smothering plants. Botaniska Notiser, 140:34–46.
KEYWORDS: Sudan, *Cyperus*, river (Nile), ecological problems

187 DRUIJFF, A.H., 1973. Possible approaches to water hyacinth control. Pesticide Abstracts, 19:219–222.
KEYWORDS: South Africa, control/management, aquatic weeds, development (land/swamp), *Eichhornia*

188 DRUIJFF, A.H., 1973. The chain scythe, a simple tool for controlling aquatic weeds in irrigation canals. Pesticide Abstracts, 19:216–218.
KEYWORDS: South Africa, control/management (weeds), *Eichhornia*, irrigation

189 DUBBERS, F.A.A., GHONEIM, S., SIEMELINK, M.E., EL GHARABLY, Z., PIETERSE, A.H. and BLOM, J.E., 1981. Aquatic weed control in irrigation and drainage canals in Egypt by means of grass carp (*Ctenopharyngodon idellus*). Proceedings of the 5th International Symposium on Biological Control of Weeds, Brisbane, 1980, pp 261–271.

KEYWORDS: control/management, aquatic weeds, fish, Egypt, development (water)

190 DUBOIS, L., 1955. La jacinthe d'eau au Congo Belge. Bulletin Agricole du Congo Belge, 46:893–900.
KEYWORDS: Zaïre, *Eichhornia*, ecological problems, river (Zaïre)
(in French)

191 DURAND, J.R. and LEVEQUE, C., 1980. Flore et Faune Aquatiques de l'Afrique Sahélo–Soudanienne. I. Office de la Recherche Scienfique et Technique Outer–Mer Collection Initiations – Documentation Techeniques No.45. Paris, 389pp.
KEYWORDS: Sudan, Cameroun, Niger, Nigeria, vegetation description, aquatic plants

192 DUVIGNEAUD, P. and HOMES, P., 1955. Les *Xyris* du Bas–Congo et du Kwango. Bulletin Société Royale Botanique Belgique (Bull. Soc. Roy. Bot.), 87:81–113.
KEYWORDS: Zaire, *Xyris*, taxonomy

193 DYE, P.J., LEE, R.E. and REAVELL, P., 1976. Plant succession on spoil dumps along the Boro River, Okavango Delta, Botswana. Tropical Ecology, 17:9–15.
KEYWORDS: Botswana, river, control/management, development (water), vegetation description (Botswana), riverine, river (Boro)

194 EDWARDS, D., 1967. A plant ecological survey of the Tugela River Basin, Natal. Memoirs of the Botanical Survey of South Africa, No. 36.
KEYWORDS: South Africa, vegetation description (South Africa), river (Tugela), floodplain (Tugela), riverine

195 EDWARDS, D., 1969. Some effects of siltation upon aquatic macrophyte vegetation in rivers. Hydrobiologia, 34:29–37.
KEYWORDS: South Africa, environmental factors, aquatic plants, river (South Africa)

196 EDWARDS, D. and BENNISON, R.H., 1973. Report by November 1972 joint Botswana–South Africa survey of the extent and degree of occurrence of *Salvinia molesta* (Kariba weed) in the Chobe – Linyati–Kwando river system. Botanical Research Institute, Pretoria.
KEYWORDS: Botswana, South Africa, *Salvinia*, river (Chobe/Linyanti/Kwando)

197 EDWARDS, D. and MUSIL, C.F., 1975. *Eichhornia crassipes* in South Africa. A general review. Journal of the Limnological Society of Southern Africa, 1:23–27.
KEYWORDS: South Africa, *Eichhornia*, control/management (*Eichhornia*), ecology

198 EDWARDS, D. and NEL, M.D.S., 1972. A reconnaissance survey of aquatic macrophytes in the Upper River catchment. Civil Engineering in South Africa, 14:85–88.
KEYWORDS: South Africa, vegetation description (South Africa), river, aquatic plants

199 EDWARDS, D. and THOMAS, P.A., 1977. The *Salvinia molesta* problem in the northern Botswana and eastern Caprivi area. Proceedings of the Second National Weeds Conference of South Africa. pp 221–237.
KEYWORDS: Botswana, Namibia, *Salvinia*, ecological problems, river (Okavango)

200 EGGELING, W.J., 1934. Notes on the flora and fauna of a Uganda swamp. Uganda Journal, 1:51–60.
KEYWORDS: Uganda, utilisation (animals), swamp, *Cyperus*, vegetation description

201 EGGELING, W.J., 1935. The vegetation of Namanve Swamp, Uganda. Journal of Ecology, 23:422–435.
KEYWORDS: Uganda, vegetation description, swamp, *Cyperus*

202 EL DIN EL ZARKA, S., 1973. Kainji Lake case history. In Man-made lakes, Their Problems and Environmental Effects. W. Ackermann *et al* (Editor). American Geophysical Union, Geophysical Monograph 17, Washington, DC.
KEYWORDS: Nigeria, lake (Kainji), ecological problems

203 EL GHARABLY, Z., KHATTAB, A.F. and DUBBERS, F.A.A., 1982. Experience with grass carps for the control of aquatic weeds in irrigation canals and drains in Egypt. Proceedings of the 2nd International Symposium on Herbiverous Fish, Novi Sad, pp 17–26.
KEYWORDS: control/management, aquatic weeds, fish, Egypt, development (water)

204 EL HADIDI, N.N., 1965. *Potamogeton trichoides* Cham. and Schlecht. in Egypt. Candollea, 20:159–165.
KEYWORDS: Egypt, *Potamogeton*

205 EL HADIDI, N.N., 1976. The riverain flora in Nubia. In: J. Rzoska (Editor), The Nile, Biology of an Ancient River. W. Junk, b.v, The Hague, pp 87–91.
KEYWORDS: Sudan, vegetation description, floodplain, swamp, river

206 ELSERAFY, Z.M., SONBOL, H.A. and ELTANTAWAY, I.M., 1980. The problem of water hyacinth in rivers and canals. I, Production of compost from plant. Soil Science and Plant Nutrition, 26:135–138.
KEYWORDS: Egypt, *Eichhornia*, utilisation (man), river, ecological problems, decomposition

207 ELSERAFY, Z.M., SONBOL, H.A. and ELTANTAWY, I.M., 1980. The problem of water hyacinths in rivers and canals. II. Physiochemical properties of humic substances occurring at various degrees of humification of the composted weed. Soil Science and Plant Nutrition, 26:399–404.
KEYWORDS: Egypt, chemistry, decomposition, *Eichhornia*, river, utilisation (man)

208 ELSTER, H.J. and VOLLENWEIDER, R., 1961. Beitrage zür Limnologie Agyptens. Archiv für Hydrobiologie, 57:241–343.
KEYWORDS: Egypt, aquatic plants, productivity
(in German)

209 ENTZ, B., 1976. Lake Nasser and Lake Nubia. In: The Nile, Biology of an Ancient River. J. Rzoska (Editor). Dr. W. Junk, b.v., The Hague, pp 271–291.
KEYWORDS: Egypt, Sudan, lake (Nasser), lake (Nubia), *Potamogeton*, *Najas*, river (Nile), development (water)

210 ENTZ, B., 1979. The appearance of a littoral fauna and flora in Lake Nasser, Nubia, a newly formed tropical man-made lake. In: J. Salauki and P. Biro (Editors), Human Impacts on Life in Fresh Waters. Akademy Kiado, Budapest, pp 201–209.
KEYWORDS: Egypt, development (water), lake, vegetation description, invertebrates

211 ENTZ, B.A.G., 1980. Ecological aspects of Lake Nasser–Nubia. The first decade of its existence, with special reference to the development of insect populations and the land and water vegetation. In: M.Kassas and S.I. Ghabbour (Editors), The Nile and its Environment. Water Supply Management, 4(1–2):67–72.
KEYWORDS: Egypt, lake, development (water), invertebrates, vegetation description, littoral

212 ERNEST, H.M., 1976. Manipulation of the Okavango Delta for increased outflow; its impacts and constraints. Proceedings of the Symposium on the Okavango Delta and its Future Utilisation. The Botswana Society, Gaborone, pp 337–346.
KEYWORDS: Botswana, river, swamp, development (water), ecological problems

213 EVANS, A.C., 1963. The grip of the water hyacinth. New Scientist, 19:666–668.
KEYWORDS: Africa, *Eichhornia*, ecological problems

214 EVERSON, T.M., 1980. Germination and nutritional studies on *Potamogeton crispus* L.. M.Sc. Thesis. University of Natal, 173pp.
KEYWORDS: *Potamogeton*, physiology, chemistry

215 EWER, D.W., 1966. Biological investigations in the Volta Lake, May 1964– May 1965. In: R.H. Lowe–McConnell (editor), Man–Made Lakes. Academic Press, London, pp 21–30.
KEYWORDS: Ghana, environmental factors, vegetation description (Ghana), lake (Volta), development (water)

216 FADEN, R.B., 1972. Some notes on *Marsilea* in East Africa. Bulletin of the East African Natural History Society, (1972):72–74.
KEYWORDS: Africa, *Marsilea*, ecology

217 FAIRHALL, N., 1981. Bio–energetics of the red–knobbed coot *Fulica crustata* on a South African estuarine lake. South African Journal of Wildlife Research, 11(1):1–4.
KEYWORDS: South Africa, utilisation (animals), birds, estuary (South Africa), aquatic plants, wildlife

218 FANSHAWE, D.B., 1972. The biology of the reed *Phragmites mauritianus* Kunth. Kirkia, 8:147–150.
KEYWORDS: Zambia, Zimbabwe, *Phragmites*, swamp (Zambia), ecology

219 FAO, 1964. Occurrence of water hyacinth in Senegal. FAO Plant Protection Bulletin, 12:93.
KEYWORDS: Sénégal, *Eichhornia*

220 FAO REPORT, 1965. Report of the FAO Conference on Quelea Bird and Water Hyacinth Control in Africa. United Nations/FAO, Rome, Report No. PL/1965/14.
KEYWORDS: Africa, *Eichhornia*, birds, control/management, wildlife

221 FAURE, J.C., 1956. Thysanoptera from papyrus in the Sudan. Journal of the Entomological Society of Southern Africa, 19:100–117.
KEYWORDS: Sudan, invertebrates, utilisation (animals), *Cyperus*

222 FEELY, J.M., 1965. Observations on *Acacia albida* in the Luangwa Valley. Puku, 3:67–70.
KEYWORDS: Zambia, river (Luangwa), *Acacia*

223 FERGUSON, A.J.D. and HARBOTT, B.J., 1982. Geographical, physical and chemical aspects of Lake Turkana. Volume 1. In: Hopson, A.J. (Editor) Lake Turkana. Overseas Development Administration, London. pp 1–107.
KEYWORDS: Kenya, lake (Turkana), chemistry

224 FERREIRA, J.C., 1974. Autecological studies of *Polygonum senegalense* Meisn. M.Sc. Thesis. University of Rhodesia.
KEYWORDS: Zimbabwe, *Polygonum*, chemistry, ecology

225 FIELD, C.R., 1970. A study of the feeding habits of the hippopotamus (*Hippopotamus amphibius* Linn.) in the Queen Elizabeth Park, Uganda, with some management implications. Zoologica Africana, 5:71–86.
KEYWORDS: Uganda, hippopotamus, utilisation (animals)

226 FLACK, T.P., 1977. Studies on the autecology of *Myriophyllum aquaticum* in the South Western Cape. MSc Thesis. University of Cape Town.
KEYWORDS: South Africa, *Myriophyllum*, ecology

227 FOTIUS, G., 1974. Problémes posés par l'évolution de la végétation liée á la baisse du lac Tchad. Cahiers d'Office de la Récherche Scientifique et Technique Outre-Mer, N'Djaména, 30pp, (Mimemeo).
KEYWORDS: Tchad, lake (Tchad), aquatic plants, swamp (Tchad), ecological problems

228 FOTIUS, G. and LEMOALLE, J., 1976. Reconnaissance de l'évolution de la végétation du lac Tchad entre Janvier 1974 et Juin 1976. Cahiers d'Office de la Récherche Scientifique et Technique Outre-Mer, Série Hydrobiologique, 13pp.
KEYWORDS: Tchad, Nigeria, vegetation description (Tchad), lake (Tchad), swamp (Tchad)

229 FREIDEL, J.W., 1978. Populations dynamik der Wasserhyazinthe (*Eichhornia crassipes* (Mart) Solms)unter besonderer Berücksichtigung des sudanesischen Befallsgebietes. 145 p.
KEYWORDS: Sudan, *Eichhornia*, productivity, control/management (*Eichhornia*), river (Nile)
(In German)

230 FRIEDEL, J.W., 1978. Aquatic weed control, Cabora Bassa, Moçambique. Technical Report, Agoa Moz/75/011, 59pp.
KEYWORDS: Moçambique, aquatic weeds, control/management (weeds), lake (Cabora Bassa), development (water)

231 FURNESS, H.D., 1981. The plant ecology of seasonally flooded areas of the Pongolo river floodplain with particular reference to *Cynodon dactylon* (L.) Pers. Ph.D Thesis. University of Natal, 2 vols:184pp.
KEYWORDS: South Africa, swamp (savanna), floodplain (Pongolo), river (Pongolo)

232 FURNESS, H.D. AND BREEN, C.M., 1980. The vegetation of seasonally flooded areas of the Pongolo River Flood Plain. Bothalia, 13:217–231.
KEYWORDS: South Africa, Moçambique, swamp (savanna), floodplain (Pongolo), river (Pongolo), vegetation description (South Africa)

233 GALLAIS, J., 1967. Le Delta intérieur du Niger; étude de géographie régionale. Memoirs de l'Institut Fondamental d'Afrique Noire, 79:1–619.
KEYWORDS: Mali, river (Niger), floodplain (Niger), development (land/swamp)

234 GAUDET, J.J., 1973. Report on the development and control of aquatic weeds in Volta Lake. In: Volta Lake Case Study. II. The Environmental Impact of Tropical Dams. Office of International and Environment Programs, Smithsonian Institute, Washington, D.C.
KEYWORDS: Ghana, aquatic weeds, *Myriophyllum*, *Pistia*, development (water), lake (Volta), control/management (weeds)

235 GAUDET, J.J., 1974. Aquatic weed problems at Lake Jipe (Tanzania–Kenya). Report to the African Fisheries Organisation, pp 1–10.
KEYWORDS: Kenya, Tanzania, fish, lake (Jipe), aquatic weeds, ecological problems

236 GAUDET, J.J., 1975. Mineral concentration in papyrus in various African swamps. Journal of Ecology, 63:483–491.
KEYWORDS: Uganda, Kenya, chemistry, swamp, *Cyperus*

237 GAUDET, J.J., 1975. Papyrus, valued from ancient times. Africana, 5(9):3–4.
KEYWORDS: Africa, *Cyperus*, utilisation (man)

238 GAUDET, J.J., 1976. Nutrient relationships in the detritus of a tropical swamp. Archiv für Hydrobiologie, 78:213–239.
KEYWORDS: Uganda, Kenya, decomposition, chemistry, swamp (Uganda), *Cyperus*

239 GAUDET, J.J., 1976. *Salvinia* infestation on Lake Naivasha in East Africa. In: C.K. Varshney and J. Rzoska (Editors), Aquatic Weeds in South East Asia. W. Junk, b.v, The Hague, 193–209.
KEYWORDS: Kenya, lake (Naivasha), *Salvinia*

240 GAUDET, J.J., 1977. Natural drawdown on Lake Naivasha, Kenya, and the formation of papyrus swamps. Aquatic Botany, 3:1–47.
KEYWORDS: Kenya, swamp (Kenya), lake (Naivasha), *Cyperus*, chemistry, littoral, decomposition

241 GAUDET, J.J., 1977. Uptake accumulation and loss of nutrients by papyrus in tropical swamps. Ecology, 58:415–422.
KEYWORDS: Kenya, Uganda, *Cyperus*, chemistry, swamp (East Africa), physiology

242 GAUDET, J.J., 1978. Control of aquatic weeds on Lake Naivasha, Kenya. Proceedings of Sixth East African Weed Science Conference, 1976. 57–65.
KEYWORDS: Kenya, lake (Naivasha), aquatic weeds, control/management (weeds)

243 GAUDET, J.J., 1978. Effect of a tropical swamp on water quality. Verhandlungen der Internationalen Vereinigung für theoretische und angewandte Limnologie (Verh. Internat. Verein. Limnol.), 20:2202–2206.
KEYWORDS: Uganda, chemistry, swamp (Uganda), *Cyperus*, decomposition

244 GAUDET, J.J., 1979. Aquatic weeds in African man-made lakes. PANS, 25:279–286.
KEYWORDS: Africa, lake, aquatic weeds

245 GAUDET, J.J., 1980. Nutrient dynamics in papyrus swamps in Africa. Published abstracts of the 21st Congress of the International Association for Theoretical and Applied Limnology, Kyoto, Japan 24–31 August, 1980. p 107.
KEYWORDS: Africa, chemistry, *Cyperus*, swamp

246 GAUDET, J.J., 1982. Nutrient dynamics of papyrus swamps. In: B. Gopal, R.E. Turner, R.G. Wetzel, and D.F. Whigham (Editors), Wetlands: Ecology and Management. Proceedings of the 1st International Wetlands Conference, New Delhi, 10–17 September, 1980. National Institute of Ecology, New Delhi, pp 305–319.
KEYWORDS: Kenya, Uganda, swamp, chemistry, *Cyperus*, decomposition

247 GAUDET, J.J. and MUTHURI, F.M., 1981. Nutrient relationships in shallow water in African lake, Lake Naivasha. Oecologia, 49(1):109–118.
KEYWORDS: Kenya, chemistry, *Cyperus*, swamp (Naivasha), lake (Naivasha), decomposition

248 GAUDET, J.J., MITCHELL, D.S. and DENNY, P., 1981. Macrophytes (Aquatic vegetation). In: Symoens, J.J., Burgis, M., and Gaudet, J.J. (Editors), The Ecology and Utilisation of African Inland Waters. United Nations Environment Programme, Nairobi, pp 27–36.
KEYWORDS: Africa, aquatic plants, review

249 GAY, P.A., 1957. Some aspects of the riverain flora of the White Nile and Bahr el Ghazal. Annual Report of the Hydrobiological Research Unit, University of Khartoum, 4:pp 10–20.
KEYWORDS: Sudan, vegetation description (Sudan), riverine, river (Nile)

250 GAY, P.A., 1958. *Eichhornia crassipes* in the Nile of the Sudan. Nature, London, 182:538.
KEYWORDS: Sudan, *Eichhornia*, river (Nile)

251 GAY, P.A., 1958. The riverain flora of the Nile. Annual Report of the Hydrobiological Research Unit, University of Khartoum, 5:pp 7–15.
KEYWORDS: Sudan, vegetation description (Sudan), riverine, river (Nile), *Eichhornia*

252 GAY, P.A., 1960. British Ecological Society Symposium Volume No. 20. In: J.L. Harper (Editor), The Biology of Weeds. British Ecologicla Society Symposium Volume 20, Blackwells, Oxford, pp 184–188.
KEYWORDS: Sudan, *Eichhornia*, river (Nile), ecological problems

253 GAY, P.A., 1960. Ecological studies of *Eichhornia crassipes* Solms in the Sudan. I. Analysis of spread in the Nile. Journal of Ecology, 48:183–191.
KEYWORDS: Sudan, *Eichhornia*, river (Nile), ecological problems

254 GAY, P.A. and BERRY, L., 1959. The water hyacinth: a new problem on the Nile. Geographical Journal, 125:89–91.
KEYWORDS: Sudan, *Eichhornia*, river (Nile), ecological problems

255 GAYED, S.K., 1963. Some observations on the distribution of the water hyacinth in the Nile between Kosti and Nimule. Annual Report of the Hydrobiological Research Unit, University of Khartoum, 9–10: pp 8–13.
KEYWORDS: Sudan, *Eichhornia*, river (Nile)

256 GELDENHUYS, J.N., 1976. Physiognomic characteristics of wetland vegetation in South African shelduck habitat. South African Journal of Wildlife Research, 6(2):75–78.
KEYWORDS: South Africa, utilisation (animals), birds, swamp (South Africa), wildlife

257 GELDENHUYS, J.N., 1982. Classification of the pans of the western Orange Free State according to vegetation structures, with reference to avifaunal communities. South African Journal of Wildlife Research, 12(2):55–62.
KEYWORDS: South Africa, utilisation (animals), birds, lake (Orange Free State), vegetation description (South Africa), checklist, wildlife

258 GERMAIN, R., 1952. Les associations végétales de la plaine de la Ruzizi (Congo Belge) en relation avec le milieu. Institut National pour l'Etude Agronomique du Congo Belge (I.N.E.A.C.), Serie Science, 52:109–118.
KEYWORDS: Zaïre, vegetation description, floodplain (Ruzizi), river (Ruzizi), swamp (savanna)
(in French)

259 GERMAIN, R., 1965. Les biotopes alluvionaires herbeux et les savannes intercalaires du Congo equatorial. Académie Royale des Sciences Outre-Mer (A.R.S.O.M.), Nouveau Serie, 15(4):237–244.
KEYWORDS: Zaïre, swamp (savanna), vegetation description (Zaïre), ésobes
(in French)

260 GERTENBACH, W.P.D., 1975. Kariba weed now in Kruger Park. Custos, 4:39–41.
KEYWORDS: South Africa, *Salvinia*

261 GIBB, SIR ALEXANDER and PARTNERS, 1956. Water resources survey of Uganda, 1954–5, applied to irrigation and swamp reclamation. Report for the Uganda Government, Kampala.
KEYWORDS: Uganda, hydrology, development (land/swamp)

262 GIBBS-RUSSELL, G.E., 1974. The vascular aquatic plants of Rhodesia. Ph.D. Thesis. University of Georgia, USA.
KEYWORDS: Zimbabwe, aquatic plants, ecology

263 GIBBS-RUSSELL, G.E., 1975. Distribution of vascular aquatic plants in Rhodesia. South African Journal of Science, 71:270–271.
KEYWORDS: Zimbabwe, aquatic plants, ecology

264 GIBBS-RUSSELL, G.E., 1975. Taxonomic bibliography of vascular aquatic plants is southern Africa. Journal of the Limnological Society of Southern Africa, 1:53–65.
KEYWORDS: Southern Africa, aquatic plants, checklist, taxonomy

265 GIBBS-RUSSELL, G.E., 1977. Keys to vascular aquatic plants in Rhodesia. Kirkia, 10:411–502.
KEYWORDS: Zimbabwe, taxonomy, checklist, aquatic plants

266 GIBBS-RUSSELL, G.E., 1977. Life forms of aquatic plants. Rhodesia Science News, 10(5):134–137.
KEYWORDS: Zimbabwe, aquatic plants, ecology

267 GIBBS-RUSSELL, G.E. and BIEGEL. H.M., 1973. Report on botanical collecting trips to Maun and the northern Okavango Delta. Botswana Notes and Records, July 1973:154–169.
KEYWORDS: Botswana, swamp (Okavango), aquatic plants

268 GIBBS-RUSSELL, G.E. and MITCHELL, D.S., 1976. Common aquatic plants on Rhodesian pans and lakes. Rhodesia Agricultural Journal, 73(1):13–17.
KEYWORDS: Zimbabwe, lake (Zimbabwe), aquatic plants, ecology

269 GILLETT, J.B., 1963. *Sesbania* in Africa (excluding Madagascar) and southern Arabia. Kew Bulletin, 17:91–153.
KEYWORDS: Africa, *Sesbania*, checklist

270 GOLDING, F.D. and GWYNN, A.M., 1938. Notes on the vegetation of the Nigerian shore of Lake Tchad. Kew Bulletin, 10:631–643.
KEYWORDS: Nigeria, Tchad, vegetation description (Tchad), lake (Tchad)

271 GOMA, L.K.H., 1958. The productivity of various mosquito breeding places in the swamps of Uganda. Bulletin of Entomological Research, 49:437–448.
KEYWORDS: Uganda, utilisation (animals), swamp (Uganda), disease, invertebrates

272 GOMA, L.K.H., 1960. The swamp-breeding mosquitoes of Uganda, records of larvae and their habitats. Bulletin of Entomological Research, 51:77–94.
KEYWORDS: Uganda, utilisation (animals), disease, swamp (Uganda), invertebrates

273 GOODMAN, P.S., 1981. Water hyacinths in Ndumu Game Reserve: past history, present status and recommendations for future control, monitoring and research. Natal Parks, Game and fish Preservation Board Report, 8pp.
KEYWORDS: South Africa, *Eichhornia*, control/management (*Eichhornia*)

274 GORDON-GRAY, K.D. and WARD, C.J., 1971. A contribution to the knowledge of *Phragmites* (Gramineae) in South Africa, with particular reference to Natal populations. Journal of South African Botany, 37:1–30.
KEYWORDS: South Africa, *Phragmites*, ecology

275 GOUVEIA, A.C.P., 1966. Vegetaçao do Cunene (Entre a Matala e Vila Rocadas). Rel. Fin do Curso dos Engenairos da Agricola da Lisboa, Instituto Sup. Agrico.
KEYWORDS: Angola, vegetation description (Angola), river (Cunene)
(mimeographed; in Portuguese)

276 GRANT, J., 1970. Feasibility study of the uses of papyrus: also of alternative raw materials for paper manufacture in Uganda. Report for the Uganda Government, United Nations International Development Organisation (UNIDO).
KEYWORDS: Uganda, utilisation (man), *Cyperus*

277 GREENWAY, P.J., 1973. A classification of the vegetation of East Africa. Kirkia, 9:1–68.
KEYWORDS: East Africa, vegetation description (East Africa), swamp (East Africa), swamp (savanna), swamp (forest), checklist

278 GREENWAY, P.J. and VESEY-FITZGERALD, D.F., 1969. The vegetation of Lake Manyara National Park. Journal of Ecology, 57:127–149.
KEYWORDS: Tanzania, vegetation description (Tanzania), lake (Manyara), swamp (savanna)

279 GREYVENSTEIN, P.S. and DE VILLIERS, C.P.M., 1975. Common reed as a reclamation plant. Farming in South Africa, Engineering Series, No. E1/1975.
KEYWORDS: South Africa, utilisation (man), *Phragmites*

280 GRIMSDELL, J.J.R. and BELL, R.H.V., 1975. Ecology of the black lechwe in the Bangweulu Basin of Zambia. Black Lechwe Research Project: Final report. Animal Productivity Research Report ARI (NCSR/TR 31), 175 pp and Appendices.
KEYWORDS: Zambia, lake (Bangweulu), swamp (savanna), floodplain (Bangweulu), vegetation description (Zambia), wildlife, lechwe

281 GRIMSDELL, J.J.R. and BELL, R.H.V., 1976. The Bangueulu Basin of Zambia. The Botswana Society, Goborone, Botswana, pp 257–262. Proceddings of a Symposium on the Okavango Delta and its Future Utilisation. pp 257–262.
KEYWORDS: Zambia, lake (Bangweulu), swamp (savanna), floodplain (Bangweulu), wildlife, lechwe, conservation

282 GROVE, A.T., 1970. Africa South of the Sahara (2nd edition). Oxford University Press.
KEYWORDS: Africa, vegetation description (Africa)

283 GUY, P.R., 1977. Notes on the vegetation types of the Zambezi Valley, Zimbabwe, between the Kariba and Mpata gorges. Kirkia, 10:543–557.
KEYWORDS: Zimbabwe, vegetation description (Zimbabwe), river (Zambezi)

284 HAINES, R.W. and LYE, K.A., 1971. Studies on African Cyperaceae. 4. *Lipocarpha* R. Br., *Hemicarpa* Neees and *Isolepis* R. Br.. Botaniska Notiser, 124:473–482.
KEYWORDS: Africa, *Lipocarpha, Hemicarpha, Isolepis*

285 HALL, J.B., 1970. Observations on the vegetation of the Lake Volta drawdown area. Volta Basin Research Project, Technical Report No.35, University of Ghana, Legon.
KEYWORDS: Ghana, lake (Volta), aquatic plants, littoral, development (water), vegetation description (Ghana)

286 HALL, J.B., 1971. Observations on *Isoetes* in Ghana. Botanical Journal of the Linnaean Society, 64:117–139.
KEYWORDS: Ghana, *Isoetes*

287 HALL, J.B., 1975. The vascular flora of Lake Kainji and its shores. In: A.M.A. Imevbore and O.S. Adegoke (Editors), The Ecology of Lake Kainji: the Transition from River to Lake. University of Ife Press, Ile–Ife, pp 149–156.
KEYWORDS: Nigeria, lake (Kainji), aquatic plants, development (water), vegetation description (Nigeria), littoral

288 HALL, J.B. and POPLE, W., 1968. Recent vegetational changes in the lower Volta River. Ghana Journal of Science, 8:24–29.
KEYWORDS: lake (Volta), Ghana, aquatic plants

289 HALL, J.B. and OKALI, D.U.U., 1974. Phenology and productivity of *Pistia stratiotes* L. on the Volta Lake, Ghana. Journal of Applied Ecology, 11:709–725.
KEYWORDS: Ghana, productivity, ecology, *Pistia*, lake (Volta)

290 HALL, J.B., LAING, E., HOSSAIN, M. and LAWSON, G.W., 1969. Observations on aquatic weeds in the Volta Basin. In: Obeng, L.E. (Editor), Man–Made lakes: The Accra Symposium. Ghana University Press, Accra, pp 331–336.
KEYWORDS: Ghana, lake (Volta), aquatic weeds

291 HALL, J.B., PIERCE, P. and LAWSON, C., 1971. Common plants of the Volta Lake. Publication of the Department of Botany, University of Ghana, Legon, 123 pp.
KEYWORDS: Ghana, aquatic plants, lake (Volta)

292 HAMDOUN, A.M., 1977. Weed control problems in the Sudan. PANS, 23(2):190–194.
KEYWORDS: Sudan, aquatic weeds, control/management (weeds)

293 HAMILTON, A.C., 1982. Environmental history of East Africa: a study of the quaternary. Acedamic Press, London, 328 pp.
KEYWORDS: East Africa, peatland, environmental factors, environmental history, afroalpine

294 HAMMERTON, D., 1972. The Nile river – a case history. In: R.T. Oglesby (Editor), River Ecology and Man. Academic Press, London.
KEYWORDS: river (Nile), riverine, swamp (Upper Nile/Sudd), hydrology, chemistry, development (water), Sudan

295 HAMMOUDA, M.A., 1968. The water outlay by *Eichhornia crassipes* and observations on the plant chemical control. Phyton, 13:97–106.
KEYWORDS: Africa, *Eichhornia*, physiology, control/management (*Eichhornia*)

296 HANCOCK, G.L.R., 1938. The mosquitoes of Namanve Swamp, Uganda. Journal of Animal Ecology, 3:204–221.
KEYWORDS: Uganda, invertebrates, disease, swamp (Namanve)

297 HARDING, D., 1960. Preliminary observations on the effects of seasonal change in water condition on the fishery in the Bangweulu area of Northern Rhodesia. Proceedings of the Third Symposium on Hydrobiology and Inland Fisheries: Major Lakes Lusaka, 1960. Conseil Scientifique pour l'Afrique au Sud du Sahara (C.S.A.), CCTA/CSA Technical and Scientific Publication, 63.
KEYWORDS: Zambia, fish, environmental factors, lake (Bangweulu), swamp (Bangweulu), hydrology

298 HARDING, D., 1965. Research on Lake Kariba 1962/63. Fisheries Research Bulletin, Zambia 1962/63, Government Printer, Lusaka, pp 38–55.
KEYWORDS: Zambia, lake (Kariba), development (water), fish, aquatic plants, hydrology

299 HARDING, D., 1966. Lake Kariba: the hydrology and development of fisheries. In: R.H. Lowe-McConnell (Editor), Man-made Lakes. Academic Press, London, pp 7–20.
KEYWORDS: Zimbabwe, Zambia, fish, lake (Kariba), development (water), hydrology, aquatic plants

300 HARRISON, A., 1964. An ecological survey of the Great Berg River. In: D. Davis (Editor), Ecological Studies in Southern Africa. Dr. W. Junk, b.v., The Hague.
KEYWORDS: South Africa, river (Great Berg)

301 HARRISON, A.D., KELLER, P. and DIMOVIC, D., 1960. Ecological studies on Olifantsvlei, near Johannesburg. Hydrobiologia, 15:89–134.
KEYWORDS: South Africa, swamp (Olifantsvlei), utilisation (animals)

302 HATTINGH, E.R., 1958. Survey of certain weed control problems in Africa south the Sahara. Proceedings of the 4th British Weed Control Conference. pp 1–11.
KEYWORDS: Uganda, Sudan, swamp, river (Nile), hydrology, utilisation (man)

303 HATTINGH, E.R., 1961. Problem of *Salvinia auriculata* Aubl. and associated aquatic weeds on Kariba Lake. Weed Research, 1:303–306.
KEYWORDS: *Salvinia*, aquatic weeds, lake (Kariba), ecological problems, Zimbabwe

304 HATTINGH, E.R., 1962. Report on investigations into the control of *Salvinia auriculata* on Lake Kariba. Part 1. Union Weedkiller Services, Johannesburg.
KEYWORDS: *Salvinia*, control/management *Salvinia*, aquatic weeds, lake (Kariba)

305 HATTINGH, E.R., 1963. Report on investigation into the control of *Salvinia auriculata* on Lake Kariba. Part 2 Field trials and recommendations. Union Weedkiller Services, Johannesburg.
KEYWORDS: *Salvinia*, control/management *Salvinia*, aquatic weeds, lake (Kariba)

306 HEDBERG, I. and HEDBERG, O. (Editors), 1968. Conservation of Vegetation in Africa South of the Sahara. Acta Phytogeographica Suecica, 54.
KEYWORDS: Africa, conservation, swamp, review, vegetation description (Africa), lake, river

307 HEDBERG, O., 1964. Features of Afro-alpine ecology. Acta Phytogeographica Suecica, 49:118–128.
KEYWORDS: East Africa, afroalpine, peatland

308 HEEG, J. and BREEN, C.M., 1982. Man and the Pongolo Floodplain, a report of the Committee for Inland Water Ecosystems, National Programme for Environmental Sciences. South African National Scientific Programmes Report No.56, Council for Scientific and Industrial Research, Pretoria, 117pp.
KEYWORDS: South Africa, floodplain (Pongolo), development (water), swamp (savanna), hydrology, chemistry, conservation, fish, invertebrates, wildlife, ecological problems, development (land/swamp)

309 HEEG, J., BREEN, C.H., COLVIN, P.M., FURNESS, W.D. and MUSIL, C.F., 1978. On the dissolved solids of the Pongolo Floodplain pans. Journal of the Limnological Society of Southern Africa, 4:59–64.
KEYWORDS: South Africa, floodplain (Pongolo), lake (Pongolo), chemistry, river (Pongolo)

310 HEEG, J., BREEN, C.M. and ROGERS, K.H., 1980. The Pongolo Floodplain: A unique ecosystem threatened. In: Bruton, M.N. and Cooper, K.H. (Editors) Studies on the Ecology of Maputaland. Rhodes University Press, Grahamstown, South Africa, pp 374–381.
KEYWORDS: South Africa, floodplain (Pongolo), conservation, swamp (savanna), development (water), ecological problems, river (Pongolo), control/management (swamp)

311 HEINEN, E.T. and AHMED, S.H., 1964. Water hyacinth control on the Nile river, Sudan. Report of the Publishing and Information Processing Centre, Department of Agriculture, Khartoum, 56pp.
KEYWORDS: Sudan, *Eichhornia*, river (Nile), control/management (*Eichhornia*)

312 HEPPER, F.N., 1966. The duckweeds of West Africa. Nigerian Field, 31:18–21.
KEYWORDS: West Africa, *Lemna*, ecology

313 HEPPER, F.N., 1970. Hovercraft on Lake Chad. Plant life on sandbanks and papyrus swamps. Geographical Magazine, 42:577–582.
KEYWORDS: Tchad, swamp (Tchad), lake (Tchad), *Cyperus*

314 HESSE, P.R., 1957. The distribution of sulphur in the muds, water and vegetation of Lake Victoria. Hydrobiologia, 11:29–39.
KEYWORDS: Uganda, decomposition, chemistry, lake (Victoria), swamp (Victoria)

315 HESSE, P.R., 1958. Fixation of sulphur in the muds of Lake Victoria. Hydrobiologia, 11:171–181.
KEYWORDS: Uganda, Kenya, Tanzania, lake (Victoria), chemistry

316 HILL, B.J., BLABER, S.J.M. and BOLTT, R.E., 1975. The limnology of Lagoon Poelela. Transactions of the Royal Society of South Africa, 41:263–271.
KEYWORDS: Moçambique, chemistry, lake (Poelela), *Potamogeton, Charophytes*

317 HIRA, P.R., 1969. Transmission of schistosomiasis in Lake Kariba, Zambia. Nature, London, 224:670–672.
KEYWORDS: Zambia, *Salvinia*, disease, aquatic plants, lake (Kariba), invertebrates

318 HOIER, R., 1950. Exploration du Parc National Albert. Institut Parcs Nationaux Congo Belge, Bruxelles, 172pp.
KEYWORDS: Zaïre, vegetation description (Zaïre), lake (Albert), swamp (savanna), swamp (Lake Albert)

319 HOLMES, R.H., 1955. Reclamation of Mumwalo Swamp, Bufumbira. Annual Report of the Department of Hydrological Survey, Uganda Government (1954), pp 11–12.
KEYWORDS: Uganda, swamp (Mumwalo), development (land/swamp)

320 HOOPER, S.S., 1972. Cyperaceae. In: F.N. Hepper (Editor), Flora of West Tropical Africa. 2nd Edition. 3(2):278–349.
KEYWORDS: West Africa, *Cyperus*, checklist

321 HOPE, C.W., 1902. The 'Sadd' of the Upper Nile: its botany compared with that of similar obstructions in Bengal and American waters. Annals of Botany, 16:495–516.
KEYWORDS: Sudan, *Cyperus*, swamp (Upper Nile/Sudd), river (Nile), ecological problems

322 HOPKINS, G.H.E., 1940. Afforestation as a method of drying up swamps. East African Medical Journal, 17:189–194.
KEYWORDS: East Africa, disease, control/management (swamp), ecological problems, hydrology, swamp (forest)

323 HORNE, A.J. and VINER, A.B., 1971. Nitrogen fixation and its significance in tropical Lake George, Uganda. Nature, London, 232:417–418.
KEYWORDS: Uganda, chemistry, lake (George), plankton

324 HOWARD-WILLIAMS, C., 1971. An ecological sampling method for swamp and marsh vegetation types. Limnological Society of Southern Africa Newsletter, 17:27–30.
KEYWORDS: Malawi, swamp (Chilwa), ecology

325 HOWARD-WILLIAMS, C., 1972. Limnological studies in an African swamp: seasonal and spatial changes in the swamps of Lake Chilwa, Malawi. Archiv für Hydrobiologie, 70:379–391.
KEYWORDS: Malawi, chemistry, swamp (Chilwa), environmental factors, littoral, aquatic plants, lake (Chilwa)

326 HOWARD-WILLIAMS, C., 1972. Preliminary botanical description of the Elephant Marsh. In: R. Hastings (Editor) Fisheries Bulletin 4. Fisheries Department, Zomba.
KEYWORDS: Malawi, vegetation description (Malawi), swamp (Elephant Marsh), aquatic plants

327 HOWARD-WILLIAMS, C., 1973. Vegetation and environment in the marginal areas of a tropical African lake (L. Chilwa, Malawi). Ph.D. Thesis. University of London.
KEYWORDS: Malawi, vegetation description (Malawi), swamp, littoral, environmental factors, chemistry, *Typha*, *Aeschynomene*, ecology, aquatic plants, lake (Chilwa), swamp (Chilwa)

328 HOWARD-WILLIAMS, C., 1975. Seasonal and spatial changes in the composition of the aquatic and semiaquatic vegetation of Lake Chilwa, Malawi. Vegetatio, 30:33–39.
KEYWORDS: Malawi, *Typha*, aquatic plants, lake (Chilwa), environmental factors, chemistry, littoral, swamp (Chilwa)

329 HOWARD-WILLIAMS, C., 1975. Vegetation changes in a shallow African lake: response of the vegetation to a recent dry period. Hydrobiologia, 47:381–398.
KEYWORDS: Malawi, *Typha*, aquatic plants, lake (Chilwa), environmental factors, chemistry, swamp (Chilwa), vegetation description (Malawi)

330 HOWARD-WILLIAMS, C., 1977. A check-list of the vascular plants of Lake Chilwa, with special reference to the influence of environmental factors on the distribution of taxa. Kirkia, 10:563–579.
KEYWORDS: Malawi, aquatic plants, lake (Chilwa), environmental factors, checklist

331 HOWARD-WILLIAMS, C., 1977. The distribution of nutrients in Swartvlei, a southern Cape coastal lake. Water SA, 3:213–217.
KEYWORDS: South Africa, chemistry, lake (Swartvlei)

332 HOWARD-WILLIAMS, C., 1978. Growth and production of aquatic macrophytes in a south temperate saline lake. Verhandlungen der Internationalen Vereinigung für theoretische und angewandte Limnologie (Verh. Internat. Verein. Limnol.), 20:1153–1158.
KEYWORDS: South Africa, productivity, lake (Chilwa), *Potamogeton*

333 HOWARD-WILLIAMS, C., 1979. Distribution, biomass and role of aquatic macrophytes in Lake Sibaya. In: B.R. Allanson (Editor), Lake Sibaya. Junk Biological Monographs, The Hague, 36:88–107.
KEYWORDS: South Africa, aquatic plants, ecology, lake (Sibaya), vegetation description (South Africa)

334 HOWARD-WILLIAMS, C., 1979. Interactions between swamp and lake. In: M. Kalk, A.J. McLachlan and C. Howard-Williams (Editors), Lake Chilwa. W. Junk, The Hague, 231–245.
KEYWORDS: Malawi, chemistry, swamp (Chilwa), lake (Chilwa), aquatic plants, productivity, decomposition, fish

335 HOWARD-WILLIAMS, C., 1979. Recommendations for aquatic plant cutting and control in Swartvlei. Institute for Freshwater Studies, Rhodes University, Grahamstown, Special report, No. 79/1:7pp.
KEYWORDS: South Africa, *Potamogeton*, control/management (weeds), lake (Swartvlei)

336 HOWARD-WILLIAMS, C., 1979. The aquatic environment. 2. Chemical and physical characteristics of the Lake Chilwa swamps. In: M. Kalk, A.J. McLachlan and C. Howard-Williams (Editors), Lake Chilwa. W. Junk, The Hague, pp 79–90.
KEYWORDS: Malawi, swamp (Chilwa), lake (Chilwa), chemistry

337 HOWARD-WILLIAMS, C., 1979. The distribution of aquatic macrophytes in Lake Chilwa: annual and long term environmental fluctuations. In: M. Kalk, A.J. McLachlan and C. Howard-Williams (Editors), Lake Chilwa. W. Junk, The Hague, 105–122.
KEYWORDS: Malawi, environmental factors, lake (Chilwa), aquatic plants, vegetation description (Malawi)

338 HOWARD-WILLIAMS, C., 1979. The use of aquatic plants as nutrient filters, especially in South African conditions– a short review. Report to the Council for Scientific and Industrial Research. Working Group for Eutrophication, March 1979, 11pp.
KEYWORDS: South Africa, chemistry, aquatic plants, control/management (Chemistry), utilisation (man)

339 HOWARD-WILLIAMS, C., 1980. Aquatic macrophyte communities of the Wilderness lakes: community structure and associated environmental conditions. Journal of the Limnological Society of Southern Africa, 6:85–92.
KEYWORDS: South Africa, *Phragmites*, *Typha*, *Potamogeton*, *Charophytes*, lake (Wilderness), environmental factors, vegetation description (South Africa)

340 HOWARD-WILLIAMS, C., 1980. Aquatic macrophytes of the coastal wetlands of Maputaland. In: M.N. Bruton and K.H. Cooper (Editors), Studies on the Ecology of Maputaland. Rhodes University Press, Grahamstown, pp 42–51.
KEYWORDS: South Africa, vegetation description (South Africa), aquatic plants, swamp (Maputaland)

341 HOWARD-WILLIAMS, C., 1981. Studies on the ability of a *Potomogeton pectinatus* community to remove dissolved nitrogen and phosphorus compounds from lake water. Journal of Applied Ecology, 18:619–637.
KEYWORDS: South Africa, chemistry, *Potamogeton*, lake (Swartvlei), physiology

342 HOWARD-WILLIAMS, C. and ALLANSON, B.R., 1978. Research Summary, Recommendations for Future Management Uses of Aquatic Macrophytes as Nutrient Filters. Institute for Freshwater Studies Special Report No. 78/2, Swartvlei Project Report, Part 1, Rhodes University, Grahamstown, South Africa, 76pp.
KEYWORDS: South Africa, chemistry, utilisation (man), aquatic plants, control/management (chemistry)

343 HOWARD-WILLIAMS, C. and ALLANSON, B.R., 1978. The Limnology of Swartvlei with Special Reference to Production and Nutrient Dynamics in the Littoral Zone. Institute for Freshwater Studies Special Report No. 78/3, Swartvlei Project Report, Part II, Rhodes University, Grahamstown, South Africa, 280pp.
KEYWORDS: South Africa, chemistry, productivity, *Potamogeton*, lake (Swartvlei)

344 HOWARD-WILLIAMS, C. and ALLANSON, B.R., 1981. An integrated study on littoral and pelagic primary production in a Southern African coastal lake. Archiv für Hydrobiologie, 92:507–534.
KEYWORDS: South Africa, productivity, lake (Swartvlei), plankton, *Potamogeton*

345 HOWARD-WILLIAMS, C. and ALLANSON, B.R., 1981. Phosphorus cycling in a dense *Potamogeton pectinatus* lake bed. Oecologia, 49:56–66.
KEYWORDS: South Africa, chemistry, *Potamogeton*, decomposition, lake (Swartvlei)

346 HOWARD-WILLIAMS, C. and DAVIES, B.R., 1979. The rates of dry matter and nutrient loss from decomposing *Potamogeton pectinatus* in a brackish south temperate coastal lake. Freshwater Biology, 9:13–21.
KEYWORDS: South Africa, decomposition, *Potamogeton*, chemistry, lake (Swartvlei)

347 HOWARD-WILLIAMS, C. and HOWARD-WILLIAMS, W.A., 1978. Nutrient leaching from the swamp vegetation of Lake Chilwa, a shallow African lake. Aquatic Botany, 4:257–267.
KEYWORDS: Malawi, decomposition, chemistry, lake (Chilwa)

348 HOWARD-WILLIAMS, C. and LENTON, G., 1975. The role of the littoral zone in the functioning of a shallow tropical lake ecosystem. Freshwater Biology, 5:445–459.
KEYWORDS: Malawi, littoral, chemistry, lake (Chilwa), swamp (Chilwa)

349 HOWARD-WILLIAMS, C. and LIPTROT, M.R.M., 1980. Submerged macrophyte communities in a brackish South African estuarine-lake system. Aquatic Botany, 9:101-116.
KEYWORDS: *Potamogeton*, *Charophytes*, vegetation description (South Africa), lake (Swartvlei)

350 HOWARD-WILLIAMS, C. and LONGMAN, T.G., 1976. A quantitative sampler for submerged aquatic macrophytes. Journal of the Limnological Society of Southern Africa, 2:31-33.
KEYWORDS: South Africa, productivity

351 HOWARD-WILLIAMS, C. and WALKER, B.H., 1974. The vegetation of a tropical African lake: classification and ordination of the vegetation of Lake Chilwa (Malawi). Journal of Ecology, 62:831-854.
KEYWORDS: Malawi, vegetation description (Malawi), lake (Chilwa), swamp (Chilwa)

352 HOWARD-WILLIAMS, C., and LIPTROT, M.R.M., 1980. Submerged macrophyte communities in a brackish South African estuarine lake-system. Aquatic Botany, 9(2):101-116.
KEYWORDS: South Africa, *Potamogeton*, lake (Swartvlei), estuary (Swartvlei)

353 HOWARD-WILLIAMS, C., DAVIES, B.R. and CROSS, R.H.M., 1978. The influence of periphyton on the surface structure of a Potamogeton pectinatus L. leaf (an hypothesis). Aquatic Botany, 5:87-91.
KEYWORDS: South Africa, *Potamogeton*, periphyton, physiology

354 HOWARD-WILLIAMS, C., FURSE, M., SCHULTEN-SENDEN, C., BOURN, D. and LENTON, G., 1972. Lake Chilwa, Malawi. Studies on a Tropical Freshwater Ecosystem. Report to the IBP/UNESCO Symposium on Synthesis of PF Results, Reading, London.
KEYWORDS: Malawi, productivity, fish, *Typha*, lake (Chilwa) (cyclostyled)

355 HUNTLY, B.J., 1978. Ecosystem conservation in Southern Africa. In: Werger, M.J.A. (Editor) Biogeography and Ecology of Southern Africa. W. Junk, The Hague, pp 1333-1384.
KEYWORDS: Africa, conservation, control/management, swamp, lake, river

356 HURST, H.E., 1933. The Sudd region of the Nile. Journal of the Royal Society of Arts, 81:721-736.
KEYWORDS: Sudan, hydrology, swamp (Upper Nile/Sudd), river (Nile)

357 HUTCHINSON, G.E., PICKFORD, G.E. and SCHUURMAN, J.F.M., 1932. A contribution to the hydrobiology of pans and other inland waters of South Africa. Archiv für Hydrobiologie, 24:1-154.
KEYWORDS: South Africa, lake (South Africa), chemistry

358 I.U.C.N., 1979. The distribution of protected areas in relation to the needs of biotic community conservation in West and Central Africa. (Draft). International Union for the Conservation of Nature and Natural Resources, Gland, Switzerland, 62 pp.
KEYWORDS: West Africa, swamp (forest), conservation

359 ILTIS, A. and LEMOALLE, J., 1983. The aquatic vegetation of Lake Chad. In: J-P. Carmouze, J-R. Durand and C. Lévêque, Lake Chad: Ecology and Productivity of a Shallow Tropical Ecosystem. Junk Monographiae Biologicae 53, pp 125-144.
KEYWORDS: Tchad, lake (Tchad), vegetation description (Tchad), swamp (Tchad)

360 ILTIS, A. and LEMOALLE. J., 1979. La Végétation Aquatique du Lac Tchad. Rapport de la Réunion de Travail sur la Limnologie Africaine, S.I.L., Nairobi, 16–23 December, 1979, Office de la Récherche Scientifique et Technique Outre-Mer, Paris, 17 pp.
KEYWORDS: Tchad, aquatic plants, lake (Tchad), swamp (Tchad), vegetation description (Tchad)

361 IMAM, M. and KOSINOVA, J., 1972. Studies on the weed flora of cultivated land in Egypt. 2. Weeds of rice fields. Botanische Jahrbuecher fuer systematik Pflanzengesichte und Pflanzengeographie (Bot. Jahrb. Syst.), 92:90–107.
KEYWORDS: Egypt, aquatic weeds

362 IMEVBORE, A.M.A., 1971. Floating vegetation of Lake Kainji, Nigeria. Nature, London, 230:599–600.
KEYWORDS: Nigeria, *Pistia*, lake (Kainji), development (water), vegetation description (Nigeria)

363 IMEVBORE, A.M.A., 1975. Floating vegetation of Lake Kainji. In: A.M.A. Imevbore and O.S. Adegoke (Editors), The Ecology of Lake Kainji: the Transition from River to Lake. University of Ife Press, Ile-Ife, Nigeria, pp 146–148.
KEYWORDS: Nigeria, *Pistia*, aquatic plants, lake (Kainji), development (water), vegetation description (Nigeria)

364 IMEVBORE, A.M.A., 1976. Aquatic weeds in Nigeria. Proceedings of 5th Conference of Weed Science Society of Nigeria, University of Ife, July 1975. pp 17–28.
KEYWORDS: Nigeria, aquatic weeds

365 IMEVBORE, A.M.A. and BAKARE, O., (In Press). A pre-impoundment study of swamps in the Kainji Lake Basin. African Journal of Tropical Hydrobiology and Fisheries.
KEYWORDS: Nigeria, swamp (Kainji), vegetation description (Nigeria), swamp (Kainji)

366 IRVING, N.S. and M.O. BESHIR, 1982. Introduction of some natural enemies of water hyacinth to the White Nile, Sudan. Tropical Pest Management, 28(1):20–26.
KEYWORDS: Sudan, *Eichhornia*, control/management (*Eichhornia*)

367 JACKSON, G., 1968. The vegetation of Malawi. Society of Malawi Journal, 21:11–19.
KEYWORDS: Malawi, aquatic plants, vegetation description (Malawi)

368 JACKSON, G., 1968–69. The grasslands of Malawi, I–III. Society of Malawi Journal, 22:7–25, 73–81.
KEYWORDS: Malawi, vegetation description (Malawi), swamp (savanna)

369 JACKSON, P.B.N. and DAVIES, B.R., 1976. Cabora Bassa in its first year: some ecological aspects and comparisons. Rhodesia Science News, 10(5):128–133.
KEYWORDS: Zimbabwe, lake (Cabora Bassa), development (water), ecological problems

370 JACOT GUILLARMOD, A., 1962. The bogs and sponges of the Basutoland mountains. South African Journal of Science, 58:179–182.
KEYWORDS: Lesotho, vegetation description (Lesotho), peatland

371 JACOT GUILLARMOD, A., 1962. The flora of Basutoland. III. Water plants and water-side plants. Basutoland Notes and Records, 3:7–15.
KEYWORDS: Lesotho, aquatic plants, checklist, riverine, vegetation description (Lesotho)

372 JACOT GUILLARMOD, A., 1963. The flora of Basutoland. III. Water plants and water side plants. South African Journal of Science, 59(4):115–118.
KEYWORDS: Lesotho, aquatic plants, checklist, riverine, vegetation description (Lesotho)

373 JACOT GUILLARMOD, A., 1968. Lesotho. In: I. and O. Hedberg (Editors), Conservation of Vegetation in Africa South of the Sahara. Acta Phytogeographica Suecica, 54:253–256.
KEYWORDS: Lesotho, vegetation description (Lesotho), aquatic plants, conservation, peatland

374 JACOT GUILLARMOD, A., 1969. Modern agricultural practice and limnology in Lesotho. Limnological Society of Southern Africa Newsletter, 13:41–44.
KEYWORDS: Lesotho, ecological problems, development (land/swamp), aquatic plants

375 JACOT GUILLARMOD, A., 1969. The effect of land usage on aquatic and semi-aquatic vegetation at high altitudes in Southern Africa. Hydrobiologia, 34:3–13.
KEYWORDS: Lesotho, development (land/swamp), aquatic plants, peatland, ecological problems

376 JACOT GUILLARMOD, A., 1970. Aquatic weeds in Southern Africa. Probe 70 (Science Students' Council Journal, University of the Witwatersrand), pp 47–49, 53.
KEYWORDS: South Africa, *Myriophyllum*, aquatic weeds, ecological problems

377 JACOT GUILLARMOD, A., 1972. The bogs and sponges of the Orange River catchment within Lesotho. Civil Engineering in South Africa, 14:84–85.
KEYWORDS: Lesotho, swamp (Lesotho), utilisation (animals)

378 JACOT GUILLARMOD, A., 1973. Report on botanical investigations at Swartvlei during September 1972: *Salvinia* and *Potamogeton*.
KEYWORDS: South Africa, *Salvinia*, *Myriophyllum*, *Potamogeton*, lake (Swartvlei)

379 JACOT GUILLARMOD, A., 1976. *Myriophyllum*, an increasing water weed menace for South Africa. Proceedings of the South African Association for the Advancement of Science, 1:93–100.
KEYWORDS: South Africa, *Myriophyllum*, ecological problems, aquatic weeds

380 JACOT GUILLARMOD, A., 1977. Emersed water milfoil/water duisendblaar, a new threat plant in Southern Africa. Eastern Cape Naturalist, 60:4–8.
KEYWORDS: Africa, *Myriophyllum*, aquatic weeds

381 JACOT GUILLARMOD, A., 1977. *Myriophyllum*, a powerful threat to our water supplies. Environment RSA, 4:(1), 3–5.
KEYWORDS: South Africa, *Myriophyllum*, development (water), aquatic weeds, ecological problems

382 JACOT GUILLARMOD, A., 1977. *Myriophyllum*, an increasing water weed menace for South Africa. South African Journal of Science, 73(3):89–90.
KEYWORDS: South Africa, *Myriophyllum*, ecological problems, aquatic weeds

383 JACOT GUILLARMOD, A., 1977. Some water weeds of the Eastern Cape Province. II. Eastern Cape Naturalist, 61:14–17.
KEYWORDS: South Africa, *Myriophyllum*, aquatic weeds

384 JACOT GUILLARMOD, A., 1977. Some water weeds of the Eastern Cape Province. III. *Salvinia*. Eastern Cape Naturalist, 62:10–13.
KEYWORDS: South Africa, *Salvinia*

385 JACOT GUILLARMOD, A., 1978. Can we come to terms with our water weeds?. Conference Proceedings S2A3 Stellenbosch. Association of Scientific Societies of South Africa, 1:409–416.
KEYWORDS: South Africa, aquatic weeds, ecological problems

386 JACOT GUILLARMOD, A., 1978. Eradication of water hyacinth. South African Journal of Science, Notes and Views, 74:122.
KEYWORDS: South Africa, *Eichhornia*, control/management (*Eichhornia*)

387 JACOT GUILLARMOD, A., 1978. Kariba weed (*Salvinia molesta*). In: C.H. Stirton (Editor), Plant Invaders – Beautiful, but Dangerous. Department of Nature and Environmental Conservation, CPA, Cape Town, pp 132–135.
KEYWORDS: South Africa, *Salvinia*

388 JACOT GUILLARMOD, A., 1978. Notes on distribution and biology of *Aponogeton ranunculiflorus*. Kew Bulletin, 32(4):781–783.
KEYWORDS: South Africa, *Aponogeton*

389 JACOT GUILLARMOD, A., 1978. Parrot's feather (*Myriophyllum aquaticum*). In: C.H. Stirton (Editor), Plant Invaders – Beautiful, but Dangerous. Department of Nature and Environmental Conservation, CPA, Cape Town, 96–99.
KEYWORDS: South Africa, *Myriophyllum*

390 JACOT GUILLARMOD, A., 1978. Some water weeds of the Eastern Cape province. IV. *Azolla*. Eastern Cape Naturalist, 63:16–18.
KEYWORDS: South Africa, *Azolla*, aquatic weeds

391 JACOT GUILLARMOD, A., 1979. Investigation of the taxonomy and distribution of *Myriophyllum* species in South Africa. Institute for Freshwater Studies Special Report, 79/2, Rhodes University, Grahamstown.
KEYWORDS: South Africa, *Myriophyllum*

392 JACOT GUILLARMOD, A., 1979. *Myriophyllum* in South Africa. In: Weeds– 3rd National Conference. pp 141–149.
KEYWORDS: South Africa, *Myriophyllum*

393 JACOT GUILLARMOD, A., 1979. Taxonomy and economic aspects of water plants in Africa. In: Taxonomic Aspects of African Economic Botany. Proceedings of the 9th plenary Meeting of A.E.T.F.A.T., Las Palmas de Gran Canaria, 18–23 March, 1978. pp 85–87.
KEYWORDS: Africa, aquatic plants, utilisation (man)

394 JACOT GUILLARMOD, A., 1979. Water weeds in Southern Africa. Aquatic Botany, 6:377–391.
KEYWORDS: Southern Africa, lake (South Africa), river (South Africa), *Eichhornia*, *Azolla*, *Myriophyllum*, aquatic weeds

395 JACOT GUILLARMOD, A., 1980. Vegetation of the Touw River and adjacent lakes area floodplain. Annual report of the Rhodes University Institute of Freshwater Studies, 1979/80, Paper No 5.
KEYWORDS: South Africa, river (Touw), floodplain (Touw), aquatic plants

396 JACOT GUILLARMOD, A. and ALLANSON, B.R., 1977. Eradication of water hyacinth. South African Journal of Science, 61:14–17.
KEYWORDS: Africa, *Eichhornia*, control/management (*Eichhornia*)

397 JARVIS, M.J.F., MITCHELL, D.S. and THORNTON, J.A., 1982. Aquatic macrophytes and *Eichornia crassipes*. In: J.A. Thornton (Editor), Lake McIlwaine: The Eutrophication and Recovery of a Tropical African Man made Lake. W. Junk, The Hague, Monographiae Biologicae, 49:137–144.
KEYWORDS: Zimbabwe, lake (McIlwaine), *Eichhornia*, *Myriophyllum*, *Phragmites*, *Typha*, *Polygonum*, ecological problems, development (water)

398 JARVIS, M.J.F., VAN DER LINGEN, M.I. and THORNTON, J.A., 1981. Water hyacinth. Zimbabwe Science News, 15(4):97–99.
KEYWORDS: Zimbabwe, *Eichhornia*

399 JENIK, J., 1970a. The pneumatophores of *Voacanga thouarsii* Roem. et Schult. (Apocynaceae)'. Bull. de L'LFan, 32(A):986–994.
KEYWORDS: West Africa, swamp (forest), physiology, structure (plant)

400 JENIK, J., 1970c. Root system of tropical trees 4. The stilted peg-roots of *Xylopia staudtii* Eng. et Diels. Preslia, 42:25–32.
KEYWORDS: West Africa, swamp (forest), physiology, structure (plant)

401 JENIK, J., 1971. Root systen of tropical trees. 7. Facultative peg-roots of *Anthocleista nobilis* G.Don. Preslia, 43:97–104.
KEYWORDS: swamp (forest)

402 JENIK, J., 1971b. Root system of tropical trees 7. The facultative peg-roots of *Anthocleista nobilis* G. Don. Preslia, 43:96–104.
KEYWORDS: West Africa, swamp (forest), physiology, structure (plant)

403 JENIK, J., 1973. Root system of tropical trees. 8. Stilt-roots and allied adaptations. Preslia, 45:250–264.
KEYWORDS: West Africa, swamp (forest), physiology, structure (plant)

404 JENIK, J. and KUBIKOVA, J., 1969. Root system of tropical trees. 3. The heterorhizis of *Aeschynomene elaphroxylon*. Preslia, 41:220–226.
KEYWORDS: Africa, *Aeschynomene*

405 JOHANSSON, D., 1976. Aquatic plants and man-made lakes in Africa. In: Ecological Studies of the Mtera Basin. SWECO, Stockholm, pp 101–113.
KEYWORDS: Tanzania, development (water), river (Great Ruaha), aquatic plants

406 JONES, M.B., 1983. Papyrus : a new fuel for the Third World. New Scientist, 11 August 1983:418–421.
KEYWORDS: Rwanda, Kenya, Sudan, Egypt, *Cyperus*, utilisation (man), productivity

407 JONES, M.B. and MILBURN, T.R., 1978. Photosynthesis in papyrus (*Cyperus papyrus* L.). 12:197–199.
KEYWORDS: Uganda, *Cyperus*, physiology

408 JONGLEI INVESTIGATION TEAM, 1954. The equatorial Nile project and its effects on the Anglo-Egyptian Sudan. Report of the Jonglei Investigation Team, Sudan Government, Khartoum, Vols 1–4.
KEYWORDS: Sudan, development (water), swamp (Upper Nile/Sudd), *Cyperus*, floodplain (Upper Nile/Sudd), hydrology, control/management (swamp)

409 JOSEPH, A.F. and MARTIN, F.J., 1902. A preliminary account of the chemistry of the Nile sudd. Journal of the Society of Chemical Industry, 29:91–94.
KEYWORDS: Sudan, chemistry, *Cyperus*, swamp (Upper Nile/Sudd)

410 JOVET-AST, S., 1968. Contribution to the study of the fresh waters of the Ennedi (French Equatorial Africa). 2. Lemnaceae. Bulletin de l'Institut Française d'Afrique Noire, Serie A, 30:830–847.
KEYWORDS: Central African Republic, *Lemna*

411 JUBB, R.A., 1979. Watch for these water weeds. South African Garden and Home, February, 1979:80–81.
KEYWORDS: South Africa, *Salvinia*

412 KALK, M. (Editor), 1970. Decline and Recovery of a Lake. Government Printer, Zomba, Malawi.
KEYWORDS: aquatic weeds, ecological problems, Malawi, lake

413 KALK, M., McLACHLAN, A.J., and HOWARD-WILLIAMS, C. (Editors), 1979. Lake Chilwa, Studies of Change in a Tropical Ecosystem. W. Junk, Monographiae Biologicae 35, The Hague, 462 pp.
KEYWORDS: aquatic plants, *Typha*, fish, review, chemistry, ecology, decomposition, floodplain, lake (Chilwa)

414 KAMAL, I.A. and LITTLE, E.C.S., 1970. The potential use of water hyacinth for horticulture in the Sudan. Pesticide Abstracts, 16:488–496.
KEYWORDS: Sudan, *Eichhornia*, utilisation (man)

415 KAMEL, El, SAYED, J., TOLBA, A. and DRUIJFF, A.H., 1978. Evaluation of some machines for mechanical control of aquatic weeds – Egypt. Proceedings of the 5th EWRS International Symposium on Aquatic Weeds.
KEYWORDS: Egypt, aquatic weeds, control/management (weeds)

416 KAPETSKI, J.M. (Editor), 1981. Seminar on River Basin Management and development. CIFA Technical Papers, 8:302pp.
KEYWORDS: Africa, river, floodplain, development (land/swamp), development (water)

417 KASKE, R., 1972. The problem of water hyacinth control of the White Nile. Berichte aus der Abteilung für Herbologie and der Universitat Hohenheim, 1972(3):27–44.
KEYWORDS: *Eichhornia*, ecological problems, river (Nile), control/management (*Eichhornia*)
(In German)

418 KASSAS, M. and GHABBOUR, S.I. (Editors), The Nile and its Environment. Water Supply Management, 4(1–2).
KEYWORDS: Sudan, Egypt, vegetation description (Sudan, Egypt), development, environmental factors, river (Nile)

419 KEDVES, M., 1978. The sporomorphae of an Angolan brown coal. Acta Bot. Acad. Sci. Hung, 24(1/2):69–90.
KEYWORDS: Angola, *Cyperus*, swamp

420 KELLEY, D.W., 1968. Fishery development in the Central Barotse floodplain. Report FAO/UNDP (TA), 2554, Rome.
KEYWORDS: Zambia, fish, floodplain (Barotse)

421 KENMUIR, D., 1978. A wilderness called Kariba: the wildlife and natural history of Lake Kariba. Wilderness Publication.
KEYWORDS: Zambia, Zimbabwe, lake (Kariba), wildlife, aquatic plants, development (water), conservation

422 KERFOOT, O., 1975. The Okavango controversy: a real cause for alarm. South African Journal of Science, 71:67.
KEYWORDS: Botswana, swamp (Okavango), conservation, wildlife,

development (land/swamp), disease, control/management (wildlife)

423 KERFOOT, O., 1975. The Okavango controversy - a follow up. South African Journal of Science, 71:196–197.
KEYWORDS: Botswana, swamp (Okavango), wildlife, conservation, development (land/swamp), disease, control/management (wildlife)

424 KEULDER, P.C., 1982. Leaf area as a measure of standing crop of floating aquatic macrophytes. Journal of the Limnological Society of Southern Africa, 8:29–32.
KEYWORDS: South Africa, *Salvinia*, *Azolla*, productivity

425 KILHAM, P., 1972. Biogeochemistry of African lakes and rivers. Ph.D. Thesis, Duke University, Durham, N.C..
KEYWORDS: Kenya, chemistry, hydrology

426 KILLICK, D.J.B., 1979. African mountain heathlands. In: R.L. Specht (Editor), Heathlands and Related Shrublands. A Descriptive Studies. Elsevier, Ecosystems of the World Series, 9A. pp 96–116.
KEYWORDS: afroalpine, peatland, ecology, classification

427 KIMPE, P. de, 1957. Le controle de la jacinthe d'eau. Bulletin Agricole du Congo Belge, 48:105–152.
KEYWORDS: Zaïre, *Eichhornia*, control/management (*Eichhornia*)

428 KIMPE, P. de, 1962. La lutte contre les jacinthes d'eau par l'administration belge au Congo de 1955 a 1960. Africa–Tervuren, 8:76–80.
KEYWORDS: Congo, Zaïre, *Eichhornia*, ecological problems, river (Zaïre) (in French)

429 KLUGE, R.L., 1977. The prospects for the biological control of water hyacinth in South Africa. Proceedings of the 2nd Entomological Congress.
KEYWORDS: South Africa, *Eichhornia*, control/management

430 KLUGE, R.L., 1978. Water hyacinth (*Eichhornia crassipes*). In: C.H. Stirton (Editor), Plant Invaders - Beautiful, but Dangerous. Department of Nature and Environmental Conservation, CPA, Cape Town, pp 68–71.
KEYWORDS: South Africa, *Eichhornia*

431 KLUGE, R.L. and ANNEKE, D.P., 1972. Prospects for the biological control of *Eichhornia crassipes* in South Africa. Workshop on Aquatic Weeds. C.S.I.R., Pretoria.
KEYWORDS: South Africa, *Eichhornia*, control/management (*Eichhornia*)

432 KNAPP, R., 1973. Die Vegetation von Afrika. Gustav Fischer Verlag, 624pp (particularly pp 235–256, 363–364, 551–553).
KEYWORDS: Africa, swamp, swamp (savanna), swamp (forest), vegetation description (Africa), review

433 KOECHLIN, G., 1961. La végétation des savannes dans le Sud de la Republique du Congo. Institut de Récherches Scientifiques au Congo Brazzaville, Office de la Récherche Scientifique et Technique Outre-Mer, 310 pp.
KEYWORDS: Zaïre, swamp (savanna), vegetation description (Zaïre), classification

434 LAING, E., 1968. Observations on the biology of *Pistia stratiotes*. Pesticide Abstracts, 8:15.
KEYWORDS: Ghana, *Pistia*, ecology

435 LAMPREY, H.F., 1975. The distribution of protected areas in relation to the needs of biotic comunity censervation in eastern Africa. I.U.C.N., Occasional Paper No.16:1–83.
KEYWORDS: Eastern Africa, conservation, swamp (savanna)

436 LAMPREY, H.F., 1975. The distribution of protected areas in relation to the needs of biotic community conservation in Eastern Africa. IUCN Occasional Paper No. 16, Morgues, Switzerland, 86 pp.
KEYWORDS: Eastern Africa, conservation, swamp (Eastern Africa)

437 LANDOLT, E., 1975. Morphological differentiation and geographical distribution of the *Lemna gibba* – *Lemna minor* group. Aquatic Botany, 1:345–363.
KEYWORDS: Africa, *Lemna*, ecology

438 LANGDALE-BROWN, I. and SPOONER, R.J., 1963. Land use prospects of Northern Bechuanaland. Developmental Study 1, Forestry and Land Use Section, Directorate of Overseas Survey, Tolworth, U.K..
KEYWORDS: Botswana, swamp (Okavango), floodplain, development (land/swamp)

439 LANGDALE-BROWN, I. OSMASTON, H.A. and WILSON, J.G., 1964. The Vegetation of Uganda and its Bearing upon Land-Use. Government Printer, Entebbe, 159 pp.
KEYWORDS: Uganda, vegetation description (Uganda), swamp (Uganda), swamp (savanna), development (land/swamp), environmental factors, control/management (swamp)

440 LAUNERT, E., 1965. A survey of the genus *Leersia* in Africa. Senck. Biol., 46(2):129–153.
KEYWORDS: *Leersia*, taxonomy

441 LAWSON, G.W., 1964. A preliminary report on the aquatic weeds in the Volta Basin. Volta Basin Research Project, University of Ghana, Legon, Report No.6, 8pp (Mimeographed).
KEYWORDS: Ghana, *Myriophyllum*, aquatic weeds, lake (Volta)

442 LAWSON, G.W., 1967. 'Sudd' formation on the Volta Lake. Bulletin de l'Institut Française d'Afrique Noire, Serie A, 29:1–4.
KEYWORDS: Ghana, aquatic weeds, ecological problems, lake (Volta), development (water)

443 LAWSON, G.W., 1970. Lessons of the Volta – a new man-made lake in tropical Africa. Biological Conservation, 2(2):90–96.
KEYWORDS: Ghana, aquatic weeds, ecological problems, lake (Volta), development (water)

444 LAWSON, G.W., HALL, J.B., LAING, E. and HOSSAIN, M., 1969. Observations on aquatic weeds in the Volta basin. In: L.E. Obeng (Editor), Man-Made Lakes: The Accra Symposium. Ghana University Press, pp 331–336.
KEYWORDS: Ghana, aquatic weeds, lake (Volta)

445 LAWTON, R.M., 1959. A pollen analysis of the lake Bangweulu peat deposits. Northern Rhodesia Journal, 4:33–43.
KEYWORDS: Zambia, peatland, lake (Bangweulu), swamp (Bangweulu), environmental history

446 LAWTON, R.M., 1963. Palaeoecological and ecological studies in the Northern Province of Northern Rhodesia. Kirkia, 3:46–76.
KEYWORDS: Zambia, vegetation description, lake (Bangweulu), peatland, swamp (Bangweulu), environmental history

447 LEBRUN, J-P., 1947. La végétation de la plaine alluviale au sud du Lac Edouard. Exploration du Parc National Albert (Mission Journal Lebrun). Institut Royal des Sciences Naturelles de Belgique (et Institut Pares Nationaux du congo Belge), Bruxelles, 800 pp.
KEYWORDS: Zaïre, floodplain (Edward), vegetation description, swamp (Edward), lake (Edward)
(in French)

448 LEBRUN, J-P., 1948. Exploration du Parc National de la Kagera (Mission Journal Lebrun, 1937–1938). Institut Royal des Sciences Naturelles de Belgique, Bruxelles.
KEYWORDS: Rwanda, vegetation description (Rwanda), river (Kagera), swamp (Kagera)
(in French)

449 LEBRUN, J-P., 1958. The present situation in respect of the problems of *Eichhornia crassipes* in Africa south of the Sahara. Pesticide Abstracts, 4:87–99.
KEYWORDS: Africa, *Eichhornia*, ecological problems

450 LEBRUN, J-P., 1959. La lutte contre le development de l'*Eichhornia crassipes.* Bulletin agricole du Congo belge, 50:251–252.
KEYWORDS: Congo, *Eichhornia*
(in French)

451 LEBRUN, J-P., 1973. Enumeration des plantes vasculaires du Sénégal. Institut d'Elévage et de Médécine Vetérinaire des Pays Tropicaux, Bruxelles.
KEYWORDS: Sénégal, vegetation description (Sénégal)

452 LEBRUN, J-P., AUDRU, J., GASTON, A. and MOSNIER, M., 1972. Catalogue des plantes vasculaires du Tchad Meridionale. Institut d'Elévage et de Médécine Vetérinaire des Pays Tropicaux.
KEYWORDS: Tchad, vegetation description (Tchad), checklist

453 LEBRUN, J. and GILBERT, G., 1954. Une classification ecologique des forêts du Congo. Institut National pour l'Etude Agronomique du Congo Belge (I.N.E.A.C.), Série Scientifique, 63, 89 pp.
KEYWORDS: classification, Zaire, swamp (forest)

454 LEISTNER, O.A., 1967. The plant ecology of the Southern Kalahari. Memoirs of the Botanical Survey of South Africa, 38:1–172.
KEYWORDS: Botswana, swamp (Okavango), aquatic plants, vegetation description (Botswana)

455 LEJOLY, J. and MANDANGO, A., 1982. L'association arbustive ripicole à *Alchornea cordifolia* dans le haut-Zaire. In: J.J. Symoens, S.S. Hooper and P. Compère, 'Sudies on Aquatic Vascular Plants'. Royal Botanical Society of Belgium, Brussels, pp 257–265.
KEYWORDS: *Alchornea*, riverine, river (Zaire), swamp (forest)

456 LEONARD, J., 1950. Botanique du Congo Belge. I. Les groupements végétaux. In: Encyclopedie du Congo belge. Cahiers d'Office de la Récherche Scientific et Technique Outre-Mer, 1:345–389.
KEYWORDS: swamp (Zaïre), aquatic plants, Zaïre, vegetation description (Zaïre)
(in French)

457 LEONARD, J., 1951. Aperçu préliminaire des groupements végétaux pioniers dans la région de Yangambi (Congo Belge). Vegetatio, 3:279–297.
KEYWORDS: Zaïre, vegetation description (Zaïre)

458 LEONARD, J., 1969. Aperçu sur la végétation aquatique. In: Monographie Hydrologique du Lac Tchad. Cahiers d'Office de la Récherche Scientifique et Technique Outre-Mer, Fort Lamy, 11 pp.
KEYWORDS: Tchad, vegetation description (Tchad), aquatic plants, lake (Tchad), swamp (Tchad)
(in French)

459 LEONARD, J., 1974. Aperçu sur la végétation de la partie est du lac Tchad. Cahiers d'Office de la Récherche Scientifique et Technique Outre-Mer, N'Djamena:14pp.
KEYWORDS: Tchad, lake (Tchad), aquatic plants, swamp (Tchad)

460 LETOUZEY, R., 1968. Etude phytogéographique de Cameroun. Encyclopedie Biologique. Paul Lechevalier, Paris, Vol. 69.
KEYWORDS: Cameroun, vegetation description (Cameroun), swamp (Cameroun)
(in French)

461 LEVEQUE, C., CARMOUZE, J.P., DEJOUX, C., DURAND, J.R., GRAS, R., ILTIS, A., LEMOALLE, J., LOUBENS, G., LAUZANNEL, L. and SAINT- JEAN, L., 1972.. A study of biomasses and production of Lake Chad. In: Z. Kajak and A. Hillbricht-Ilkowska (Editors), Productivity Problems of Freshwaters. Proceedings of the IBP-UNESCO Symposium, Kazimierz Dolny, Poland, 6-12 May 1970. Polish Scientific Publishers, Warsaw, pp 165-181.
KEYWORDS: Tchad, Nigeria, productivity, lake (Tchad)

462 LIND, E.M., 1956. Studies in Uganda swamps. Uganda Journal, 20:166-176.
KEYWORDS: Uganda, chemistry, swamp, vegetation description (Uganda)

463 LIND, E.M. and BEADLE, L.C., 1959. Research on the swamps of Uganda. Makerere College, University of East Africa, Kampala (mimeographed).
KEYWORDS: Uganda, swamp (Uganda), invertebrates, vegetation description (Uganda)

464 LIND, E.M. and MORRISON, M.E.S., 1974. East African Vegetation. Longmans, London.
KEYWORDS: East Africa, vegetation description (East Africa), swamp (East Africa), swamp (forest), swamp (savanna), lake, ecology, review

465 LIND, E.M. and VISSER, S.A., 1962. A study of a swamp at the north end of Lake Victoria. Journal of Ecology, 50:599-613.
KEYWORDS: Uganda, chemistry, swamp (Victoria), lake (Victoria)

466 LIPTROT, M.R.M., 1977. The ecology of Swartvlei Estuary. 1. The effect of mouth closure on submerged vegetation and physico-chemical characteristics of the water column during 1976. Institute for Freshwater Studies Special Report, 77/3, Grahamstown, 28 pp.
KEYWORDS: South Africa, aquatic plants, estuary (Swartvlei), environmental factors, chemistry

467 LIPTROT, M.R.M. and ALLANSON, B.R., 1978. Swartvlei Project Report. Part III:Community metabolism and phosphorus dynamics in the Swartvlei Estuary. Institute for Freshwater Studies Special Report, 78/3, Rhodes University, Grahamstown, 125pp.
KEYWORDS: South Africa, chemistry, environmental factors, aquatic plants, estuary (Swartvlei)

468 LISOWSKI, S., MALAISSE, E. and SYMOENS, J.J., 1975. Plantes nouvelles pour la flore du Zaïre. Bulletin du Jardin Botanique National de Belgique, 45:383-389.
KEYWORDS: Zaïre, *Cyperus*

469 LITTERICK, M.R., GAUDET, J.J., KALEF, J. and MELACK, J.M., 1980. The limnology of an African lake, Lake Naivasha, Kenya. 73pp.
KEYWORDS: Kenya, swamp, lake (Naivasha), chemistry, hydrology, aquatic plants

470 LITTLE, A.M. and PHELAN, A.J., 1978. St Lucia Estuary: Evaluation of Natural Resources. South African Journal of Science, 74:8–11.
KEYWORDS: South Africa, estuary (St Lucia), swamp (South Africa), development (land/swamp)

471 LITTLE, E.C.S., 1965. Occurrence of *Salvinia auriculata* Aubl. on the Congo River. Nature, London, 208:1111–1112.
KEYWORDS: Zaïre, *Salvinia*, river (Zaïre)

472 LITTLE, E.C.S., 1966. The invasion of man-made lakes by plants. In: R.H. Lowe-McConnell (Editor), Man-made Lakes. Academic Press, London, pp 75--86.
KEYWORDS: Africa, *Myriophyllum*, lake, development (water), ecological problems, aquatic weeds

473 LITTLE, E.C.S., 1967. Progress report on transpiration of some tropical water weeds. Pesticide Abstracts, 13:127–132.
KEYWORDS: Africa, *Eichhornia*, physiology, hydrology

474 LITTLE, E.C.S. (Editor), 1968. Handbook of Utilisation of Aquatic Plants. UN/FAO, Rome, 123pp.
KEYWORDS: Africa, aquatic weeds, utilisation (man), aquatic plants, control/management (weeds)

475 LITTLE, E.C.S., 1968. The control of water weeds. Weed Research, 8:75–105.
KEYWORDS: Africa, *Eichhornia*, *Pistia*, *Salvinia*, *Potamogeton*, *Vossia*, *Cyperus*, control/management (weeds)

476 LITTLE, E.C.S., 1969. Weeds and man-made lakes. In: L.E. Obeng (Editor), Man-Made Lakes: The Accra Symposium. Ghana University Press, Accra, Ghana, 284–291.
KEYWORDS: Africa, *Eichhornia*, lake, development (water), aquatic weeds

477 LITTLE, E.C.S., 1979. Handbook of utilisation of aquatic plants (second edition). FAO Fisheries Technical Paper No.187, 176pp.
KEYWORDS: Africa, aquatic weeds, utilisation (man), aquatic plants, control/management (weeds)

478 LIVERSIDGE, R., 1955. A preliminary account of the vegetation of the Rondevlei bird sanctuary. B.Sc. Thesis. University of Capetown, Bolus Herbarium.
KEYWORDS: South Africa, vegetation description (South Africa), swamp (Rondevlei), lake (Rondevlei)

479 LOCK, J.M., 1973. The aquatic vegetation of Lake George, Uganda. Phytocoenologia, 1:250–262.
KEYWORDS: Uganda, *Cyperus*, swamp, lake (George), aquatic plants, vegetation description

480 LOCK, J.M., 1977. The vegetation of Ruwenzori National Park, Uganda. Botanische Jahrbuecher fuer Systematik Pflanzengesichte und Pflanzengeographie (Bot. Jahrb. Syst.), 98(3):372–448.
KEYWORDS: Uganda, vegetation description (Uganda), checklist

481 LONGMAN, K.A. and JENIK, J., 1974. Tropical forest and its environment. Longman, 196 pp.
KEYWORDS: swamp (forest)

482 LOUIS, J., 1947. L'origine et la végétation des i s du fleuve dans la region

de Yangambi. Comptes rendus de la semaine Agricole de Yangambi. Publications de l'Institut National pour l'Etude Agronomique du Congo belge (INEAC), 26:2–5 924–933.
KEYWORDS: Zaïre, vegetation description (Zaïre), swamp (Yangambi) (in French)

483 LOVELESS, A.R., 1969. The possible role of pathogenic fungi in local degeneration of *Salvinia auriculata* Aublet on Lake Kariba. Annals of Applied Biology, 63:61–69.
KEYWORDS: Zambia, Zimbabwe, *Salvinia*, control/management (*Salvinia*), lake (Kariba)

484 LOWE–McCONNELL, R.H. (Editor), 1966. Man–Made Lakes. Academic Press, London.
KEYWORDS: Africa, development (water), aquatic weeds, aquatic plants

485 LUBKE, R.A., RAYNHAM, G.L. and REAVELL, P.E., 1981. Reassessment of plant succession on spoil heaps along the Boro river, Okavango Delta, Botswana. South African Journal of Science, 77:21–23.
KEYWORDS: Botswana, river (Okavango), ecological problems, development (water)

486 MACEDO, J. de A, 1974. Vegetaçao aquatica em Cabora Bassa Alguns problemas futuros. Instituto da Investigaçao de Agricole de Moçambique Memoir No, 5:1–86.
KEYWORDS: Moçambique, aquatic plants, lake (Cabora Bassa), development (water), ecological problems

487 MAGADZA, C.H.D., 1969. The Shoreline Vegetation. Nuffield Lake Kariba Research Station, Sinamwenda, Report 1962–1968, pp 33–34.
KEYWORDS: Zimbabwe, lake (Kariba), littoral, vegetation description (Zimbabwe)

488 MAGADZA, C.H.D., 1970. A preliminary survey of the vegetation of the shore of Lake Kariba. Kirkia, 7:253–267.
KEYWORDS: Zimbabwe, vegetation description (Zimbabwe), lake (Kariba), littoral

489 MAGADZA, C.H.D., 1977. Observations on the behaviour of macrophytes following impoundment of the Kafue River at Kafue Gorge (1974–1976). Transactions of the Rhodesia Scientific Association, 58(3):17–23.
KEYWORDS: Zambia, river (Kafue), floodplain (Kafue), aquatic plants, development (water), hydrology, *Vossia*, *Typha*, *Aeschynomene*

490 MAKIN, M.J., KINGHAM, T.J., WADDAMS, A.E., BIRCHALL, C.J. and EAVIS, B.W., 1976. Prospects for irrigation development around Lake Zwai, Ethiopea. Land Resources Division, Ministry of Overseas Development, Tolworth, U.K., Land Resource Study 26:316pp.
KEYWORDS: Ethiopea, lake (Zwai), swamp (Zwai), development (land/swamp), irrigation

491 MALAISSE, F., 1968. Etude écologique de la rivière Luanza et de son bassin. Ph.D. Thesis. University of Lubumbashi, Zaïre, 473pp.
KEYWORDS: Zaïre, ecology, river (Luanza), aquatic plants, vegetation description (Zaïre)

492 MALAISSE, F., 1975. Carte de la végétation du bassin de la Luanza. Exploration Hydrobiologique du Bassin du Lac Bangweolo et du Luapula, 18(2):1–41.
KEYWORDS: Zambia, Zaïre, river (Luanza), vegetation description (Zaïre), swamp (savanna)

493 MALAISSE, F., 1976. Ecologie de la Rivière Luanza. Exploration Hydrobiologique du Bassin du Lac Bangweolo et du Luapula. Cercle Hydrobiologique de Bruxelles, Belgique, 17:151pp.
KEYWORDS: Zambia, Zaïre, ecology, river (Luanza), aquatic plants, vegetation description (Zaïre)
(in French)

494 MANN, O., 1977. The Jonglei Canal: Environmental and Social aspects. Report for the Environmental Liaison Centre, Nairobi, Kenya, 70pp.
KEYWORDS: Sudan, swamp (Upper Nile/Sudd), development (water), swamp (savanna), utilisation (man)

495 MARSHALL, B.E., 1978. Lake McIlwaine after twenty-five years. 12(3):79–82.
KEYWORDS: Zimbabwe, lake (McIlwaine), aquatic plants

496 MARSHALL, B.E., 1979. The current status of *Salvinia molesta* in Lake Kariba. Lake Kariba Fisheries Research Insitute, Project Report, 33:1–15.
KEYWORDS: Zambia, Zimbabwe, lake (Kariba), *Salvinia*, ecological problems

497 MARSHALL, B.E. and JUNOD, F.J.R., 1981. The decline of *Salvinia molesta* on Lake Kariba. Hydrobiologia, 83:477–484.
KEYWORDS: Zambia, Zimbabwe, *Salvinia*, control/management (*Salvinia*), lake (Kariba)

498 MARTIN, A.R.H., 1960. The ecology of Groenvlei, a South African fen, I: The primary communities. Journal of Ecology, 48:55–71.
KEYWORDS: South Africa, peatland, vegetation description (South Africa), swamp (Groenvlei)

499 MARTIN, F.R.M., 1960. The ecology of Groenvlei, a South African Fen II: The secondary communities. Journal of Ecology, 48:307–339.
KEYWORDS: South Africa, peatland, swamp (Groenvlei), vegetation description (South Africa)

500 MAUVE, A.A., 1965. The South African species of *Potamogeton*. Limnological Society of Southern Africa Newsletter, 2(1):27–28.
KEYWORDS: South Africa, *Potamogeton*

501 MAVUTI, K.M., 1982. Wetlands and shallow water bodies in Kenya: characteristics and general status of knowledge. In: D.O. Logofet and N.K. Lukyanov (Editors), Ecosystem Dyanamics in Freshwater, Wetlands and Shallow Water Bodies. Proceedings of a SCOPE/UNEP International Scientific Workshop, Minsk–Pinsk–Tskhaltoubo, 12–26 July 1981. Centre of International Projects GKNT, Moscow, pp 188–204.
KEYWORDS: Kenya, classification, *Cryperus*, fish

502 McLACHLAN, A.J., 1969. The effect of aquatic macrophytes on the variety and abundance of benthic fauna in a newly created lake in the tropics (Lake Kariba). Archiv für Hydrobiologie, 66:212–231.
KEYWORDS: Zimbabwe, aquatic plants, utilisation (animals), invertebrates, fish, lake (Kariba)

503 McLACHLAN, A.J., 1975. The role of aquatic macrophytes in the recovery of the benthic fauna of a tropical lake after a dry phase. Limnology and Oceanography, 20:54–63.
KEYWORDS: Malawi, utilisation (animals), aquatic plants, lake (Chilwa), invertebrates, fish, environmental factors

504 McLACHLAN, A.J. and McLACHLAN, S.M., 1971. Benthic fauna and sediments in the newly created Lake Kariba (Central Africa). Ecology, 52:800–809.
KEYWORDS: Zimbabwe, *Salvinia*, lake (Kariba), invertebrates, fish

505 McVEIGH, S., 1981. Control of water weeds. Tight Lines, 22(12):42–43.
KEYWORDS: South Africa, aquatic weeds, control/management (weeds)

506 MENAUT, J–C, 1983. The vegetation of Africa savannas. In: Boulière, F. (Editor), Tropical Savannas. Elservier "Ecosstems of the World' series, 13. pp 109–149.
KEYWORDS: classification, swamp (savanna)

507 MENAUT, J–C. and CESEAR, J., 1983. The structure and dynamics of a West African savanna. Springer–Verlag, pp 80–100. In: B.J. Huntley and B.H. Walker (Editors), Ecology of Tropical Savannas. pp 80–100.
KEYWORDS: West Africa, swamp (savanna), control/management (swamp), ecological problems, palms

508 MENDONÇA, A., 1958. Etat actuel du problème de l'*Eichhornia crassipes* au Moçambique et en Angola. Bulletin Agricole du Congo Belge, 49:1362–1363.
KEYWORDS: Moçambique, Angola, *Eichhornia*, ecological problems (in French)

509 MICHELMORE, A.P.G., 1939. Observations on tropical African grasslands. Journal of Ecology, 27:282–312.
KEYWORDS: Eastern Africa, swamp (savanna)

510 MICKLEBURGH, D., 1980. From floating menace to cattle feed (*Eichhornia crassipes*). Farmer's Weekly, August 13, 1980:30–31.
KEYWORDS: South Africa, *Eichhornia*, utilisation (animals)

511 MIGAHID, A.M., 1947. An ecological study of the 'sudd' swamps of the Upper Nile. Proceedings of the Egyptian Academy of Science, 3:57–86.
KEYWORDS: Sudan, swamp (Upper Nile/Sudd), *Cyperus*, ecology, river (Nile)

512 MIGAHID, A.M., 1948. Report on a Botanical Excursion to the 'Sudd' Region. Cairo University Press, Cairo, 159pp.
KEYWORDS: Sudan, *Cyperus*, vegetation description, swamp (Upper Nile/Sudd), river (Nile)

513 MIGAHID, A.M., 1952. Further observations on the flow and loss of water in the sudd swamps of the Upper Nile. Bulletin of the Faculty of Science, Cairo University, 16pp.
KEYWORDS: Sudan, hydrology, swamp (Upper Nile/Sudd), *Cyperus*, river (Nile)

514 MIGAHID, A.M., 1952. Velocity of water current and its relation to swamp vegetation in the sudd region of the Upper Nile. Bulletin of the Faculty of Science, Cairo University, 76pp.
KEYWORDS: Egypt, *Cyperus*, hydrology, swamp (Upper Nile/Sudd), river (Nile)

515 MIGAHID, A.M. and AMER, F.A., 1951. Three types of transpiration curves. Proceedings of the Egyptian Academy of Science, 8:92–113.
KEYWORDS: Egypt, *Eichhornia*, *Lemna*, hydrology, physiology

516 MIGAHID, A.M., ADBEL RAHMAN, A.A. and EL SHAFEI ALI M., 1955. An ecological study of a swampy island in the Upper Nile. Proceedings of the Arabic Congress, 1965. pp 677–718.
KEYWORDS: Sudan, vegetation description (Sudan), swamp (Upper Nile/Sudd), river (Nile)

517 MILLARD, N.A.H. and SCOTT, K.M.F., 1953. A contribution to the hydrobiology of pans and other inland waters of South Africa. Transactions of the Royal Society of South Africa, 34:278–324.
KEYWORDS: South Africa, aquatic plants, lake (South Africa)

518 MILLS, W.R., 1953. Fertilizer trials in Nyanza Swamp. Annual Report of the Uganda Government Department of Agriculture, Kampala, pp 55–64.
KEYWORDS: Uganda, swamp (Nyanza), development (land/swamp), control/management (swamp)

519 MILNE, G., 1947. 'A soil reconnaisance journey through parts of Tanganyika territory, December 1935 to February 1936'. C. Gillman (editor). Journal of Ecology, 35:129–265.
KEYWORDS: classification, Tanzania, swamp (savanna)

520 MILSTEIN, P. and ZALOUMIS, E.A., 1976. Threatened wetland habitat and associated species. Proceedings of a Symposium on Endangered Wildlife in Southern Africa. Published by Endangered Wildlife Trust, Johannesburg, pp 105–115.
KEYWORDS: Southern Africa, birds, conservation, wildlife, utilisation (animals)

521 MITCHELL, D.S., 1960. Report on incidence of *Salvinia auriculata* on Lake Kariba. Federal Science Teachers' Journal, 1:39–46.
KEYWORDS: Zimbabwe, *Salvinia*, lake (Kariba)

522 MITCHELL, D.S., 1964. The biological explosion of *Salvinia auriculata* Aublet on Lake Kariba. Journal of the Institute of Sewerage Purification, 3:252 (Summary).
KEYWORDS: Zambia, Zimbabwe, *Salvinia*, ecological problems, lake (Kariba)
(Summary)

523 MITCHELL, D.S., 1965. Research on *Salvinia auriculata*. Proceedings of the Lake Kariba Fisheries Research Institute Symposium, Kariba, June 1965. Kariba, Zimbabwe, pp 51–56.
KEYWORDS: Zambia, Zimbabwe, *Salvinia*, ecological problems, lake (Kariba)

524 MITCHELL, D.S., 1967. A survey of *Salvinia auriculata* in the Chobe River System, March 1967. Report to the Government of Botswana, Gaborone, Botswana.
KEYWORDS: Botswana, *Salvinia*, river (Chobe)

525 MITCHELL, D.S., 1967. *Salvinia auriculata* in the estuaries of Lake Kariba. Proceedings of the First Rhodesian Science Congress. p 64.
KEYWORDS: Zimbabwe, *Salvinia*, ecological problems, lake (Kariba)
(Abstract)

526 MITCHELL, D.S., 1968. Ecological studies of *Salvinia auriculata* with particular reference to Lake Kariba (Rhodesia–Zambia) and the Chobe River (Botswana). Proceedings of the International Hydrological Decade Symposium on "Ecology and control of aquatic vegetation", December, 1968. UNESCO House, Paris.
KEYWORDS: Botswana, *Salvinia*, lake (Kariba), river (Chobe), ecological problems

527 MITCHELL, D.S., 1969. Autecology of *Salvinia auriculata*. Report of the Nuffield Lake Kariba Research Station, 1962–1968, Sinamwenda, Zimbabwe, pp 20–22.
KEYWORDS: Zimbabwe, *Salvinia*, ecology, lake (Kariba)

528 MITCHELL, D.S., 1969. The development of the shoreline vegetation. Report of the Nuffield Lake Kariba Research Station, 1962–1968, Sinamwenda, pp 23–24.
KEYWORDS: Zimbabwe, littoral, vegetation description (Zimbabwe), lake (Kariba)

529 MITCHELL, D.S., 1969. The ecology of vascular hydrophytes on Lake Kariba. Hydrobiologia, 34:448–464.
KEYWORDS: Zimbabwe, *Salvinia*, lake (Kariba), ecology, aquatic plants

530 MITCHELL, D.S., 1970. Autecological studies of *Salvinia auriculata* Aubl. Ph.D. Thesis, University of London. London, U.K..
KEYWORDS: Zimbabwe, productivity, chemistry, utilisation (animals), *Salvinia*, ecology

531 MITCHELL, D.S., 1971. Factors influencing the explosive growth of floating aquatic weeds in tropical areas. Rhodesia Science News, 5:17–19.
KEYWORDS: Africa, *Salvinia*, *Eichhornia*, ecological problems

532 MITCHELL, D.S., 1972. Aquatic weeds. Rhodesia Science News, 6:105–106.
KEYWORDS: Africa, *Eichhornia*, aquatic weeds

533 MITCHELL, D.S., 1972. The Kariba weed: *Salvinia molesta*. British Fern Gazette, 10:251–252.
KEYWORDS: Zimbabwe, *Salvinia*

534 MITCHELL, D.S., 1973. Aquatic weeds in man-make lakes. In: W.C. Ackermann, G.F. White and E.B. Worthington (Editors), Man-Made lakes: Their problems and Environmental Effects. American Geophysical Union, Washington, D.C., pp 606–611.
KEYWORDS: Zimbabwe, *Salvinia*, chemistry, lake (Kariba)

535 MITCHELL, D.S. (Editor), 1974. Aquatic vegetation and its use and control. UNESCO, Paris, 135 pp.
KEYWORDS: Africa, aquatic plants, aquatic weeds, review, utilisation (man), control/management (weeds), *Salvinia*, *Eichhornia*, ecological problems

536 MITCHELL, D.S., 1977. Water weed problems in irrigation systems. In: E.B. Worthington (Editor), Arid Land Irrigation in Developing Countries. Pergamon Press, Oxford, pp 317–328.
KEYWORDS: Africa, aquatic weeds, control/management (weeds), development (land/swamp), ecological problems

537 MITCHELL, D.S., 1978. Freshwater plants. In: M.J.A. Werger and A.C. Van Bruggen (Editors), Biogeography and Ecology of Southern Africa. W. Junk, b.v., The Hague, pp 1113–1138.
KEYWORDS: Southern Africa, vegetation description (Southern Africa), aquatic plants, lake (Southern Africa), river (Southern Africa), floodplain, swamp (Southern Africa)

538 MITCHELL, D.S. and MARSHALL, B.E., 1974. Hydrobiological observations on three Rhodesian reservoirs. Freshwater Biology, 4:61–72.
KEYWORDS: Zimbabwe, lake, development (water), aquatic weeds

539 MITCHELL, D.S. and ROSE, D.J.W., 1979. Factors affecting fluctuations in extent of *Salvinia molesta* on Lake Kariba. PANS, 25:171–177.
KEYWORDS: Zambia, Zimbabwe, *Salvinia*, environmental factors, lake (Kariba)

540 MITCHELL, D.S. and TUR, N.M., 1975. The rate of growth of *Salvinia molesta* (*S. auriculata* Auct.) in laboratory and natural conditions. Journal of Applied Ecology, 12:213–225.
KEYWORDS: Zimbabwe, productivity, *Salvinia*

541 MOGG, A.O.D., 1967. Comments on the flora on Inhaca Island, Moçambique. South African Journal of Science, 63:440–442.
KEYWORDS: Moçambique, vegetation description (Moçambique), swamp (Inhaca)

542 MOGHRABY, A.I. el, 1975. Some effects of *Eichhornia crassipes* (Mart.) Solms on the productivity of the White Nile. In: Aquatic Weeds in the Sudan. National Council for Research, Khartoum, pp 133–150.
KEYWORDS: Sudan, productivity, *Eichhornia*, river (Nile), ecological problems

543 MOGHRABY, A.I. el, 1982. The Jonglei Canal – needed development or potential ecodisasters?. Environmental Conservation, 9:141–148.
KEYWORDS: Sudan, development (water), *Cyperus*, *Vossia*, *Phragmites*, *Typha*, aquatic weeds, ecological problems, river (Nile)

544 MOHAMED, B.F., 1975. The ecology of water hyacinth in the White Nile, Sudan. Hyacinth Control Journal, 13:39–43.
KEYWORDS: Sudan, river (Nile), ecology, *Eichhornia*, control/management (*Eichhornia*)

545 MOHAMED, B.F. and BEBAWI, F.F., 1973. Burning as a supporting management in the control of water hyacinth in the Sudan. II. Back burning. Hyacinth Control Journal, 11:34–37.
KEYWORDS: Sudan, *Eichhornia*, control/management (*Eichhornia*)

546 MOHAMED, B.F. and BEBAWI, F.F., 1973. Burning as a supporting treatment in controlling water hyacinth in the Sudan. I. Routine burning. Hyacinth Control Journal, 11:31–34.
KEYWORDS: Sudan, *Eichhornia*, control/management (*Eichhornia*)

547 MOLL, F.J., 1968. A note on the aquatic vegetation of Midmar Dam, Howick, Natal. Lammergeyer, 9:37–41.
KEYWORDS: South Africa, aquatic plants, lake (Midmar), development (water)

548 MONTEIRO, J.F. de S., 1965. Resgate dos Machongos do Sul do Save. Primeiras Jornadas de Engenharia de Moçambique, Communicaçao, 60:34pp.
KEYWORDS: Moçambique, swamp (Moçambique), development (land/swamp) (in Portuguese)

549 MOORE, H.E., 1973. Palms in the tropical forest ecosystems of Africa and South America. In: B.J. Meggers, E.S. Ayensu and W.D. Duckworth (Editors), Tropical Forest Ecosystems in Africa and South America: a Comparative Review. Smithsonian Institution, Washington, pp 63–88.
KEYWORDS: palms, swamp (savanna), swamp (forest)

550 MORGAN, N.C. and BOY, V., 1982. An ecological survey of standing waters in northwest Africa. 1. Rapid survey and classification. Biological Conservation, 24(1):5–44.
KEYWORDS: classification, lake (Northwest Africa), wildlife, birds, vegetation description (Northwest Africa), checklist, Morocco, Tunisia, Algeria

551 MORRISON, M.E.S., 1968. Vegetation and climate in the uplands of south–western Uganda during the later pleistocene period. I. Muchoya Swamp, Kigezi District. Journal of Ecology, 56:363–384.
KEYWORDS: Uganda, swamp (Muchoya), peatland, vegetation description (Uganda), environmental factors, environmental history

552 MOSS, B., 1979. The Lake Chilwa ecosystem – a limnological overview. In: Kalk, M., McLachlan, A.J. and Howard–williams, C. (Editors), Lake Chilwa. Dr W. Junk, b.v., The Hague, pp 401–415.
KEYWORDS: Malawi, lake (Chilwa), chemistry, hydrology

553 MUKHTAR, A.M.S., 1967. A preliminary study on the chemical composition of water hyacinth (*Eichhornia crassipes*) in the Sudan. Sudan Journal of Veterinary Science, 8:124–126.
KEYWORDS: Sudan, *Eichhornia*, chemistry

554 MULLIGAN, H.F., 1974. *Pistia stratiotes* dans le lac de Kossou, Cote d'Ivoire. Bulletin Phytosanitaire, UN/FAO, 22:7–10.
KEYWORDS: Ivory Coast, *Pistia*, lake (de Kossou), ecological problems (in French)

555 MUSIL, C.F., 1972. Ecological studies on the aquatic macrophytes of the Pongolo river floodplain pans. M Sc Thesis, University of Natal, Pietermaritzburg, Vols I and II:216pp.
KEYWORDS: South Africa, aquatic plants, floodplain (Pongolo), ecology, vegetation description (South Africa), lake (Pongolo), swamp (savanna), river (Pongolo)

556 MUSIL, C.F., 1973. Water plants of Natal. Wildlife Protection and Conservation Society of South Africa, Durban, 62pp.
KEYWORDS: South Africa, aquatic plants, conservation, checklist

557 MUSIL, C.F., 1977. The application of growth kinetics for the control of *Eichornia crassipes* and *Salvinia molesta* through nutrient removal. Proceedings of the 2nd National Weeds Conference of South Africa, pp 239–247.
KEYWORDS: South Africa, *Eichhornia*, *Salvinia*, productivity, control/management (*Eichhornia*), chemistry, control/management (*Salvinia*), chemistry

558 MUSIL, C.F. and BREEN, C.M., 1977. The application of growth kinetics to the control of *Eichornia crassipes* (Mart.) Solms, through nutrient removal by mechanical harvesting. Hydrobiologia, 53:165–171.
KEYWORDS: South Africa, productivity, *Eichhornia*, control/management (*Eichhornia*), chemistry

559 MUSIL, C.F. and BREEN, C.M., 1977. The influence of site and position in the plant community on the nutrient distribution in, and content of *Eichhornia crassipes* (Mart.) Solms. Hydrobiologia, 53:67–72.
KEYWORDS: South Africa, chemistry, environmental factors, *Eichhornia*

560 MUSIL, C.F., BORNMAN, C.H. and GRUNOW, J.O. , 1976. Some observed interrelationships between the cover of aquatic vegetation and various physical properties of the water medium. Journal of South African Botany, 42:157–169.
KEYWORDS: South Africa, environmental factors, aquatic plants, productivity

561 MUSIL, C.F., GRUNOW, J.O. and BORNMAN, C.H., 1973. Classification and ordination of aquatic macrophytes in the Pongolo River pans, Natal. Bothalia, 11:181–190.
KEYWORDS: South Africa, vegetation description (South Africa),

classification, lake (Pongolo), aquatic plants, river (Pongolo)

562 MWANYIKA, N.A., 1982. Human use of wetlands in Tanzania: the case of Lower Rufiji floodplain. In: D.O. Logofet and N.K. Lukyanov (Editors), Ecosystem Dynamics in Freshwater Wetlands and Shallow Water Bodies. Proceedings of a SCOPE/UNEP International Scientific Workshop, Minsk–Pinsk–Tskhaltoubo, 12–26 July, 1981. Centre of International Projects GKNT, Moscow, pp 357–379.
KEYWORDS: Tanzania, utilisation (man), development (land/swamp), floodplain (Rufiji), river (Rufiji)

563 MYRE, M., 1963. Algumas plantas nocivas pelo sen caracter infestante (contribuiçao para o seu estudo). Instituto de Investigacao Cientifica de Moçambique, 32pp.
KEYWORDS: Moçambique, *Eichhornia*, control/management (weeds)
(in Portuguese)

564 MYRE, M., 1964. A vegetaçao do extremo sul de Moçambique (contribuicao para o seu estudo). Junta Invest. do Ultr, 145pp.
KEYWORDS: Moçambique, vegetation description (Moçambique), swamp (Moçambique)
(in Portuguese)

565 NAPPER, D.M., 1965. Cyperaceae of East Africa III. Journal of the East African Natural History Society, XXV:1(110):1–27.
KEYWORDS: East Africa, *Cyperus*, checklist

566 NAPPER, D.M., 1971. Typhaceae. In: E.Milne-Redhead and R.M. Pohill (Editors), Flora of Tropical East Africa. pp 1–6.
KEYWORDS: East Africa, *Typha*, checklist

567 NATAL PARKS BOARD, 1977. Water weed control: NPB succeeds with carp. South Africa Waterbulletin, 2(4):1–2.
KEYWORDS: South Africa, aquatic weeds, control/management (weeds), fish

568 NIELSEN, M.S., 1965. Introduction to the flowering plants of West Africa. University of London Press, 246 pp.
KEYWORDS: aquatic plants, vegetation description (West Africa), West Africa, taxonomy

569 NOEL, A.R.A., 1959. The vegetation of the freshwater swamps of Inhaca Island. Journal of South African Botany, 25:189–205.
KEYWORDS: Moçambique, swamp (Inhaca), vegetation description (Moçambique)

570 NORDENSTAM, B., 1970. Notes on the flora and vegetation of Etosha Pan, South West Africa. Dinteria, 5:3–18.
KEYWORDS: Namibia, vegetation description (Namibia), checklist, lake (Etosha), swamp (savanna)

571 NORDLINDH, T. and WEIMARCK, H., 1934. Beitrage zür Kenntnis der flora von Sud Rhodesia II. Botaniska Notiser, 1934:64–110.
KEYWORDS: Zimbabwe, vegetation description (Zimbabwe)

572 NORDLINDH, T. and WEIMARCK, H., 1948. Beitrage zür Kenntnis der Flora von Sud Rhodesia VIII. Botaniska Notiser, 1948:17–40.
KEYWORDS: Zimbabwe, vegetation description (Zimbabwe)

573 NTETA, D.N. and HERMANS, J. (Editors), 1976. The Okavango Delta and Future Utilisation. Proceedings of the Okavango Delta Symposium, 30 August – 2 September, 1976, Gaborone, Botswana, Botswana Society, Gaborone, 350 pp.
KEYWORDS: Botswana, vegetation description (Botswana), aquatic plants, swamp (Okavango), wildlife, development, chemistry, hydrology, utilisation (man), control/management (swamp)

574 OBEID, M. (Editor), 1975. Aquatic weeds in the Sudan: With special reference to the water hyacinth. National Council for Research and Agricultural Research Council, Khartoum.
KEYWORDS: Sudan, chemistry, aquatic weeds, *Eichhornia*

575 OBEID, M. and CHADWICK, M.J., 1964. Some factors affecting the growth of two aquatic weed species of the Nile, water hyacinth (*Eichhornia crassipes* Solms) and water lettuce(*Pistia stratiotes* L.). Proceedings of the Seventh British Weed Control Conference. pp 548–552.
KEYWORDS: Egypt, Sudan, *Eichhornia*, *Pistia*, environmental factors, productivity, river (Nile)

576 OBEID, M. and TAG el SEED, M., 1976. Factors affecting dormancy and germination of seeds of *Eichhornia crassipes* (Mart) Solms from the Nile. Weed Research, 16:71–80.
KEYWORDS: Sudan, *Eichhornia*, physiology, river (Nile)

577 OBENG, L.E., 1969. The invertebrate fauna of aquatic plants of the Volta Lake in relation to the spread of helminth parasites. In: L.E. Obeng (Editor), Man–Made Lakes: the Accra Symposium. Ghana Universities Press, Accra, Ghana.
KEYWORDS: Ghana, invertebrates, utilisation (animals), aquatic plants, lake (Volta), disease

578 OBERMEYER, A.A., 1964. The South African species of *Lagarosiphon*. Bothalia, 8:139–146.
KEYWORDS: South Africa, *Lagarosiphon*, checklist

579 ODEI, M., 1973. Observations of some weeds of malacological importance in the Volta Lake. Bulletin de l'Institut Française d'Afrique Noire, Serie A, 35:57–66.
KEYWORDS: Ghana, *Myriophyllum*, lake (Volta), disease, invertebrates

580 OKAFOR, L.I., 1980. Weeds of Lake Chad and their potential problems in the South Chad Irrigation Project. Weed Science Society of Nigeria, Tenth Annual Conference, Benin City, Nigeria.
KEYWORDS: aquatic weeds, Tchad, lake (Tchad), irrigation

581 OKAFOR, L.I., 1982. A preliminary survey on the major aquatic weeds of Lake Chad, Nigeria. Proceedings of the 6th International Symposium on Aquatic Weeds, Novisad, pp10–18.
KEYWORDS: Nigeria, Tchad, aquatic weeds, lake (Tchad)

582 OKALI, D.U.U. and HALL, J.B., 1974. Die–back of *Pistia stratiotes* on Volta Lake, Ghana. Nature, London, 248:452–453.
KEYWORDS: Ghana, *Pistia*, productivity, lake (Volta), control/management (*Pistia*)

583 OKALI, D.U.U. and HALL, J.B., 1974 b. Colonization of *Pistia stratiotes* mats by *Scirpus cubensis* on the Volta Lake. Ghana Journal of Agricultural Science, 7:31–36.
KEYWORDS: Ghana, lake (Volta), *Pistia*, *Scirpus*, ecology

584 OLIFF, W., 1959. Hydrobiological studies on the Tugela River system. Part 1. The main Tugela River. Hydrobiologia, 14:281–385.
KEYWORDS: South Africa, river (Tugela), chemistry

585 OOSTHUIZEN, C.J. and WALTERS, M.M., 1961. Control water fern with diesoline. Farming in South Africa, 37:35–37.
KEYWORDS: South Africa, *Azolla*, control/management (*Azolla*)

586 OPPERMANN, TESSA, 1982. Water hyacinth – pest or provider?. Environment RSA, 9(2):1–2.
KEYWORDS: South Africa, *Eichhornia*, utilisation (man), ecological problems

587 OSBORNE, P.L., BLACKMORE, S. and DUDLEY, C.O., 1980. A preliminary report on the distribution of potential aquatic weeds in the Shire River, Malawi. Unpublished report, 10pp.
KEYWORDS: Malawi, aquatic weeds, river (Shire), checklist

588 OSMAN, H.E., EL HAG, G.A. and OSMAN, M.M. , 1975. Studies on the nutritive value of waterhyacinth (*Eichhornia crassipes* Mart. Solms). In: M. Obeid (Editor), Aquatic Weeds in the Sudan with Special Reference to Water Hyacinth. National Council for Research, Agricultural Research Council, Khartoum, pp 104–127.
KEYWORDS: chemistry, Sudan, aquatic weeds, *Eichhornia*, utilisation(animals)

589 OTTO, N.E., 1969. Aquatic vegetation problems of the Okavango Swamp. United Nations Development Project /BOT/67/501, Technical Note 3, Food and agriculture Organisation of the United Nations, Rome.
KEYWORDS: Botswana, swamp (Okavango), *Cyperus*, aquatic plants, ecological problems

590 PAPERNA, I., 1969. Aquatic weeds, snails and transmission of bilharzia in the new man-made Volta Lake in Ghana. Bulletin de l'Institut Française d'Afrique Noire, Serie A, 31:487–499.
KEYWORDS: Ghana, utilisation (animals), disease, aquatic weeds, lake (Volta), development (water), invertebrates

591 PAPERNA, I., 1969. Snail vector of human schistosomiasis in the newly-formed Volta Lake. In: L.E.Obeng (Editor), Man-Made Lakes: the Accra Symposium. Ghana Universities Press, Accra, 326–330.
KEYWORDS: Ghana, utilisation (animals), invertebrates, disease, aquatic weeds, lake (Volta), development (water)

592 PARKER, C., 1980. Symposium on Weed Research in Sudan, Wad Medani, Sudan, November 1979. Conference reports. Tropical Pest Management, 26:216–217.
KEYWORDS: Sudan, *Eichhornia*, aquatic weeds

593 PATTERSON, L., 1976. An introduction of the ecology and zoogeography of the Okavango Delta. In: Proceedings of the Symposium on the Okavango Delta and its Future Utilisation. Published by the Botswana Society, Gaborone, Botswana, pp 55–60.
KEYWORDS: Botswana, river (Okavango), swamp (Okavango), ecology, wildlife

594 PENMAN, H.L., 1963. Vegetation and Hydrology. Commonwealth Agricultural Bureaux, Technical Communication 53, pp 64–65.
KEYWORDS: Africa, hydrology, swamp, development (land/swamp)

595 PERERA, N.P., 1982. Ecological considerations in the management of the wetlands of Zambia. In: B. Gopal, R.E. Turner, R.G. Wetzel and D.F. Whigham (Editors), Wetlands: Ecology and Management. National Institute of Ecology and International Scientific Publications, Jaipur, India, pp 21–30.
KEYWORDS: Zambia, floodplain (Kafue), conservation, wildlife, vegetation description (Zambia), swamp (savanna), control/management (swamp)

596 PERERA, N.P., 1983. Ecology of wetlands of Zambia and their evaluation for agriculture: a model for the management of wetlands in sub-humid Eastern and Southern Africa. International Journal of Ecology and Environmental Sciences, 8(1):27–38.
KEYWORDS: Zambia, swamp (savanna), floodplain (Kafue), wildlife, development (land/swamp), control/management (swamp), dambos

597 PETER, A., 1928. Wasserpflanzen und Sumpfgewachse in Deutsch Ostafrika. Weidemannsche Buchhandlung, Berlin.
KEYWORDS: Tanzania, aquatic plants, vegetation description (Tanzania) (in German)

598 PETR, T., 1968. Population changes in aquatic invertebrates living on two water plants in a tropical man-made lake. Hydrobiologia, 32:449–485.
KEYWORDS: Ghana, invertebrates, *Pistia*, utilisation (animals), lake (Volta)

599 PETR, T., 1973. Some factors limiting the distribution of *Povilla adusta* Navas (Ephemeroptera, Polymitarcidae) in African Lakes. In: E.J. Brill (Editor), Proceedings of the First International Conference on Ephemeroptera. Leiden, pp 223–230.
KEYWORDS: Africa, invertebrates, *Pistia*, *Cyperus*, utilisation (animals), lake

600 PETR, T., 1975. Limnology and fisheries of the Nyumba ya Mungu, a man-made lake in Tanzania. East African Journal of Tropical Hydrobiology and Fisheries, 4(1) Special Issue 3:39–50.
KEYWORDS: Tanzania, fish, chemistry, lake (Nyumba ya Mungu), development (water)

601 PETR, T., 1976. Limnology of the Great Ruaha River above Kidatu. In: Ecological Studies of the Mtera Basin. Great Ruaha Power Project (Tanzania). SWECO, Stockholm, Sweden, pp 87–99.
KEYWORDS: Tanzania, river (Great Ruaha), chemistry, aquatic plants

602 PETTET, A., 1964. Seedlings of *Eichhornia crassipes*: a possible complication to control measures in the Sudan. Nature, London, 201:516–517.
KEYWORDS: Sudan, *Eichhornia*, control/management (*Eichhornia*)

603 PETTET, A. and PETTET, S.J., 1970. Biological control of *Pistia stratiotes* L. in Western State, Nigeria. Nature, London, 226:282.
KEYWORDS: Nigeria, *Eichhornia*, control/management (*Pistia*)

604 PEYRE, de FABREGUES, B. and LEBRUN, J-P., 1976. Catalogue des plantes Vasculaires du Niger. Institut d'Elevage et de Médécine Veterinaires des Pays Tropicaux.
KEYWORDS: Niger, vegetation description (Niger), checklist

605 PIAGET, J. and SCHLIEMANN, G., 1973. Weeds choking Western Cape rivers – appearance of *Myriophyllum* spp in the Boland. Deciduous Fruit Grower, 23:176–179.
KEYWORDS: South Africa, *Myriophyllum*, river (Boland), ecological problems

606 PICKWORTH, G., 1969. Water hyacinths in farm dams – a serious problem. Farming in South Africa, 45:13–15.
KEYWORDS: South Africa, *Eichhornia*, lake (South Africa), ecological problems

607 PIELOU, E.C., 1952. Notes on the vegetation of the Rukwa Rift Valley, Tanganyika. Journal of Ecology, 40:383–392.
KEYWORDS: Tanzania, vegetation description (Tanzania), swamp (savanna)

608 PIERCE, P., 1971. Aquatic weed development, impact and control at Volta Lake. Report for the U.S.A.I.D. Volta Lake Technical Assistance Project 641/11/190/C.28.
KEYWORDS: Ghana, *Myriophyllum*, *Vossia*, lake (Volta), control/management (*Myriophyllum*), control/management (*Vossia*)

609 PIERCE, P. and OPOKU, A., 1971. Summary of aquatic weed survey and control data for Volta Lake during 1969. Hyacinth Control Journal, 9:49–56.
KEYWORDS: Ghana, aquatic weeds, control/management (*Myriophyllum*), *Myriophyllum*, lake (Volta)

610 PIERCE, S.M., 1979. The contribution of *Spartina maritima* (Curtis) Fernald to the primary production of the Swartkops estuary. M.Sc. Thesis, Rhodes University. Grahamstown, South Africa, 135pp.
KEYWORDS: South Africa, productivity, *Spartina*, estuary (Swartkops)

611 PIERCE, S.M., 1982. What is *Spartina* doing in our estuaries?. South African Journal of Science, 78:229–230.
KEYWORDS: South Africa, *Spartina*, estuary (South Africa)

612 PLATENKAMP, J.D.M., 1978. The Jonglei Canal: its impact on an integrated system in the southern Sudan, Instituut voor Culturele Antropologie en Sociologie der Niet–WesterseVolken, Leiden, The Netherlands, Publication no.26:126 pp.
KEYWORDS: Sudan, swamp (Upper Nile/Sudd), development (water), ecological problems

613 POLE EVANS, I.B., 1948 pp. A reconnaissance trip through the eastern portion of the Bechuanaland Protectorate , April 1931, and expedition to Ngamiland. Memoirs of the Botanical Survey of South Africa, 21pp.
KEYWORDS: Botswana, swamp (Okàvango), vegetation description (Botswana)

614 PORTER, R.N., 1969. A vegetation ecological study of the Sinamwenda area. Report of the Nuffield Lake Kariba Research Station, 1962–1968, Sinamwenda, pp 24–25.
KEYWORDS: Zimbabwe, vegetation description (Zimbabwe), littoral, lake (Kariba)

615 PORTERES, R., 1951. Les variations des ceintures hydrophytiques et gramino–helophytiques des eaux vives du systeme lagunaise de la Cote d'Ivoire. (in French). Bulletin de l'Institut Française d'Afrique Noire, Serie A.
KEYWORDS: Ivory Coast, vegetation description (Ivory Coast), aquatic plants

616 PROCTOR, J., 1976. Vegetation. In: Shepherd, C.J., Investigation into fish productivity in a shallow freshwater lagoon in Malawi, Report prepared by the Ministry of Overseas Development, London, for the Malawi Government, pp 22–29.
KEYWORDS: Malawi, aquatic plants, fish, lake (Bangula Lagoon), vegetation description (Malawi), swamp (Bangula Lagoon)

617 PROCTOR, J., 1980. The macrophyte vegetation of Bangula Lagoon, Malawi. Kirkia, 12:141–149.
KEYWORDS: Malawi, lake (Bangula Lagoon), aquatic plants, vegetation description (Malawi)

618 RANTZIEN, H.H., 1951. Certain aquatic plants collected by Dr. J.T. Baldwin Jr. in Liberia and the Gold Coast. Botaniska Notiser, 124:368–398.
KEYWORDS: Liberia, Ghana, aquatic plants, vegetation description (Liberia), checklist

619 RANTZIEN, H.H., 1952. Notes on some tropical African species of *Najas* in the Kew Herbarium. Kew Bulletin, 6:29–40.
KEYWORDS: Africa, *Najas*, checklist

620 RATCLIFFE, C., 1972. The fishery of the lower Shire River area. Malawi Fisheries Bulletin No.3.
KEYWORDS: Malawi, river (Shire), fish, utilisation (animals)

621 RAVEN, P.H., 1963. The Old World species of *Ludwigia* (including *Jussiaea*) with a synopsis of the genus (Onagraceae). Reinwardtia, 6:327–427.
KEYWORDS: Africa, *Ludwigia*, checklist

622 RAYNAL-ROQUES, A., 1980. Les plantes aquatiques (plantes à fleurs et fougères). In: Durand, J.R. and Lévêque, C., Flore et Faune Aquatiques de l'Afrique Sahélo-Soundanienne. I. Cahiers d'Office de la Récherche Scientifique et Technique Outre-Mer, Serie Hydrobiologique, pp 63–152.
KEYWORDS: Sudan, aquatic plants, checklist, vegetation description (Sudan), Tchad, Nigeria, Niger

623 REES, W.A., 1978. The ecology of the Kafue lechwe: soils, water levels and vegetation. Journal of Applied Ecology, 15:163–176.
KEYWORDS: Zambia, wildlife, utilisation (animals), vegetation description, floodplain (Kafue), swamp (savanna), control/management (wildlife), lechwe

624 RICHARDS, P.W., 1939. Ecological studies on the rain forest of Southern Nigeria. 1. The structure and floristic composition of the primary forest. Journal of Ecology, 27:1–61.
KEYWORDS: swamp (forest)

625 RIJKS, D.A., 1969. Evaporation from a papyrus swamp. Quarterly Journal of the Royal Meteorological Society, 95:643–649.
KEYWORDS: Uganda, *Cyperus*, swamp (Kabanyolo), hydrology

626 ROBB, F., DAVIES, B.R., CROSS, R., KENYON, C. and HOWARD-WILLIAMS, C., 1979. Cellulolytic bacteria as primary colonisers of *Potamogeton pectinatus* L. (sago pond weed) from a brackish south temperate coastal lake. Microbial Ecology, 5:167–177.
KEYWORDS: South Africa, decomposition, *Potamogeton*, lake (Swartvlei), plankton

627 ROBERTY, G., 1946. Les associations végétales de la vallée moyenne du Niger. Veroeffentlichungen des Geobotonischen Institutes Rübel in Zürich (Veroff. Geobot Inst. Rubel.), 22:1–168.
KEYWORDS: Mali, river (Niger), floodplain (Niger), vegetation description (Mali), aquatic plants

628 ROBYNS, W., 1936. Contribution á l'étude des formations herbeuses du district forestier central du Congo Belge. Mémoires de L'Institut Royale Colonial Belge, Section des Sciences Naturelles et Médicales, 5(1).
KEYWORDS: Zaïre, swamp (forest), vegetation description (Zaïre), swamp (savanna), ésobes

629 ROBYNS, W., 1947–1955. Flore des spermatophytes du Parc National Albert, vols 1–3. Institut des Parcs Nationale du Congo Belge, Bruxelles.
KEYWORDS: classification, lake (Albert), lacustrine, Zaire, swamp, conservation

630 ROGERS, F.E.J., 1981. Studies on the epiphyton of *Potamogeton crispus* L. MSc thesis, University of Natal, 172pp.
KEYWORDS: South Africa, *Potamogeton*, decomposition, chemistry

631 ROGERS, K.H., 1980. The vegetation of Pongolo floodplain: distribution and utilization. In: Bruton M.N. and Cooper, K.H. (Editors), Studies on the Ecology of Maputaland. Rhodes University, Grahamstown and The Natal Branch of the Wildlife Society of Southern Africa, pp 69–77.
KEYWORDS: South Africa, vegetation description (South Africa), floodplain (Pongolo), development (land/swamp)

632 ROGERS, K.H. and BREEN, C.M. , 1980. Growth and reproduction of *Potamogeton crispus* in a South African lake. Journal of Ecology, 68:561–571.
KEYWORDS: South Africa, *Potamogeton*, productivity, lake (Swartvlei), nutrients

633 ROGERS, K.H. and BREEN, C.M., 1981. The effects of epiphyton on *Potamogeton crispus* L. leaves. Microbial Ecology, 7:351–363.
KEYWORDS: South Africa, *Potamogeton*, decomposition, periphyton

634 ROGERS, K.H. and BREEN, C.M., 1982. Decomposition of *Potamogeton crispus* L.: the effects of drying on the pattern of mass and nutrient loss. Aquatic Botany, 12:1–12.
KEYWORDS: South Africa, *Potamogeton*, decomposition, chemistry

635 RUDESCU, L., 1969. Der Einfluss der Umweltbedingungen auf die produktivitat (Photosynthese Intensitat) des Schilfrohres (*Phragmites communis* Trin.) aus dem Donaudelta, Dnepr–und Wolgadelta, Mesopotamien, Pakistan (*P. Karka*) und Nildelta. Hidrobiologia, Bucurest, 10:77–86.
KEYWORDS: Egypt, productivity, *Phragmites*, river (Nile)
(in German)

636 RUSSEL–SMITH, A. and RUCKERT, E., 1981. The effects of aerial spraying of endosulfan for tsetse fly control on aquatic invertebrates in the Okavango Swamps, Botswana. Environmental Pollution, Series A, 24(1):57–73.
KEYWORDS: Botswana, aquatic plants, invertebrates, swamp (Okavango), disease, control/management (Schistosomiasis), ecological problems

637 RZOSKA, J., 1966. The Upper Nile swamps as an example of integrated, astatic ecology. Verhandlungen der Internationalen Vereinigung für theoretische und angewandte Limnologie (Verh. Internat. Verein. Limnol.), 16:561–562.
KEYWORDS: Sudan, swamp (Upper Nile/Sudd), ecology
(Abstract)

638 RZOSKA, J., 1974. The Upper Nile swamps, a tropical wet–land study. Freshwater Biology, 4:1–30.
KEYWORDS: Sudan, *Cyperus*, utilisation (animals), ecology, swamp (Upper Nile/Sudd)

639 RZOSKA, J. (Editor), 1976. The Nile, Biology of an Ancient River. W. Junk, b.v. Monographiae Biologieae 29, The Hague, pp197–214, 315–320, 417.
KEYWORDS: Uganda, Sudan, Egypt, swamp (Upper Nile/Sudd), river (Nile)

640 SANDER, F., 1955. Some effects of swamp drainage on the distribution of birds in the Lagos area. Nigerian Field, 20:4–15.
KEYWORDS: Nigeria, birds, development (land/swamp), wildlife

641 SANOKHO, A., 1976. *Salvinia nymphellula* Desv. au Senegal. Bulletin de l'Institut Française d'Afrique Noire, Serie A, 38:16–19.
KEYWORDS: Sénégal, *Salvinia*
(in French)

642 SCHAFER, F., 1977. Is the water-hyacinth a curse or blessing. Farmer's Weekly, March 2:19–24.
KEYWORDS: South Africa, ecological problems, *Eichhornia*

643 SCHELPE, E.A.C., 1961. The ecology of *Salvinia auriculata* and associated vegetation on Kariba Lake. Journal of South African Botany, 27:181–187.
KEYWORDS: Zambia, Zimbabwe, *Salvinia*, aquatic plants, lake (Kariba), ecology

644 SCHMITZ, A., 1963. Aperçu sur les groupments végétaux du Katanga. Bulletin de la Société Royale de Botanique de Belgique, 96:233–447.
KEYWORDS: Zaïre, vegetation description (Zaïre), swamp (Shaba)

645 SCHMITZ, A., 1971. La végétation de la plaine de Lubumbashi (Haut–Katanga). Institut National pour l'Etude Agronomique du Congo Belge (I.N.E.A.C.), Serie Science, 113:1–390.
KEYWORDS: Zaïre, vegetation description (Zaïre), floodplain (Shaba), swamp (savanna)

646 SCHOONBEE, H.J., SWANEPOEL, J.H. and VAN GRAAN, J.J.M., 1975. The use of the isotope 32P in the study of some ecological aspects of a laboratory stream ecosystem. (In Afrikaans). Tydskrif vir Natuurwetenskappe, 15(2):48–66.
KEYWORDS: South Africa, chemistry, *Potamogeton*, physiology

647 SCHUTTE, K.H. and ELSWORTH, J.F., 1954. The significance of large pH fluctuations observed in some South African vleis. Journal of Ecology, 48:55–71, 307–329.
KEYWORDS: South Africa, chemistry, swamp (South Africa)

648 SCOTCHER, J.S.B. , 1974. A quantitative assessment of the food preferences of *Hippopotamus amphibius* L. in the Ndumu Game Reserve, Tongaland. MSc thesis, University of Natal, 190pp.
KEYWORDS: South Africa, wildlife, utilisation (animals), swamp (savanna), swamp (Ndumu), hippopotamus

649 SCOTT, W.E., ASHTON, P.J. and STEYN, D.J., 1979. Chemical control of the water hyacinth on Hartbeespoort Dam. Water Research Commission, Pretoria, 84pp.
KEYWORDS: South Africa, *Eichhornia*, control/management (*Eichhornia*), lake (Hartebeespoort)

650 SCOTT, W.E., SEAMAN, M., WALMSLEY, R.D. and TOERIEN, D.F., 1975. An investigaton of Lake Liambezi and its surroundings. Part II. Project Report No.3: Advisory Services and Ad Hoc Studies: File No. W 6/151/3Project No. 151 Code No. 6201/6151, The Council for Scientific and Industrial Research (CSIR), National Institute for Water Research, Pretoria, South Africa.
KEYWORDS: Namibia, vegetation description (Namibia), lake (Liambezi), utilisation (man)

651 SEAGRIEF, S.C., 1962. The Lukanga Swamps of Northern Rhodesia. Journal of South African Botany, 10:251–252.
KEYWORDS: Zambia, swamp (Lukanga), vegetation description (Zambia)

652 SEAMAN, M.T., SCOTT, W.E., WALMSLEY, R.D., VAN DER WAAL, B.C.W. and TOERIEN, D.F., 1978. A limnological investigation of Lake Liambezi, Caprivi. Journal of the Limnological Society of Southern Africa, 4:129–144.
KEYWORDS: Namibia, chemistry, aquatic plants, lake (Liambezi)

653 SHAHEEN, A.H. and YOSEF, S.F., 1980. Physio-chemical conditions, fauna and flora of Lake Manzala, Egypt. In: M. Kassas and S.I. Ghabbour (Editors), The Nile and its Environment. Water Supply Management, 4(1–2):103–113.
KEYWORDS: Egypt, chemistry, aquatic plants, fish, invertebrates, lake (Manzala)

654 SHEPHERD, C.J. (Editor), 1976. Investigation into fish productivity in a shallow freshwater lagoon in Malawi. Report prepared by the Ministry of Overseas Development, London, for the Malawi Government , 90pp.
KEYWORDS: Malawi, fish, lake (Bangula Lagoon), swamp (Malawi)

655 SHEPPE, W. and OSBORNE, T., 1971. Patterns of use of a floodplain by Zambian mammals. Ecological Monographs, 41:179–205.
KEYWORDS: Zambia, wildlife, lechwe, utilisation (animals), swamp (savanna), floodplain (Kafue)

656 SHILLINGLAW, S.N., 1981. Dissolved oxygen depletion and nutrient uptake in an impoundment infested with *Eichhornia crassipes* (Mart.) Solms. Journal of the Limnological Society of Southern Africa, 7(2):63–66.
KEYWORDS: South Africa, chemistry, *Eichhornia*, ecological problems

657 SHILLINGLAW, S.N., 1981. The war on water weeds. African Wildlife, 35(3):31–34.
KEYWORDS: South Africa, aquatic weeds, control/management (weeds)

658 SIDIQUI, A.Q., 1977. Lake Naivasha (Kenya, East Africa) fishery and its management together with a note on the food habits of fishes. Biological Conservation, 12:217–228.
KEYWORDS: Kenya, lake (Naivasha), fish, utilisation (animals)

659 SIMPSON, C.D., 1975. A detailed vegetation study on the Chobe River in northeast Botswana. Kirkia, 10:185–227.
KEYWORDS: Botswana, river (Chobe), swamp (Chobe), vegetation description (Botswana), aquatic plants

660 SITA, P., 1970. Etude de la végétation de l'ile M'bamou (Stanley-pool). Office de la Recherche Scientifique et Technique Outre-Mer, Congo-Brazzaville, 94pp.
KEYWORDS: Congo, Zaïre, swamp (m'bamou), vegetation description (Congo), aquatic plants
(in French)

661 SMITH, P.A., 1969. A herbicide spraying experiment and other measures to control the *Salvinia* infestation at Shaile on the Linyanti River in Northern Botswana. Report to the Government of Botswana.
KEYWORDS: Botswana, *Salvinia*, control/management (*Salvinia*), river (Linyanti)

662 SMITH, P.A., 1969. A search for the aquatic weed, *Salvinia auriculata* Aublet in the Kwando, Linyanti and Savuti Rivers and the Selinda Spillway. Report of the Department of Veterinary Services and Tsetse Control, Botswana.
KEYWORDS: Botswana, *Salvinia*, river (Linyanti/Kwando/Savuti/Selinda)

728 VAN MEEL, L., 1953. Contribution a l'Étude du lac Upemba. A. Le milieu physico-chimique. Exploration du Parc National de l'Upemba. Institut Royal des Sciences Naturelles de Belgique, Bruxelles.
KEYWORDS: Zaïre, chemistry, swamp (Upemba), lake (Upemba), vegetation description (Zaïre)
(in French)

729 VAN RENSBERG, J., 1971. The ecology of the Okavango delta. FAO Technical Notes No.19, Rome.
KEYWORDS: Botswana, river (Okavango), swamp (Okavango), development (land/swamp)

730 VAN RENSBURG, H.J., 1968. Multipurpose survey of the Kafue River Basin, Zambia. Vol.4. The ecology of the Kafue Flats. UN/FAO, SF:35/ZAM:138 pp.
KEYWORDS: Zambia, river (Kafue), development (water), swamp (savanna), wildlife, floodplain (Kafue)

731 VAN RENSBURG, H.J., 1971. Fire: its effect on grasslands including swamps. Proceedings of the Tall Timbers Fire Ecology Conference. Tall Timbers Research Station, Florida, U.S.A., 11:175–199.
KEYWORDS: Zambia, swamp (savanna), development (land/swamp), ecological problems, control/management (swamp)

732 VAN WYK, H.R., 1972. The problem of aquatic weeds at Hartebeespoort Dam. Proceedings of the Workshop for Aquatic Weeds. C.S.I.R., Pretoria, South Africa.
KEYWORDS: South Africa, aquatic weeds, ecological problems, lake (Hartebeespoort), development (water)

733 VAN ZINDEREN BAKKER, E.M., 1955. A preliminary survey of the peat bogs of the alpine belt of Northern Basutoland. Acta Geographica, 14:413–422.
KEYWORDS: peatland, Lesotho, vegetation description (Lesotho)

734 VAN ZINDEREN BAKKER, E.M., 1965. Uber Moorvegetation und den Aufbau der Moore in Sud-und Ostafrika. Botanische Jahrbuecher fuer Systematik Pflanzengesichte und Pflanzengeographie (Bot. Jahrb. Syst.), 84:215–231.
KEYWORDS: Southern Africa, peatland, vegetation description (Lesotho)
(in German)

735 VAN ZINDEREN BAKKER, E.M. (Editor), 1974. Orange River Project, Progress report. The University of the Orange Free State, Blomfontein.
KEYWORDS: South Africa, environmental factors, *Azolla*, aquatic weeds, river (Orange)

736 VAN ZINDEREN BAKKER, E.M. and WERGER, M.J.A., 1974. Environment, vegetation and phytogeography of the high-altitude bogs of Lesotho. Vegetatio, 29:37–49.
KEYWORDS: Lesotho, peatland, vegetation description (Lesotho)

737 VANDEN BERGHEN, C., 1982. Premier aperçu sur la végétation aquatique de la Basse Casamance (Sénégal Méridional). In: J.J. Symoens, S.S. Hooper, and P. Compére (Editors), Studies on Aquatic Vascular Plants: Proceedings of the International Colloquium on Aquatic Vascular Plants, Brussels, 23–25 January, 1981. Société Royale de Botanique de Belgique, Brussels, pp 266–271.
KEYWORDS: Sénégal, vegetation description (Sénégal), aquatic plants, *Utricularia*

738 VERBOOM, W.C., 1971. *Salvinia auriculata* (Kariba Weed). Department of Wildlife Chilanga - 8th–10th Dec., Ecol/1 S. 842/71. 3pp.
KEYWORDS: Zambia, *Salvinia*

739 VERBOOM, W.C., 1975. List of plant species in the main vegetation types of the Bangweulu Basin. In: Grimsdell and Bell (Editors), Black Lechwe Research Project Final Report. Lusaka, Zambia. Report ARI–NCSR/TR 31, Animal Production Research, Lusaka, Zambia.
KEYWORDS: Zambia, checklist, lake (Bangweulu), aquatic plants

740 VERMAAK, J.F., SWANEPOEL J.H. and SCHOONBEE H.J., 1982. The phosphorus cycle in Germiston Lake. II. The *in vitro* and *in situ* absorption of 32P by *Potamogeton pectinatus* L.. Water SA, 8(1):52–57.
KEYWORDS: South Africa, chemistry, *Potamogeton*, lake (Germiston)

741 VERMAAK, J.F., SWANEPOEL, J.H. and SCHOONBEE, H.J., 1976. Absorption and accumulation of 32P by *Oedogonium* and some aquatic macrophytes. Water SA, 2:7–12.
KEYWORDS: South Africa, chemistry, aquatic plants, *Potamogeton*

742 VERMAAK, J.F., SWANEPOEL, J.H. and SCHOONBEE, H.J., 1981. The phosphorus cycle in Germiston Lake. I. Investigational objectives and aspects of the limnology of the lake. Water SA, 7:160–165.
KEYWORDS: South Africa, chemistry, lake (Germiston), aquatic plants

743 VERMAAK, J.H., SWANEPOEL, J.H. and SCHOONBEE, H.J., 1982. The phosphorus cycle in Germiston Lake. III. Seasonal patterns in the absorption, translocation and release of phosphorus by *Potamogeton pectinatus* L.. Water SA, 8:138–141.
KEYWORDS: South Africa, *Potamogeton*, chemistry, lake (Germiston), environmental factors

744 VESEY-FITZGERALD, D.F., 1963. Central African grasslands. Journal of Ecology, 51:243–274.
KEYWORDS: Central Africa, swamp (savanna), floodplain (Central Africa), vegetation description (Tanzania), vegetation description (Zambia), dambos

745 VESEY-FITZGERALD, D.F., 1963a. Annotated list of grasses collected in the Congo drainage basin of Northern Rhodesia and Tanganyika. Kirkia, 3:78–111.
KEYWORDS: Zambia, Tanzania, swamp (savanna), checklist, rever (Zaire)

746 VESEY-FITZGERALD, D.F., 1970. The origin and distribution of valley grasslands in East Africa. Journal of Ecology, 58:51–75.
KEYWORDS: East Africa, swamp (savanna), floodplain (East Africa), vegetation description (East Africa)

747 VESEY-FITZGERALD, D.F., 1973. East African Grasslands. East African Publishing House. Nairobi, Kenya, 95pp.
KEYWORDS: East Africa, swamp (savanna), vegetation description (East Africa), classification

748 VILJOEN, L., 1976. Uses of *Phragmites australis*. Hand. Weidingsveren. South Africa, 11:19–22.
KEYWORDS: South Africa, *Phragmites*, utilisation (man)

749 VINER, A.B., 1974. The supply of minerals to tropical rivers and lakes (Uganda). In: Olson, G. (Editor), An Introduction to Land-water Relationships. Academic Press, New York, pp 227–261.
KEYWORDS: Uganda, chemistry, swamp (Uganda)

750 VINER, A.B., 1977. The influence of sediment upon nutrient exchanges in tropical lakes. In: H.L. Golterman (Editor), Interactions Between Sediments and Freshwaters. Proceedings of the SIL–UNESCO Symposium, Amsterdam, 10–16 September, 1976.

KEYWORDS: Uganda, decomposition, chemistry

751 VINER, A.B., 1982. Nitrogen fixation and denitrification in sediments of two Kenyan lakes. Biotropica, 14:91–98.
KEYWORDS: Kenya, lake (Naivasha), decomposition, chemistry

752 VISSER, S.A., 1962. Chemical investigations into a system of lakes, rivers and swamps in S.W. Kigezi, Uganda. East African Agricultural and Forestry Journal, 28:81–86.
KEYWORDS: Uganda, lake (Kigezi), river (Kigezi), swamp (Kigezi), chemistry

753 VISSER, S.A., 1963. Gas production in the decomposition of *Cyperus papyrus*. Journal of the Water Pollution Control Federation, 6:973–988..
KEYWORDS: Uganda, decomposition, *Cyperus*, utilisation (man), swamp (Uganda)

754 VISSER, S.A., 1964. A study in the decomposition of *Cyperus papyrus* in the swamps of Uganda, in natural peat deposits, as well as in the presence of various additives. East African Agricultural and Forestry Journal, 29:268–287.
KEYWORDS: Uganda, *Cyperus*, decomposition, peatland, swamp (Uganda)

755 VISSER, S.A., 1964. The occurrence of micro-organisms in tropical swamps. Biochemical Journal, 89:83–84.
KEYWORDS: Africa, decomposition, swamp (Uganda), *Cyperus*

756 VON RICHTER, W. AND OSTERBERG, R., 1977. The nutritive values of some major food plants of lechwe, puku and waterbuck along the Chobe River, Botswana. East African Wildlife Journal, 15(2):91–97.
KEYWORDS: Botswana, utilisation (animals), wildlife, swamp (savanna), swamp (Chobe), chemistry, river (Chobe), control/management (wildlife), lechwe, puku, waterbuck

757 WAGER, V.A., 1928. The structure and life history of South African Lagarosiphons and notes and descriptions of a few uncommon water plants. Transactions of the Royal Society of South Africa, 16:191–204.
KEYWORDS: South Africa, *Lagarosiphon*, aquatic plants, checklist

758 WALMSLEY, R.D., BREEN, C.M. and KYLE, E., 1973. Aspects of the fern-alga relationship in *Azolla filiculoides*. Limnological Society of Southern Africa Newsletter, 20:13–21.
KEYWORDS: Southern Africa, *Azolla*, chemistry, physiology

759 WALTER, H., 1971. Ecology of Tropical and Subtropical Vegetation. Oliver and Boyd, Edinburgh.
KEYWORDS: Africa, vegetation description, aquatic plants, ecology, swamp

760 WARD, C.J., 1976. Peripheral vegetation and submerged macrophytes. In: A.E.F. Heydorn (Editor), Proceedings of St. Lucia Scientific Advisory Council Meeting, Charters Creek, February 1976. Oceanographic Research Institute, Durban.
KEYWORDS: South Africa, littoral, vegetation description (South Africa), aquatic plants, estuary (St Lucia)

761 WARNIMONT, F.J., 1965. La problème de la jacinthe d'eau (*Eichhornia crassipes*) dans le basin hydrographique de Congo. Directions de Voies Navigables, Léopoldville, 23pp.
KEYWORDS: Zaïre, *Eichhornia*, ecological problems

762 WASAWO, D.P.S., 1963. Some problems of Uganda Swamps. Abstracts of the Ninth Technical Meeting of the I.U.C.N. Publications (N.S.), 4:193–204.
KEYWORDS: Uganda, utilisation (animals), invertebrates, swamp (Uganda)

763 WATTS, J.C.D., 1968. Effect of Endrox on aquatic flora. Bulletin de l'Institut Française d'Afrique Noire, Serie A, 30:469–470.
KEYWORDS: Africa, *Eichhornia*, aquatic weeds, control/management (*Eichhornia*)

764 WEBER, N.A., 1942. A biocoenose of papyrus heads (*Cyperus papyrus*). Ecology, 23:115–119.
KEYWORDS: Sudan, invertebrates, *Cyperus*, utilisation (animals)

765 WEIMARCK, H., 1946. Studies in Juncaceae with special reference to the species in Ethiopia and the Cape. Svensk Botanisk Tidskrift, 40:141–178.
KEYWORDS: Ethiopea, South Africa, Juncaceae, checklist

766 WEINTROUB, D., 1933. A preliminary account of the aquatic and subaquatic vegetation flora of the Witwatersrand. Journal of Ecology, 21:44–57.
KEYWORDS: South Africa, aquatic plants, vegetation description (South Africa), swamp (Witwatersrand)

767 WEISSER, P.J. and HOWARD-WILLIAMS, C., 1982. The vegetation of the Wilderness Lakes System and the macrophyte encroachment problem. Bontebok, 12:19–40.
KEYWORDS: South Africa, vegetation description (South Africa), lake (Wilderness), ecological problems, aquatic plants

768 WEISSER, P.J. and R.J. PARSONS, 1981. Monitoring *Phragmites australis* increases from 1937 to 1976 in the Siyi Lagoon (Natal, South Africa) by means of air photo interpretation. Bothalia, 13(3/4):553–556.
KEYWORDS: South Africa, *Phragmites*, control/management (*Phragmites*)

769 WELCOMME, R.L., 1972. A brief review of the floodplain fisheries of Africa. FAO Committee for the Floodplain Fisheries of Africa, Report CIFA/72/S.17, and African Journal of Tropical Hydrobiology and Fisheries Special Issue I, pp.67–76.
KEYWORDS: Africa, floodplain, fish, swamp

770 WELCOMME, R.L., 1975. The fisheries ecology of African floodplains, UN/FAO/CIFA, Rome, 51pp.
KEYWORDS: Africa, utilisation (animals), fish, floodplain, swamp

771 WELCOMME, R.L. (Editor), 1978. Symposium on River and Floodplain Fisheries in Africa. FAO/CIFA Technical Papers, 5:378 pp.
KEYWORDS: Africa, utilisation (animals), fish, swamp, floodplain, river

772 WELCOMME, R.L., 1979. Fisheries Ecology of Floodplain Rivers. Longmans, London, 317pp.
KEYWORDS: Africa, utilisation (animals), fish, swamp, floodplain, river, vegetation description, ecological problems, development (water), control/management

773 WELDON, L.W. and BLACKBURN, R.D., 1967. Water lettuce: nature, problem and control. Weeds, 15:5–9.
KEYWORDS: *Pistia*, ecological problems, control/management (*Pistia*)

774 WELLINGTON, J.H., 1952. Niger and Okavango: physical and human factors in their development. South African Geographical Journal, 34.
KEYWORDS: Botswana, Nigeria, Niger, development (land/swamp), utilisation (man), swamp (Okavango), swamp (Niger), floodplain (Niger), floodplain (Okavango)

775 WELSH, R. and DENNY, P., 1978. The vegetation of Nyumba ya Mungu reservoir, Tanzania. Biological Journal of the Linnaean Society, 10:67–92.
KEYWORDS: Tanzania, vegetation description (Tanzania), aquatic plants, lake (Nyumba ya Mungu)

776 WERGER, M.J.A. (Editor), Biogeography and Ecology of Southern Africa. W. Junk b.v., Monographiae Biologicae 31, The Hague, 1439 pp.
KEYWORDS: Southern Africa, review, vegetation description (Southern Africa), conservation

777 WERGER, M.J.A. and ELLENBROEK, G.A., 1980. Water resource management and floodplain ecology: an example from Zambia. Tropical Ecology and Development, 1980:693–702.
KEYWORDS: Zambia, floodplain (Kafue), river (Kafue), swamp (savanna), vegetation description (Zambia), development (water)

778 WHITE, E., 1973. Zambia's Kafue hydroelectric scheme and its biological problems. In: Ackerman (Editor), Man-made Lakes, Their Problems and Environmental Effects. U.S. Geophysical Union, Washington D.C., Geophysical Monograph 17.
KEYWORDS: Zambia, development (water), ecological problems, floodplain (Kafue), river (Kafue), utilisation (man)

779 WHITE, F., 1983. The vegetation of Africa : a descriptive memoir to accompany the Unesco/AETFAT/UNSO vegetation map of Africa. UNESCO, Paris, 356 pp.
KEYWORDS: Africa, review, swamp (savanna), swamp (forest), dambos, classification, checklist, vegetation description (Africa), swamp, lacustrine, riverine, floodplain

780 WHITFIELD, A.K., 1981. A quantitive assessment of the trophic relationships of the dominant fish species associated with the aquatic macrophytes of Swartvlei. Report of the Institute of Freshwater Studies, Rhodes University, 1980–81, 2, 10pp.
KEYWORDS: South Africa, fish, aquatic plants, productivity, estuary (Swartvlei)

781 WHITFIELD, A.K. and BLABER, S.J.M., 1979. Feeding ecology of piscivorous birds at Lake St Lucia. 2. Wading birds. Ostrich, 50(1):1–9.
KEYWORDS: South Africa, utilisation (animals), fish, birds, estuary (St Lucia), wildlife

782 WICKENS, G.E., 1976. Jebel Marra. Her Majesty's stationary Office (HMSO), London, 368 pp.
KEYWORDS: Sudan, vegetation description (Sudan)

783 WILD, H., 1956. Water hyacinth (*Eichhornia crassipes*) (Mart.) Solms, in Southern Rhodesia. Proceedings of the Second Symposium on African Hydrobiology and Inland Fisheries. CSA/CTAS Publication, 25:127–129.
KEYWORDS: Zimbabwe, *Eichhornia*

784 WILD, H., 1961. Harmful aquatic plants in Africa and Madagascar. CCTA/CSA Project 14; reprinted from Kirkia, 2:1–66.
KEYWORDS: Africa, aquatic plants, aquatic weeds, checklist, control/management (weeds)

785 WILD, H. and BARBOSA, L.A. GRANDVAUX, 1968. Vegetation map of the Flora Zambesiaca area. Flora Zambesiaca Supplement, M.O. Collins, Salisbury, 71pp.
KEYWORDS: Zimbabwe, Moçambique, vegetation description (Zimbabwe), vegetation description (Moçambique), swamp (Zimbabwe), swamp (Moçambique)

786 WILD, H. and MITCHELL, D.S., 1970. Der Einfluss von Bayluscid auf den Wasserfarn *Salvinia auriculata* und andere Wasserpflanzen. Hofchenbr. Bayer. Pflanzenschutz Nachrichten, 23:112–117.
KEYWORDS: Africa, *Salvinia*, *Eichhornia*, control/management (*Salvinia*)

787 WILSON, B.H., 1974. Some natural and man-made changes in the channels of the Okavango Delta. Botswana Notes and Records, 1973. 22pp.
KEYWORDS: Botswana, *Cyperus*, hydrology, swamp (Okavango), control/management (swamp)

788 WILSON, V.H. and DINCER, T., 1976. An introduction to the hydrology and hydrography of the Okavango Delta. Proceedings of the Symposium on the Okavango Delta and its Future Utilisation, The Botswana Society, Gaborone, pp 33–47.
KEYWORDS: Botswana, hydrology, river (Okavango), swamp (Okavango), control/management (swamp)

789 WISS, H.J., 1981. Wasserpflanzer in Südwestrafrika. S.W.A. Wissenschaftliiche Gesellschaft. Botanische mitteilunger, 71:239–253.
KEYWORDS: Namibia, aquatic plants, vegetation description (Namibia)

790 WOOD, R.D., 1976. Key to the Charophytes of South Africa. Journal of the Limnological Society of Southern Africa, 2:47–49.
KEYWORDS: South Africa, *Charophytes*, checklist

791 WOODALL, P.F., 1979. Food of the redbilled teal in Rhodesia. South African Journal of Wildlife Research, 9(1/2):9–11.
KEYWORDS: Zimbabwe, utilisation (animals), birds, aquatic plants, wildlife

792 WORTHINGTON, E.B., 1976. The conservation of wetlands in Africa. Proceedings of the Symposium on the Okavango Delta and its Future Utilisation . Botswana Society, Gaborone, p 61–66.
KEYWORDS: Botswana, conservation, swamp (Okavango), development

793 YEO, R.R. and BAILEY, W., 1976. Aquatic weed control, report of travel to Egypt and Sudan, 1976. National Research Council., Washington, D.C., 46pp.
KEYWORDS: Egypt, Sudan, aquatic weeds, control/management (weeds)

794 ZAKI, S., 1960. Density distribution of rooted hydrophytes in Nozha Hydrome. Notes Mem. Hydrobiology Department of United Arab Republic.
KEYWORDS: Egypt, aquatic plants, ecology

Author index*

* This index does not cover Chapter 10: the cross-indexed bibliography of African wetland plants and vegetation.

Subject index*

* The alphabetical arrangement of this index is word-by-word. *et seq.*, *(et sequentia)* after a reference number denotes that the subject of the heading is continued in the succeeding pages. *Passim* used after a sequence of page references denotes that the subject of the heading is referred to not continuously, but in scattered passages, throughout those pages of the text. Chapter 10 is not covered by this index.

Reference map of Africa indicating the approximate position of countries. Some names may have changed since the preparation of the map.

Geographical index*

LAKES

* The geographical index has three sections: Lakes, Rivers and General. We have included latitude, longitude and altitude where appropriate, but it must be emphasised that these are for guidance only and can be used in conjunction with the endpaper maps. The index does not cover Chapter 10: The cross-indexed bibliography of African wetland plants and vegetation. Whilst every care has been taken to maximise accuracy, errors and omissions may well occur, for which apologies are offered.

RIVERS

GENERAL

Systematic index*

* Synonyms are included in the index with no implication to the present 'taxonomic correctness' of a name. This index does not cover Chapter 10: The cross-indexed bibliography of African wetland plants and vegetation. Please note: where authors have used the generic name without explicit reference to the species the entry has been indexed under the genus.

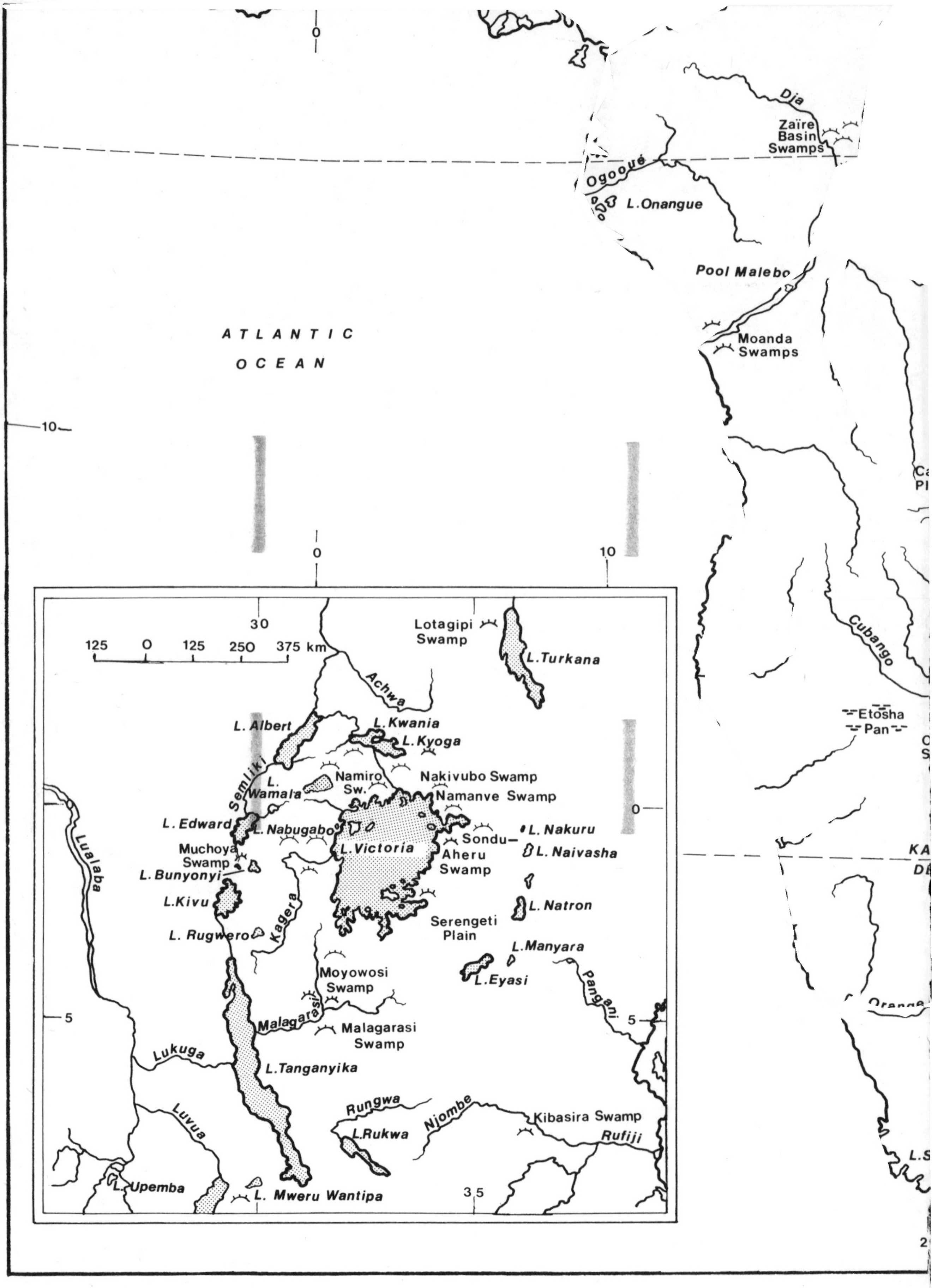

ATLANTIC
OCEAN
Dja
Zaïre
Basin
Swamps
Ogooué
L.Onangue
Pool Malebo
Moanda
Swamps
Cubango
Etosha
Pan
Orange
125 0 125 250 375 km
30
35
0
5
10
Lotagipi
Swamp
L.Turkana
Achwa
L. Albert
L. Kwania
L. Kyoga
Semliki
L.
Wamala
Namiro
Sw.
Nakivubo Swamp
Namanve Swamp
L. Edward
L. Nabugabo
L. Victoria
Sondu–
Aheru
Swamp
L. Nakuru
L. Naivasha
Lualaba
Muchoya
Swamp
L. Bunyonyi
L.Kivu
Kagera
L. Natron
Serengeti
Plain
L. Rugwero
L.Manyara
L.Eyasi
Pangani
Moyowosi
Swamp
Malagarasi
Malagarasi
Swamp
Lukuga
L.Tanganyika
Rungwa
Njombe
Kibasira Swamp
Rufiji
Luvua
L.Rukwa
L.Upemba
L. Mweru Wantipa